Arctic Marine Governance

Elizabeth Tedsen · Sandra Cavalieri
R. Andreas Kraemer
Editors

Arctic Marine Governance

Opportunities for Transatlantic Cooperation

 Springer

Editors
Elizabeth Tedsen
Ecologic Institute Washington DC
Washington DC
USA

Sandra Cavalieri
R. Andreas Kraemer
Ecologic Institute
Berlin
Germany

ISBN 978-3-642-38594-0   ISBN 978-3-642-38595-7   (eBook)
DOI 10.1007/978-3-642-38595-7
Springer Heidelberg New York Dordrecht London

Library of Congress Control Number: 2013942467

Printed on acid-free paper

Springer is part of Springer Science+Business Media (www.springer.com)

Foreword

Nowhere is climate change more visible than in the Arctic, which is a vital and vulnerable component of the Earth's environment and climate system. The melting of Arctic sea ice is progressing rapidly and affecting ecosystems as well as the traditional livelihoods of indigenous peoples. As climate change and economic development accelerate in the Arctic region, the European Union is stepping up its engagement with its Arctic partners to jointly meet the challenge of safeguarding the environment while ensuring the sustainable development of the Arctic region.

The EU sponsored the Arctic TRANSFORM project (<http://arctic-transform.org>), the results of which form a basis for this volume. The goal of Arctic TRANSFORM was to develop transatlantic policy options for supporting adaptation in the marine Arctic environment, placing a special emphasis on involving a broad range of stakeholders to address the major climate issues facing the region. Key project objectives included:

- To promote mutual exchange among EU and US policymakers and stakeholders on policies and approaches in the Arctic in stakeholder working groups;
- To provide a comparative analysis of existing policies and make recommendations with substantial buy-in as to how to strengthen cooperation between the EU and US; and
- To encourage dialogue and thus improve conditions for further transatlantic policy development and more effective protection of the Arctic marine environment.

Today, as the impacts of climate change become apparent, the need for good collaboration on preserving the Arctic marine environment is even more pressing than it was at the time of the project. Enhanced transatlantic cooperation can take advantage of emerging opportunities for improving protection of the Arctic marine environment. Through formal cooperation on policy strategies as well as informal channels of information sharing, research exchanges, and stakeholder networks, a strong transatlantic partnership can create a foundation for improved knowledge and efficient and timely addressing of the challenges ahead.

Helga Schmid
Deputy Secretary General
European External Action Service

Preface

The Arctic region plays an important role in regulating the world's climate and also is highly impacted by climate change, with average temperatures rising almost twice as fast as the rest of the world and sea ice melting much faster than previously predicted. These rapid changes will have significant impacts on human activity in the region and on the Arctic marine environment.

Recognizing the importance of these changes, the Arctic TRANSFORM project was developed to explore the roles of the EU and US in light of the changing climate and the region's political and legal complexities. The project sought to promote exchange between EU and US policymakers and stakeholders on approaches in the Arctic, provide comparative analyses of existing policies, make recommendations as how to strengthen cooperation between the EU and US, encourage dialogue, and thus improve conditions for further transatlantic policy development and more effective protection of the Arctic marine environment. The project was funded by the European Commission (formerly DG External Relations, now European External Action Service) and led by Ecologic Institute, along with the Arctic Center, the Netherlands Institute for the Law of the Sea, and the Heinz Center.

Four years later, this book begins where Arctic TRANSFORM left off, addressing the new and significant changes and developments in the marine Arctic with updates, new policy recommendations, and additional topics to better reflect the current governance environment.

Within four years, the Arctic region has undergone substantial changes. The impacts of climate change are swiftly altering the landscape and ushering in new activity and actors. Likewise, the policy environment has also undergone transformations, including modifications to adapt existing frameworks to the changed and changing Arctic conditions, as described in detail throughout this book.

A look to the future leaves little doubt that this rapid pace of change will continue—for the environment and policy. For instance, in May 2013—after the time of writing of this volume—the Arctic Council Ministerial Meeting in Kiruna is slated to take up a wide range of critical governance issues, including a legally binding agreement on marine oil preparedness and response, updating PAME's Arctic Marine Strategic Plan, and releasing reports for the Arctic Ocean Review and Arctic Biodiversity Assessment. Most important in the transatlantic context, the meeting will also address the EU's application to become a permanent observer of the Arctic Council. In addition, in 2015, the US will assume chairmanship of the Arctic Council.

While such developments may continue to further alter the Arctic governance landscape, we believe that the background, principles, and discussion presented herein, do and will continue to promote better understanding of marine Arctic governance needs and options and help to lay a foundation for future effective environmental governance.

March 2013   E. Tedsen
                                                                                 S. Cavalieri
                                                                        R. Andreas Kraemer

Acknowledgments

The editors thank and recognize the contributions of the Arctic TRANSFORM project to this book. The Arctic TRANSFORM project was funded by the **European Commission** (formerly DG External Relations, now European External Action Service) and led by **Ecologic Institute**, along with the **Arctic Centre**, the **Netherlands Institute for the Law of the Sea**, and the **Heinz Center**. The project was enriched by discussions and outcomes from five thematic working groups, which engaged more than 50 experts from multiple disciplines on Arctic indigenous peoples, environmental governance, fisheries, offshore hydrocarbon activities, and shipping. In particular, the authors recognize the contribution of the Co-Chairs of each working group as follows:

- **Environmental Governance Working Group**: Dr. Stuart Chapin (University of Alaska, Fairbanks) and Dr. Neil Hamilton (WWF Arctic International Programme).
- **Fisheries Working Group**: Ambassador David Balton (US Department of State, Bureau of Oceans, Environment and Science) and Kjartan Hoydal (North East Atlantic Fisheries Commission).
- **Indigenous Peoples Working Group**: Patricia Cochran (Inuit Circumpolar Conference) and Dr. Mark Nuttall (University of Alberta, University of Oulu).
- **Offshore Hydrocarbon Working Group**: Dr. Cutler Cleveland (Boston University) and Kevin O'Carroll (OSPAR Offshore Industry Committee).
- **Shipping Working Group**: Dr. Lawson Brigham (US Arctic Research Commission) and Rene Piil Pedersen (Danish Shipowners' Association).

The editors thank Lucy Smith and Erika Sulik at Ecologic Institute for their careful attention and support in finalizing this book.

Contents

Part I   The Arctic Environment

1   **Introduction to the Arctic**.   3
Erik J. Molenaar, Timo Koivurova, Elizabeth Tedsen,
Andrew Reid, and Kamrul Hossain

2   **The Arctic Marine Environment**   21
Arne Riedel

3   **Environmental Governance in the Marine Arctic**   45
Susanah Stoessel, Elizabeth Tedsen, Sandra Cavalieri, and Arne Riedel

4   **Arctic Indigenous Peoples and the Challenge of Climate Change**   71
Adam Stepien, Timo Koivurova, Anna Gremsperger, and Henna Niemi

Part II   Impacts and Activities in the Marine Arctic

5   **Status and Reform of International Arctic Fisheries Law**   103
Erik J. Molenaar

6   **Status and Reform of International Arctic Shipping Law**   127
Erik J. Molenaar

7   **Understanding Risks Associated with Offshore Hydrocarbon
Development**   159
Kamrul Hossain, Timo Koivurova, and Gerald Zojer

Part III Improving Marine Governance

8 Impact Assessments and the New Arctic Geo-Environment 179
Pamela Lesser and Timo Koivurova

**9 Pan-Arctic Marine Spatial Planning: An Idea Whose
Time Has Come** 199
Charles N. Ehler

**10 Marine Protected Areas as a Tool to Ensure Environmental
Protection of the Marine Arctic: Legal Aspects**.................. 215
Ingvild Ulrikke Jakobsen

Part IV Opportunities for Transatlantic Cooperation

11 EU–US Cooperation to Enhance Arctic Marine Governance 237
Elizabeth Tedsen and Sandra Cavalieri

Annex: List of Relevant Treaties, Instruments, and Agreements........ 263

Abbreviations

ABA	Arctic Biodiversity Assessment
ACAP	Arctic Contaminants Action Program (working group)
ACIA	Arctic Climate Impact Assessment
ACS	Arctic Council System
ADHR	Arctic Human Development Report
AECO	Association of Arctic Expedition Cruise Operators
AEPS	Arctic Environmental Protection Strategy
AMAP	Arctic Monitoring and Assessment Programme (working group)
AMSA	Arctic Marine Shipping Assessment
AOR	Arctic Ocean Review
APM	Associated Protective Measure
ARR	Arctic Resilience Report
ATS	Antarctic Treaty System
AWPPA	Arctic Waters Pollution Prevention Act
BEAC	Barents Euro-Arctic Cooperation
BEAR	Barents Euro-Arctic Region
BePOMAr	Best Practices in Ecosystem-Based Oceans Management in the Arctic
BFR	Brominated Flame Retardant
BWM	Convention for the Control and Management of Ships' Ballast Water and Sediments
CAFF	Conservation of Arctic Flora and Fauna (working group)
CBD	Convention on Biological Diversity
CBMP	Circumpolar Biodiversity Monitoring Programme
CCAC	Climate and Clean Air Coalition
CDEM	Construction, Design, Equipment, and Manning
CFP	Common Fisheries Policy
CITES	Convention on International Trade in Endangered Species
CLCS	Commission on the Limits of the Continental Shelf
CMS	Convention on the Conservation of Migratory Species
COP	Conference of Parties
CPAN	Circumpolar Protected Areas Network
DDE	Dichlorodiphenyldichloroethylene
DDT	Dichlorodiphenyltrichloroethane

DE	Maritime Safety Committee Sub-Committee on Design and Equipment
DOI	United States Department of the Interior
EA	Environmental Assessment
EAF	Ecosystem Approach to Fisheries
EBM	Ecosystem-Based Management
EBSA	Ecologically or Biologically Significant Area
EC	European Community
EEA	European Environment Agency
EEAS	European External Action Service
EEC	European Economic Community
EEZ	Exclusive Economic Zone
EFTA	European Free Trade Agreement
EIA	Environmental Impact Assessment
EPPR	Emergency Prevention, Preparedness, and Response (working group)
ESA	Endangered Species Act
EU	European Union
FAO	Food and Agriculture Organization
FMP	Fishery Management Plan
FPZ	Fisheries Protection Zone
GAIRAS	Generally Accepted International Rules and Standards
HFC	Hydrofluorocarbon
HFO	Heavy Fuel Oil
IACS	International Association of Classification Societies
IAEA	International Atomic Energy Agency
IAIA	International Association for Impact Assessment
IASC	International Arctic Science Committee
ICAO	International Civil Aviation Organization
ICC	Inuit Circumpolar Council
ICCAT	International Commission for the Conservation of Atlantic Tunas
ICES	International Council for the Exploration of the Sea
ICJ	International Court of Justice
ICRW	International Convention on the Regulation of Whaling
ICS	International Chamber of Shipping
IEA	International Energy Agency
IHO	International Hydrographic Organization
ILO	International Labour Organization
IMO	International Maritime Organization
IMP	Integrated Maritime Policy
IPCC	Intergovernmental Panel on Climate Change
IPHC	International Pacific Halibut Commission
IRF	International Regulators Forum
ISA	International Seabed Authority
IUCN	International Union for Conservation of Nature

IUU	Illegal, Unreported, and Unregulated
JAMP	Joint Assessment Monitoring Programme
JIP	Joint Industry Programme
LME	Large Marine Ecosystem
LNG	Liquid Natural Gas
LOS	United Nations Convention on the Law of the Sea
LRTAP	Convention on Long-range Transboundary Air Pollution
MAP	Marine Advisory Programme
MARPOL	International Convention for the Prevention of Pollution from Ships
MEPC	Marine Environment Protection Committee
MOPPR	Marine Oil Pollution Preparedness and Response
MOU	Memorandum of Understanding
MPA	Marine Protected Area
MSC	Maritime Safety Committee
MSFD	Marine Strategy Framework Directive
MSP	Marine Spatial Planning
MSY	Maximum Sustainable Yield
NAFO	Northwest Atlantic Fisheries Organization
NAMMCO	North Atlantic Marine Mammal Commission
NASCO	North Atlantic Salmon Conservation Organization
NAV	Maritime Safety Committee Sub-Committee on Navigation
NDP	Northern Dimension Policy
NEAFC	North East Atlantic Fisheries Commission
NEPA	National Environmental Policy Act
NGO	Non-Governmental Organization
NIRB	Nunavut Impact Review Board
NM	Nautical Mile
NMFS	United States National Marine Fisheries Service
NOAA	United States National Oceanic and Atmospheric Administration
NORDREG	Northern Canada Vessel Traffic Services Regulations
NPAFC	North Pacific Anadromous Fish Commission
NPFMC	North Pacific Fishery Management Council
NTNU	Norwegian University of Science and Technology
OATRC	Oden Arctic Technology Research Cruise
OIC	Offshore Industry Committee
OPRC	International Convention on Oil Pollution Preparedness, Response, and Cooperation
OSPAR	Convention for the Protection of the Marine Environment of the North-East Atlantic
PAME	Protection of the Arctic Marine Environment (working group)
PBSG	Polar Bear Specialist Group
PCB	Polychlorinated Biphenyl
PFO	Perfluorooctane Sulfonate
PICES	North Pacific Marine Science Organization
PM	Particulate Matter

POP	Persistent Organic Pollutant
PSC	Pacific Salmon Commission
PSC	Port State Control
PSSA	Particularly Sensitive Sea Area
RAIPON	Russian Association of Indigenous Peoples of the North
REIO	Regional Economic Integration Organization
RFMO	Regional Fisheries Management Organization
SAMCoT	Sustainable Arctic Marine and Coastal Technology
SAO	Senior Arctic Official
SAR	Search And Rescue
SDWG	Sustainable Development Working Group
SEA	Strategic Environmental Assessment
SLCP	Short-Lived Climate Pollutant
SOLAS	International Convention for the Safety of Life at Sea
SPRS	Swedish Polar Research Secretariat
SRS	Ship Reporting System
STCW	Standards of Training, Certification, and Watchkeeping
TAC	Total Allowable Catch
TEA	Transboundary Environmental Assessment
TEK	Traditional Ecological Knowledge
TFEU	Treaty on the Functioning of the European Union
UN	United Nations
UNECE	United Nations Economic Commission for Europe
UNGA	United Nations General Assembly
UNFCCC	United Nations Framework Convention on Climate Change
UNTS	United Nations Treaty Series
US	United States of America
USGS	United States Geological Survey
VOC	Volatile Organic Compound
VTS	Vessel Tracking Service
WCPFC	Western and Central Pacific Fisheries Commission
WSSD	Word Summit on Sustainable Development

Part I
The Arctic Environment

Chapter 1
Introduction to the Arctic

**Erik J. Molenaar, Timo Koivurova, Elizabeth Tedsen,
Andrew Reid, and Kamrul Hossain**

Abstract Climate change is occurring rapidly in the Arctic, bringing new economic opportunities alongside challenges for environmental governance in the region. Evaluating these changes, and options for effectively addressing them, requires an understanding of existing institutions and frameworks. This chapter provides a foundation for the book and an introduction to Arctic marine governance and transatlantic cooperation, setting the scene with sections on the spatial scope of the Arctic marine area, the law of the sea in the Arctic marine area, the Arctic Council, and the respective Arctic and marine policies of the European Union and the United States.

Based on Molenaa EJ., Corell R, Koivurova T, Cavalieri S. (2008) Introduction to the background papers. Arctic TRANSFORM. 8 Sept 2008.

E. J. Molenaar (✉)
Netherlands Institute for the Law of the Sea (NILOS), Utrecht University,
and University of Tromsø, Tromsø, Norway
e-mail: E.J.Molenaar@uu.nl

T. Koivurova · K. Hossain
Northern Institute for Environmental and Minority Law, Arctic Centre,
University of Lapland, Rovaniemi, Finland
e-mail: timo.koivurova@ulapland.fi

K. Hossain
e-mail: kamrul.hossain@ulapland.fi

E. Tedsen
Ecologic Institute, 1600 Connecticut, Ave NW, Suite 300,
Washington D.C 20009, USA
e-mail: elizabeth.tedsen@ecologic.eu

A. Reid
Ecologic Institute, Pfalzburger Strasse 43/44, Berlin 10717, Germany
e-mail: andrew.reid@ecologic.eu

E. Tedsen et al. (eds.), *Arctic Marine Governance*,
DOI: 10.1007/978-3-642-38595-7_1, © Springer-Verlag Berlin Heidelberg 2014

3

1.1 Arctic Marine Area

As there is no generally accepted definition of either the 'Arctic' or the 'marine Arctic', the spatial scope of this book has been determined as the marine areas included within the area agreed by the Arctic Monitoring and Assessment Programme (AMAP) working group of the Arctic Council. These are the marine areas north of the Arctic Circle (66°32′N), and north of 62°N in Asia and 60°N in North America, modified to include the marine areas north of the Aleutian chain, Hudson Bay, and parts of the North Atlantic Ocean, including the Labrador Sea. For the purpose of this book, these marine areas are referred to as the 'Arctic marine area' or 'marine Arctic'.

Figure 1.1 shows the AMAP area, as well as the borders of the Arctic according to the Arctic Circle (the parallel of latitude that runs approximately 66.56083° north of the Equator), and to certain scientific parameters (10 °C July isotherm, treeline, marine, and vegetation).

Similarly, there is no universally accepted definition for the 'Arctic Ocean'. However, it is generally accepted that the five coastal states to the Arctic Ocean are Canada, Denmark (in relation to Greenland and the Faroe Islands), Norway, the Russian Federation, and the United States (US).

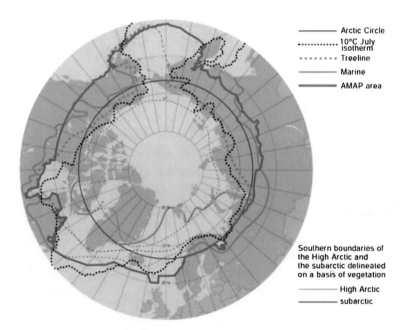

Fig. 1.1 Arctic monitoring and assessment programme (AMAP) boundary (*Source* AMAP (1997))

1.2 Law of the Sea in the Arctic Marine Area

The international law of the sea is made up of a multitude of global, regional, and bilateral instruments, decisions by international (intergovernmental) organizations, and international rules from other sources, including customary international law. The cornerstones of the current international law of the sea are the United Nations (UN) Law of the Sea (LOS) Convention (1982) and its two implementation agreements, the Part XI Deep-Sea Mining Agreement (1994) and the Fish Stocks Agreement (1995). The LOS Convention's overarching objective is to establish a universally accepted, just, and equitable legal order—or 'Constitution'—for the oceans that lessens the risk of international conflict and enhances stability and peace in the international community. As of January 2013, the LOS Convention has 165 parties, the Part XI Deep-Sea Mining Agreement 144 parties, and the Fish Stocks Agreement 80 parties.

All eight Arctic states (Canada, Denmark, Finland, Iceland, Norway, the Russian Federation, Sweden, and the US) are parties to these three treaties, except the US is not a party to either the LOS Convention or the Part XI Deep-Sea Mining Agreement (Division for Ocean Affairs and Law of the Sea 2013). The US is therefore, among other things, not subject to the LOS Convention's Part XV on dispute settlement.[1] The European Union (EU) is party to all three treaties. This is important in view of the fact that Denmark, Finland, and Sweden are Member States of the EU and Iceland and Norway are parties to the EEA Agreement (1993).

All of the global instruments that are part of the law of the sea apply to the marine environment of the entire globe, including therefore the entire marine Arctic, however defined. The mandate of the global bodies associated with these instruments has the same geographical scope. The perception that there is an international law vacuum in the Arctic, which only became a matter of attention following the melting of Arctic ice and the planting of a Russian flag on the seabed of the North Pole in August 2007, is therefore incorrect.

The most basic distinction between marine areas made by the LOS Convention is between the maritime zones of coastal states—also referred to as 'areas within national jurisdiction'—and the commons seaward thereof—also referred to as 'areas beyond national jurisdiction'. The maritime zones of coastal states can consist of: internal waters, archipelagic waters, territorial sea, contiguous zone, exclusive economic zone (EEZ), and the continental shelf. As clarified below, the EEZ includes the continental shelf, but in some cases, there is also an 'outer' continental shelf that extends seaward of the EEZ. The two marine commons are the high seas—usually seaward of the EEZ (where established)—and the so-called 'Area'—seaward of the EEZ or outer continental shelf. The Area is defined as "the sea-bed and ocean floor and subsoil thereof, beyond the limits of national jurisdiction" (LOS Convention 1982, art. 1(1)(1)).

---

[1] In the domain of straddling and highly migratory fish stocks, however, the US is subject to Part XV of the LOS Convention due to its being a party to the Fish Stocks Agreement. See also Reagan (1983).

Except for archipelagic waters,[2] all of the maritime zones recognized in the LOS Convention also exist in the marine Arctic: internal (marine) waters, territorial seas, contiguous zones, EEZs, (outer) continental shelves, the so-called 'Area' (the deep sea-bed beyond continental shelves) and the high seas. There are four high seas pockets (or enclaves) in the marine Arctic, namely the so-called 'Banana Hole' in the Norwegian Sea, the so-called 'Loop Hole' in the Barents Sea, the so-called 'Donut Hole' in the central Bering Sea, and the so-called 'Central Arctic Ocean'. There may be two or more pockets of the Area that could remain in the Arctic Ocean. Some region-specific maritime zones exist as well, as Norway has—instead of a regular EEZ—established a Fishery Zone around Jan Mayen and a Fisheries Protection Zone around Svalbard.

The outer limits of the maritime zones of coastal states are measured from baselines drawn in accordance with several provisions of the LOS Convention. The normal baseline is the low-water line along the coast (*inter alia* LOS Convention 1982, arts. 5–7, 9–14). It should be noted here that sea level rise, a consequence of climate change, could in many situations mean that new baselines may have to be drawn landward of the older ones and, as a consequence, the high seas and the Area will increase in size. In certain situations, the LOS Convention also allows coastal states to draw straight baselines. The straight baselines drawn by Canada around its Arctic Archipelago are regarded by the US and certain EU Member States as inconsistent with international law (Roach and Smith 1996) (Fig. 1.2).

Internal waters lie landward of the baselines. The maximum breadth of the territorial sea is 12 nautical miles (nm; 1 nm = 1,852 m) measured from the baselines, 24 nm is the maximum breadth for the contiguous zone, and 200 nm for the EEZ. Article 76 of the LOS Convention also recognizes that in certain circumstances, the continental shelf extends beyond 200 nm from the baselines. This is the so-called 'outer continental shelf'. Coastal states that take the view that they have an outer continental shelf must submit information on its outer limits on the basis of the criteria in Article 76 to the Commission on the Limits of the Continental Shelf (CLCS). The planting of a Russian flag on the seabed of the North Pole in August 2007 actually took place during the gathering of scientific data for this process. In 2002, the CLCS recommended that the Russian Federation gather and submit new data to complement the data submitted in 2001, which the Russian Federation is still expected to do.

The limits of the outer continental shelf established by the coastal state "on the basis of" the recommendations of the CLCS "shall be final and binding" (LOS Convention 1982, art. 76(8)). So far, only the Russian Federation and Norway have made submissions to the CLCS in relation to their outer continental shelves that lie within the marine Arctic. The CLCS has issued recommendations in relation to both submissions. Canada, Denmark (in relation to Greenland), and the

---

[2] Terms such as the 'Canadian Arctic Archipelago' and the 'Spitsbergen Archipelago', even if used consistently by Canada and Norway, do not imply that these states qualify—or claim to qualify—as archipelagic states in the domain of the international law of the sea.

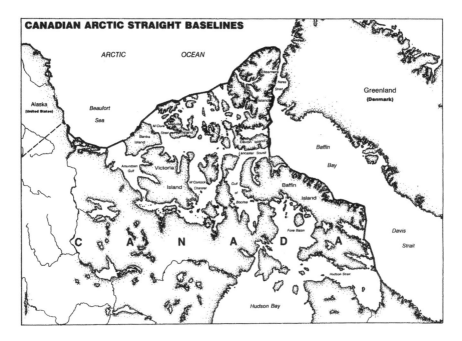

Fig. 1.2 Canadian Arctic straight baselines (*Source* United States Department of State (1992))

US are all engaged in activities to enable them to make submissions to the CLCS, despite the fact that the US is not yet a party to the LOS Convention. Canada has to make its submission before November 2013 and Denmark before November 2014 (Cf. LOS Convention 192, art. 4, Annex II).

The LOS Convention recognizes the sovereignty of a coastal state over its internal waters, archipelagic waters, and territorial sea, the airspace above, and its bed and subsoil. Sovereignty entails exclusive access to and jurisdiction over all resources (living and non-living; e.g., fish and oil) as well as full jurisdiction over all human activities, unless states have in one way or another consented to restrictions on these. The LOS Convention also recognizes that a coastal state has economic and resource-related sovereign rights and jurisdiction in its EEZ and outer continental shelf (if present). These sovereign rights give coastal states exclusive access to and jurisdiction over all resources in the EEZ and over all non-living and some living resources of the continental shelf. Nevertheless, other states have navigational rights or freedoms within the maritime zones of coastal states and, with respect to their EEZ and—where present—outer continental shelf, also the freedoms of overflight and laying of submarine cables and pipelines.

In the high seas, all states have the freedoms already mentioned above as well as the freedom to construct artificial islands and other installations, the freedom of fishing, and the freedom of scientific research. These freedoms are all subject to certain conditions and obligations. The Area and its resources are considered the common heritage of mankind and the International Seabed Authority (ISA) is charged with organizing and controlling all activities of exploration for, and exploitation of, the resources of the Area.

Maritime delimitation is always necessary for adjacent coastal states, but opposite coastal states are only required to do this if the maximum widths of their maritime zones cannot be reached. Many maritime boundaries between adjacent and opposite coastal states in the marine Arctic have already been agreed upon, but some are still under negotiation. By means of the bilateral Murmansk Treaty (2010), Norway and the Russian Federation managed to finalize their lengthy negotiations on a maritime boundary in the Barents Sea and the Arctic Ocean. Enhanced access to the marine Arctic caused by climate change is likely to have been an incentive towards finalizing these negotiations.

Even though the Arctic and the Antarctic are both polar regions, they are radically different from the perspective of international law. The Arctic is not subject to a comparably fundamental disagreement on sovereignty over territory that exists in the Antarctic. The only dispute on title to land territory in the Arctic is that between Canada and Denmark/Greenland on the tiny Hans Island in the Nares Strait.[3] News reports in April 2012 suggested that the dispute would soon be resolved by means of dividing the island in two (see e.g., Humphreys 2012), but by mid-September 2012 this had not (yet) happened.

It is true that there are several unresolved maritime boundaries in the marine Arctic and that most of the outer limits of the continental shelves still have yet to be established. But that is true for most regions in the world. There are nevertheless two international law of the sea disputes that are Arctic-specific. First, the disagreement between Canada and the Russian Federation on the one hand and the US and other states—including several EU Member States—on the other hand, on the legal status of the Northwest Passage and other waters within the Canadian Arctic Archipelago and the waters within the Northern Sea Route, respectively (Bartenstein 2011). As regards Canada, this disagreement is related to Canada's straight baselines around the Canadian Arctic Archipelago (see above). Second is the disagreement between Norway and several other parties to the Spitsbergen Treaty (1920) as to whether or not the treaty also applies seaward of the territorial sea (Molenaar 2012).

Apart from these disputes however, the marine Arctic is not fundamentally different from most other marine areas or oceans; except, of course, the waters adjacent to the Antarctic continent. As all other states have rights under the international law of the sea in all coastal state's maritime zones, the coastal states to the marine Arctic do not have full jurisdiction and control over these areas. In other words: they cannot just do as they please because they have to respect the rights of others.

---

[3] See the Joint Statement by the then Canadian Minister of Foreign Affairs Pettigrew and then Danish Minister of Foreign Affairs Møller made in New York on 19 Sept 2005 and the short article by P.E.D. Kristensen, then Ambassador of Denmark to Canada, published in the *Ottawa Citizen* on 28 July 2005, which place the dispute in the proper perspective of the good and ongoing cooperation between the two states.

1.3 Arctic Council

The first stage of Arctic-wide cooperation started with the 1991 Arctic Environmental Protection Strategy (AEPS), which was adopted in Rovaniemi by the eight Arctic States (AEPS 1991). In the Strategy, six high-priority environmental problems facing the Arctic were first identified (persistent organic contaminants, radioactivity, heavy metals, noise, acidification, and oil pollution) as well as international environmental protection treaties that apply in the region, and specific actions to counter the threats. Interestingly, the AEPS stated that "The implementation of the Strategy will be carried out through national legislation and in accordance with international law, including customary international law as reflected in the [LOS Convention]" (AEPS 1991). As part of the environmental protection action by the eight Arctic States, four environmental protection working groups were established: Conservation of Arctic Flora and Fauna (CAFF), Protection of the Arctic Marine Environment (PAME), Emergency Prevention, Preparedness, and Response (EPPR), and AMAP. Three ministerial meetings (following the signing of the Rovaniemi Declaration (1991) and the AEPS) were held in this first phase of Arctic cooperation, generally referred to as 'AEPS cooperation' (Koivurova and VanderZwaag 2007).

The Arctic Council was established between the eight Arctic states through the Ottawa Declaration in 1996 to enhance Arctic cooperation. The establishment of the Arctic Council broadened the mandate of cooperation to all common issues facing the Arctic (excluding matters related to military security), especially those relating to environmental protection and sustainable development. The four environmental protection working groups of the Strategy were integrated into the structure of the Council, and one new working group was established (the Sustainable Development Working Group (SDWG)). In the absence of a permanent secretariat—although by means of the Nuuk Declaration in 2011 the Minsters agreed to establish a permanent secretariat in Tromsø, Norway—the work of the Arctic Council has been heavily influenced by the priorities that the chair-state lays out for its two-year chair period, at the end of which a ministerial meeting is organized. Senior Arctic Officials (SAOs), a group of high-level officials, guide the work of the Council in between the ministerial meetings. The Arctic Council has also adopted new programmes related to environmental protection, such as the Arctic Council Action Plan to Eliminate Pollution in the Arctic (ACAP), which was recently turned into a sixth working group (re-titled the Arctic Contaminants Action Program), and commissioned the Arctic Climate Impact Assessment (ACIA; ACIA 2005).

One unique aspect of the Arctic Council is the role it gives to the region's indigenous peoples: They are normally accorded the status of non-governmental organizations (NGOs) in different intergovernmental organizations and forums, but the Arctic Council defines them as 'permanent participants', a distinct category of membership between members proper and observers, whom the Arctic Council member states must consult prior to any consensus decision making. The group of observers is large, and consists of intergovernmental and non-governmental organizations as well as states that are active in the Arctic region (Koivurova and VanderZwaag 2007).

The Arctic Council is engaged in various kinds of activities related to the Arctic marine area, especially through its AMAP and PAME working groups, but to some extent also through CAFF, which has marine projects. The Council has produced many important scientific assessments following the dramatic findings of the ACIA, which have played important roles in governing the Arctic marine area. It conducted a comprehensive assessment of Arctic marine shipping, which led to the Arctic Marine Shipping Assessment (AMSA) that was adopted in the 2009 Ministerial Meeting and contains policy recommendations. One of these recommendations urged the Council members to act in concert to push for comprehensive, stringent, and mandatory rules on shipping in extreme polar conditions. The work to convert the non-binding Polar Shipping Guidelines (2009) into mandatory measures is now in progress under the aegis of the International Maritime Organization (IMO).[4] The Council has also sponsored the making of an Oil and Gas Assessment and is in the process of completing the Arctic Biodiversity Assessment, which will play a role in evaluating the effectiveness of conservation policies. Moreover, an instrument that was adopted during the AEPS, the Arctic Council's Offshore Oil & Gas Guidelines (PAME 2009), has been updated already two times and the most recent version endorsed at the 2009 Ministerial Meeting. Currently, the Arctic Ocean Review process is examining gaps in Arctic marine governance and will make recommendations on routes to be followed by member states and other stakeholders in their future Arctic marine policy and law.

Until recently, Arctic cooperation functioned for over fifteen years in a fairly similar and consistent mode of operation. Yet in response to alarming climate change, the Council has recently strengthened the way it functions. In May 2011, the ministers decided to establish a permanent secretariat and adopted the first ever legally binding instrument, the Arctic Search and Rescue (SAR) Agreement (2011), which marks a change in the Council, using treaties as ways of reaching policy goals in marine areas. The agreement is meant to strengthen search and rescue coordination and cooperation efforts in the Arctic by allocating responsibilities to each Arctic state and by establishing procedures for states to cooperate in cases of emergency. There is also an ongoing process to conclude an Agreement on Cooperation on Marine Oil Pollution Preparedness and Response (MOPPR), which is scheduled to be signed during the May 2013 Ministerial Meeting. Like the SAR Agreement, the special scope of the agreement goes beyond the Arctic

---

[4] To be more specific, AMSA's recommendation was for "updating and mandatory application of relevant parts of the Guidelines for Ships Operating in Arctic Ice-covered Waters (Arctic Shipping Guidelines)". Those were adopted in 2002, and since then the AMSA report was published in April 2009 and the 'Guidelines for ships operating in polar waters' were adopted by the 26th IMO Assembly in November–December 2009); the AMSA recommendation referred to the earlier 2002 Guidelines.

Ocean; the MOPPR Agreement will likely apply not only in Arctic Ocean waters, but also in the Baltic Sea (Gulf of Bothnia), and may also have a few legally non-binding appendices (e.g., a manual on emergency response). Both treaties are firmly anchored in broader agreements already in existence.

Although the Arctic Council is not a treaty-based organization, it seems to have gradually strengthened its ways of conducting policy, which many describe as move from a policy *shaping* body to policy *making* one. The Council continues to do important scientific work also in relation to marine areas, but is getting stronger in terms of its institutional structure and ways of doing policy via legally binding agreements, all adding to the Council's capacity to respond effectively to the challenges of climate change in the region.

1.4 Arctic Policies of the EU and US

1.4.1 EU Arctic Policy and Competences

As global attention has turned towards the environmental and geopolitical changes in the Arctic, the EU has taken an increasing interest and more active role in developing its own Arctic policy. Beginning in 2007 with its Integrated Maritime Policy (European Commission 2007a) and Action Plan for Integrated Marine Policy (European Commission 2007b), the European Commission drew attention to Arctic issues and called for preparation of a report on Arctic Ocean strategic issues. In March 2008, a paper from the European Commission and the High Representative called for the development of an EU Arctic policy, highlighting the increasing geopolitical importance of the Arctic resulting from the melting of the Arctic sea ice and increased accessibility of Arctic waters (High Representative and European Commission 2008).

More concrete Arctic policy development was initiated with the European Parliament's resolution in October 2008 on Arctic governance (European Parliament 2008) and the November 2008 European Commission Communication on 'The European Union and the Arctic Region', which set out proposals for a coordinated Arctic approach for the EU (European Commission 2008). In 2009, the Council of the European Union adopted 'Council conclusions on Arctic issues' which welcomed the gradual formulation of an EU Arctic policy based upon effective mitigation of climate change, multilateral governance, the LOS Convention, maintaining the Arctic as an area of peace and stability, and formulating EU actions and policies that respect the sensitivities of Arctic ecosystems and biodiversity and the rights of indigenous peoples (Council of the European Union 2009). The January 2011 European Parliament's resolution on 'Sustainable EU policy for the High North' affirmed the EU's Arctic interests and commitment to developing policies based on best scientific knowledge, and stressed a need for a coordinated EU policy (European Parliament 2011).

Most recently, in June 2012, the EU Commission and the High Representative for Foreign Affairs and Security Policy adopted a Joint Communication on 'Developing a European Union Policy towards the Arctic Region: progress since 2008 and next steps'(European Commission and High Representative 2012). The Joint Communication sets out the case for a refined policy and increased EU engagement in Arctic issues based on knowledge, responsibility to achieve sustainable development, and engagement with Arctic states, indigenous peoples, and other partners. The Communication provides a continuation of the 2008 Communication, emphasizing the importance of combating climate change, funding research, supporting indigenous peoples, maritime safety, sustainable economic development, and multilateral cooperation. The EU considers the Arctic Council to be the region's primary forum for international cooperation and considers the LOS Convention to be a basis for the management of the marine Arctic.

As the development of an Arctic EU policy has gradually unfolded, certain areas have been consistently highlighted:

- Promoting sustainable development and resource use,
- Protecting and preserving the Arctic in unison with its population,
- Contributing to enhanced Arctic multilateral governance and international cooperation,
- Supporting research and knowledge, and
- Commitment to combating climate change

In December 2008, the European Commission applied to become a permanent observer to the Arctic Council. The decision was postponed in 2009—a move attributed to Canada in response to the EU's ban on seal products, as well as reluctance by Russia. Following the adoption of new criteria for the admission of observers in May 2011, updated information was submitted by the Commission in December 2011 and a decision will be made regarding the EU's status as an observer at the 2013 Ministerial Meeting in Kiruna.

The fact that none of the current EU Member States are coastal states with respect to the Arctic marine area, as defined in this book, is clearly a major feature and constraint of EU policy regarding the Arctic marine area. Currently, three Arctic states—Denmark, Finland, and Sweden—are members of the EU. Iceland applied to become a member of the EU in June 2009 and is now a candidate country. Greenland and the Faroe Islands are not a part of the EU, although Greenland is a member of the Overseas Countries and Territories Association (Koivurova et al. 2010). It should nevertheless be noted that EU law has considerable impact on contracting parties to the EEA Agreement (1993), which includes Iceland and Norway and requires implementation of certain EU legislation related to the common market. Up until now, Norway has decided not to extend the applicability of the EEA Agreement to Svalbard (Koivurova et al. 2010).

While neither the EU nor its Member States can act as coastal states with respect to the Arctic marine area, they can still act in a wide range of other capacities: for instance as flag states, port states, market states, or with respect to their natural and legal persons. In a flag state capacity, EU Member States are able to exercise their rights and discharge their obligations with respect to the Arctic marine area, most notably the freedoms of the high seas in the high seas pockets in the Arctic marine area, the navigational rights and freedoms in the maritime zones of coastal states to the Arctic marine area, and obligations relating to marine living resources and the marine environment connected to these rights and freedoms.

The competence of the EU and its Member States regarding the Arctic marine area is determined by general international law as well as by EU law. It goes without saying that EU Member States cannot confer more extensive competence to the EU then they themselves possess in accordance with international law. Competence between the EU and its Member States is distributed based upon first the Treaty on the Functioning of the European Union (TFEU 2008), the EU Treaty (2010), and other treaties concluded within the framework of the EU.

Most of the EU's sectoral competences that are relevant to the marine Arctic fall under the shared competence between the EU and its Member States. 'Transport' and 'environment' are among the areas listed in Article 4(2) of the TFEU where the EU and its Member States share competence. One of the clearest changes brought by the TFEU is that energy is now also explicitly listed as a shared competence between the EU and Member States. The main exception to these shared competences is the conservation of marine biological resources under the common fisheries policy, an exclusive competence for the EU under Article 3(1) of the TFEU.

1.4.2 US Arctic Policy

For the US, like the EU, the loss of sea ice and prospects of increased activity in the Arctic have heightened interest in the region's future. Unlike the EU, however, the US, by virtue of the state of Alaska, is an Arctic Ocean coastal state.

The US's current Arctic policy was adopted in January 2009 (NSPD-66 2009). The 2009 Presidential Directive, under President George W. Bush, was the US's first official statement on the Arctic since 1994. The 2009 Arctic Policy highlights the national and homeland security interests of the US foremost, but also awards significant attention to protection of the Arctic environment, sustainable development, and regional cooperation. It sets forth the interests of the US in the region and a six-point policy to:

- Meet national security and homeland security needs relevant to the Arctic,
- Protect the Arctic environment and biological resources,
- Ensure sustainable natural resource and economic development,
- Strengthen institutions for cooperating among the Arctic nations,
- Involve indigenous communities in decision making, and
- Enhance scientific monitoring and research into local, regional, and global environmental issues.

Other issues such as international governance, extended continental shelf and boundary issues, scientific cooperation, maritime transportation, and economic and energy issues are also covered. In its Arctic Policy, the US articulates a commitment to cooperation with other Arctic actors through bilateral and other institutional mechanisms. The Arctic Council is viewed as an important institution for Arctic governance, albeit with the intention that the Council remain largely within its current mandate rather than be transformed into an international organization.

The US is not a party to the LOS Convention, thus limiting its participation under the treaty. The 2009 Arctic Policy recommends Senate ratification of the LOS Convention to serve US security and environmental interests. Despite not being a party to the LOS Convention, in a May 2008 meeting of the 'Arctic five' coastal states (Canada, Denmark/Greenland, Norway, the Russian Federation, and the US) in Ilulissat, Greenland, the US, along with the other states, affirmed a commitment to the law of the sea framework for determining continental shelf claims (Ilulissat Declaration 2008). The US has also taken steps to determine the outer limits of its extended continental shelf in accordance with the LOS Convention in preparation for future accession.

Although the US has not formally updated its Arctic policy since 2009, it is currently developing an implementation plan for the National Ocean Policy that pays special attention to the Arctic as a priority region (see Sect. 1.5.2).

1.5 EU and US Marine Policy

1.5.1 EU Maritime Policy

Ocean and coastal management is an active area of EU policy development. While the economic, social, and cultural importance of marine waters to the EU is well established, there is also an understanding that the intensity and scope of its maritime activities is expanding due to advances in technology and increased demand. In recent years, the EU has clearly identified that it is "at a crossroads in our relationship with the ocean" (European Commission 2007a). Simultaneously, the last several decades have witnessed a growing recognition of the limitations of sector-based maritime policies, owing to the strong interlinkages of matters related to the marine environment. In this context, the EU is implementing a new system of integrated and holistic maritime policies, most notably through the Integrated Maritime Policy (IMP; European Commission 2007a) and its associated environmental pillar, the Marine Strategy Framework Directive (MSFD 2008).

Adopted in 2007, the IMP aims to improve coordination of marine policies among EU Member States so as to overcome the compartmentalization and incoherence that occur across sector-based policies. Its objective is to produce a more integrated and holistic approach to governing EU marine waters that will "enhance Europe's capacity to face the challenges of globalization and competitiveness, climate change, degradation of the marine environment, maritime safety and security, and energy security and sustainability" (European Commission 2007a). As the IMP encompasses a broad spectrum of policy areas, it has been called "the most comprehensive policy ever adopted by the EU" (Koivurova 2009). The IMP has a dual focus on economic development while maintaining environmental sustainability.

Perhaps the most significant instrument implemented under the IMP is the MSFD. The MSFD, adopted in 2008, is a legally binding instrument designed to establish a policy framework within which EU Member States will maintain or achieve "good environmental status" of their marine environment by 2020. The MSFD states that marine policies will use an ecosystem-based approach to the management of human activities, with Member States formulating their own national marine strategies with regards to distinct geographic regions, such as the Baltic Sea. It is important to note that the MSFD is not a harmonizing measure intended to produce a uniform set of standards across all Member States with regards to what constitutes good environmental status. Rather, Member States are required to establish their own marine strategies for their own marine waters. Member States determine what constitutes good environmental status, as well as the optimal methods for achieving that target.

1.5.2 US Ocean Policy

Oceans and coasts play a critical role for the US economy and the quality of life of its citizens. Increased attention is being paid to the need for ocean and coastal management to be integrated and scientifically-based, including using an ecosystem approach, and with the involvement of stakeholders at all stages of development and implementation. In light of this, the US has recently adopted a new National Ocean Policy, an initiative aiming to implement an integrated and holistic approach to maritime policy (Exec. Order No. 13547 2000).

Many of the challenges the US faces in sustainably developing ocean and coastal ecosystems and economies stem from mismatches between the way natural systems work and the way the activities affecting them are managed. Management has been fragmented by an outdated and disjointed collection of laws, institutions, and jurisdictions: At the federal level alone, oceans and coasts are managed under more than 140 different federal laws implemented by a wide range of agencies.

Federal law generally applies to areas of the ocean beyond the 3 nm jurisdiction from shore that most states and territories hold within the 200 nm US EEZ. Coastal lands generally fall under the jurisdiction of states. An important foundation for federal management of oceans and coasts in the US is the public trust doctrine. Under the public trust doctrine, the bottom and water column resources seaward of the land are held in trust by the government, which has a duty to ensure that public interests in those lands are protected.

Attempts to improve US ocean governance coordinating structures have met with varying successes. Both President George W. Bush and President Barack Obama have made concerted efforts to address the need for an overarching national ocean policy. Those efforts recently culminated on July 2010 when the Interagency Ocean Policy Task Force released its final recommendations on a new ocean policy (Council on Environmental Quality 2010). On the same day, President Obama signed Executive Order 13547 establishing a National Policy for the Stewardship of the Ocean, Our Coasts, and the Great Lakes (Exec. Order No. 13547 2010). The Executive Order adopts most of the final recommendations in the Interagency Task Force report and directs executive agencies to implement those recommendations under the guidance of a National Ocean Council. It establishes a national policy to ensure the protection, management, and conservation of US ocean and coastal ecosystems and resources, to respond to climate change and ocean acidification through adaptive management, and to coordinate with national security and foreign policy interests. The Order also provides for the development of coastal and marine spatial plans that build upon existing federal, state, tribal, local, and regional decision making and planning processes, which are intended to pave the way for a more integrated, ecosystem-based, flexible, and proactive approach to managing sustainable multiple use of the oceans and coasts.

Subsequently, in 2012, a draft implementation plan for the National Ocean Policy was released and opened to public comment (National Ocean Council 2012). The implementation plan lays out initial steps required to achieve the Policy's objectives, focusing on specific actions and nine priority objectives. One of these priority areas is on "Changing Conditions in the Arctic" and "addressing environmental stewardship needs in the Arctic Ocean and adjacent coastal areas in the face of climate-induced and other environmental changes". The draft plan focuses on the impacts of climate change in the Arctic region, highlighting opportunities and challenges presented by rapidly diminishing sea ice and resulting human and environmental changes and needs. The plan outlines potential actions and outcomes for improving Arctic environmental response management, observing and forecasting sea ice changes, establishing a network of biological observatories, improving Arctic communication networks and architecture, advancing Arctic marine mapping and charging, and improving (national and international) coordination on Arctic issues.

Submerged lands extending beyond three miles from the US coast are subject to federal government jurisdiction, led by the US Department of the Interior (DOI). DOI develops five-year framework leasing programmes for the outer continental shelf (Outer Continental Shelf Lands Act 1975, sec. 18). Marine resources and oceans are governed by a variety of other laws including the Coastal Zone Management Act (CZMA 1972), the Marine Protection, Research, and Sanctuaries Act (1972), and the Clean Water Act (1972). The Alaska Coastal Management Act (ACMA 1977) was passed pursuant to the federal Coastal Zone Management Act and created the Alaska Coastal Management Program. Prior to the passage of ACMA, more than 60 % of Alaska's coastal area was controlled by federal agencies.

Compared to many other countries, state governments in the US play prominent roles. State and regional governance are critical elements in managing sectors like marine fisheries and areas such as coastal zones. Localized efforts to improve ocean and coastal management and coordination have developed in a number of

US coastal states as well as multi-state initiatives in regions that share important ocean and coastal ecosystems.[5]

1.6 Conclusion

Climate change is occurring rapidly in the Arctic region, bringing new economic opportunities alongside challenges for environmental governance in the region. Evaluating these changes, and options for effectively addressing them, requires an understanding of existing institutions and frameworks. From the foundation established here, this book further explores the Arctic policy landscape, changing governance needs, and ways to promote a sustainable future.

References

ACIA (2005) Arctic Climate Impact Assessment. Cambridge University Press, New York
ACMA (1977) Alaska Coastal Management Act. AS 46.40
AEPS (1991) Arctic Environmental Protection Strategy, 14 Jan 1991, 30 I.L.M. 1624
AMAP (1997) Arctic Pollution Issues: A State of the Arctic Environment Report. Arctic Monitoring and Assessment Programme, Oslo
AMSA (2009) Arctic Marine Shipping Assessment 2009 Report. Arctic Council, Apr 2009
Arctic SAR Agreement (2011) Agreement on Cooperation on Aeronautical and Maritime Search and Rescue in the Arctic, 12 May 2011, 50 I.L.M. 1119 (2011). Entered into force on 19 Jan 2013
Arctic Shipping Guidelines (2002) Guidelines for Ships Operating in Arctic Ice-covered Waters, IMO MSC/Circ. 1056, MEPC/Circ. 399, 23 Dec 2002
Bartenstein K (2011) The "Arctic Exception" in the Law of the Sea Convention: A Contribution to Safer Navigation in the Northwest Passage? Ocean Dev Int Law 42:22–52
Clean Water Act (1972) 33 U.S.C. 1251 et seq
Coastal Zone Management Act (1972) 16 U.S.C. 1451 et seq
Council of the European Union (2009) Council conclusions on Arctic issues, 2985th Foreign Affairs Council Meeting (8 Dec 2009)
Council on Environmental Quality (2010) Final Recommendations of the Interagency Ocean Policy Task Force. 19 July 2010
Division for Ocean Affairs and Law of the Sea (2013) United Nations. <http://www.un.org/Depts/los/index.htm>. Accessed 31 Jan 2013
EEA Agreement (1993) Agreement on the European Economic Area, 17 Mar 1993. Entered into force 1 Jan 1994
European Commission (2007a) Communication from the Commission to the European Parliament, The Council, the European Economic and Social Committee and the Committee of the Regions: an Integrated Maritime Policy for the European Union, 10 Oct 2007, COM (2007) 575 final

---

[5] Notable state initiatives include: Marine Life Protection Act (California), Territorial Sea Plan (Oregon), Puget Sound Partnership and Marine Waters Planning and Management Act (Washington), Ocean and Great Lakes Ecosystem Conservation Act (New York), Massachusetts Ocean Act (Massachusetts), and the Rhode Island Ocean Special Area Management Plan (Rhode Island).

European Commission (2007b) Accompanying Document to the Integrated Maritime Policy, 10 Oct 2007, SEC (2007) 1278

European Commission (2008) Communication from the Commission to the European Parliament and the Council: the European Union and the Arctic Region, 20 Nov 2008, COM (2008) 763

European Commission and High Representative (2012) Joint Communication to the European Parliament and the Council. Developing a European Union Policy towards the Arctic Region: progress since 2008 and next steps, European Commission and High Representative of the European Union for Foreign Affairs and Security Policy, 26 June 2012, JOIN (2012) 19 final

European Parliament (2008) Arctic Governance. European Parliament resolution of 9 Oct 2008 on Arctic governance, P6_TA(2008)0474

European Parliament (2011) A sustainable EU policy for the High North. European Parliament resolution of 20 Jan 2011 on a sustainable EU policy for the High North (2009/2214(INI)), 20 Jan 2011, P7_TA(2011)0024

Exec. Order No. 13547 (2010) Stewardship of the Ocean, Our Coasts, and the Great Lakes. 10 July 2010, 75 Fed. Reg. 43,023

Fish Stocks Agreement (1995) Agreement for the Implementation of the Provisions of the United Nations Convention on the Law of the Sea of 10 December 1982 relating to the Conservation and Management of Straddling Fish Stocks and Highly Migratory Fish Stocks, 4 Aug 1995, 2167 U.N.T.S. 3. Entered into force 11 Dec 2001

High Representative and European Commission (2008) Paper from the High Representative and the European Commission to the European Council. Climate Change and International Security, 14 Mar 2008, S113/08

Humphreys A (2012) New proposal would see Hans Island split equally between Canada and Denmark. National Post, 11 Apr 2012. Available at: <http://news.nationalpost.com/2012/04/11/new-proposal-would-see-hans-island-split-equally-between-canada-and-denmark/>

Ilulissat Declaration (2008) Arctic Ocean Conference. Ilulissat, Greenland. 27 May 2008, 48 I.L.M. 382 (2009)

Koivurova T (2009) A note on the European Union's integrated maritime policy. Ocean Dev Int Law 40:171–183

Koivurova T, VanderZwaag DL (2007) The Arctic Council at 10 Years: Retrospect and Prospects. Univ British Columbia Law Rev 40:121–194

Koivurova T, Kokko K, Duyck S, Sellheim N, Stepien A (2010) EU Competencies Affecting the Arctic: Directorate General for External Policies of the Union. Directorate B Policy Department Study

LOS Convention (1982) United Nations Convention on the Law of the Sea, 10 Dec 1982, 1833 U.N.T.S. 396. Entered into force 16 Nov 1994

Marine Protection, Research, and Sanctuaries Act (1972) 16 U.S.C. 1401 et seq

Marine Strategy Framework Directive (2008) Directive 2008/56/EC of the European Parliament and the Council of 17 June 2009 establishing a framework for community action in the field of marine environmental policy, June 17 2009, 2008 O.J. (L 164)

Molenaar EJ (2012) Fisheries Regulation in the Maritime Zones of Svalbard. Int J Mar Coast Law 27:3–58

Murmansk Treaty (2010) Treaty between the Kingdom of Norway and the Russian Federation concerning Maritime Delimitation and Cooperation in the Barents Sea and the Arctic Ocean, 15 Sept 2010, U.N.T.S. Reg. No. 49095. Entered into force 7 July 2011

National Ocean Council (2012) Draft National Ocean Policy Implementation Plan, 12 Jan 2012. <http://www.whitehouse.gov/sites/default/files/microsites/ceq/national_ocean_policy_draft_implementation_plan_01-12-12.pdf>. Accessed 5 Feb 2013

NSPD-66 (2009) National Security Presidential Directive and Homeland Security Presidential Directive. Arctic Region Policy, 9 Jan 2009, NSPD-66/HSPD-25

Nuuk Declaration (2011) Nuuk Declaration on the occasion of the Seventh Ministerial Meeting of the Arctic Council. 12 May 2011

Ottawa Declaration (1996) Declaration on the Establishment of the Arctic Council, 19 Sept 1996, 35 I.L.M. 1382 (1996)

Outer Continental Shelf Lands Act (1975) 43 U.S.C. 1331 et seq

PAME (2009) Arctic Offshore Oil and Gas Guidelines. Last updated 29 Apr 2009

Part XI Deep-Sea Mining Agreement (1994) Agreement relating to the Implementation of Part XI of the United Nations Convention on the Law of the Sea of 10 Dec 1982, 28 July 1994. 1836 U.N.T.S. 3. Entered into force 28 July 1996

Polar Shipping Guidelines (2009) Guidelines for Ships Operating in Polar Waters, IMO Assembly Resolution A.1024(26), 2 Dec 2009

Reagan PR (1983) President Ronald Reagan, 'Statement on United States Ocean Policy' 19 Weekly Comp Pres Doc 383. 10 Mar 1983

Roach JA, Smith RW (1996) United States Responses to Excessive Maritime Claims, 2nd edn. Martinus Nijhoff Publishers, The Hague

Rovaniemi Declaration (1991) Declaration on the Protection of the Arctic Environment, 14 June 1991, 30 I.L.M. 1624

Spitsbergen Treaty (1920) Treaty concerning the Archipelago of Spitsbergen, 9 Feb 1920, 2 L.N.T.S. 7. Entered into force 14 Aug 1925

TFEU (2008) Treaty on the Functioning of the European Union. Consolidated Version. O.J. C 115/47

Treaty on European Union (2010) Consolidated Version of the Treaty on European Union, 2010 O.J. C 83/01

United States Department of State (1992) Limits in the Seas. No. 112. United States Responses to Excessive Maritime Claims. Available at <http://www.state.gov/documents/organization/58381.pdf>. Accessed 5 Feb 2013

Chapter 2
The Arctic Marine Environment

Arne Riedel

Abstract The following sets the scene for subsequent chapters of this book: It presents a descriptive overview of the Arctic environment, demonstrating the scope of this term and highlighting unique environmental features of the circumpolar region, while focusing on the marine environment in general and the transatlantic region in particular. The global and regional threats for the Arctic environment are presented to establish a basic understanding of the evolving and increasing risks that this relatively pristine area encounters already, and those yet to come. Next to the primary global threat of climate change—bringing with it increasing sea ice loss, ocean acidification, thawing permafrost, and melting glaciers—developments in the areas of pollutants and chemicals, natural resources, shipping, fisheries, tourism, and military activities show increasing impacts on the Arctic environment.

2.1 Introduction

The Arctic is a vast, ice-covered ocean, surrounded by tree-less, frozen ground, that teems with life, including organisms living in the ice, fish and marine mammals, birds, land animals and human societies (NOAA 2012).

This quote by the United States (US) National Administration for Oceanic and Atmospheric Research (NOAA), describes the Arctic environment succinctly: The Arctic environment is unique. Additionally, distinctive and fragile Arctic ecosystems are increasingly threatened, but what exactly is at risk? What are these risks? And why does it matter?

This chapter is based on previous publications by Ecologic Institute within the Arctic TRANSFORM project and EU Arctic Footprint and Policy Assessment (Cavalieri et al. 2010).

A. Riedel (✉)
Ecologic Institute, Pfalzburger Strasse 43/44, Berlin 10717, Germany
e-mail: arne.riedel@ecologic.eu

E. Tedsen et al. (eds.), *Arctic Marine Governance*,   21
DOI: 10.1007/978-3-642-38595-7_2, © Springer-Verlag Berlin Heidelberg 2014

The Arctic marine area is a place where local communities fight to maintain traditional livelihoods, as well as where the adjacent circumpolar states—and now others—have vested interests. Although the environmental conditions of the Arctic still exhibit extreme, and often challenging, variations in light and temperature, the quickly diminishing ice cover in the polar region is opening up new and increased opportunities for economic activities, particularly during extended periods of daylight in the long northern summers. This includes activities such as hydrocarbon exploration and development, shipping across the (at this time) two possible shipping transport routes (the Northern Sea Route and Northwest Passage), and the possibility of increased fisheries activity. As considered throughout this book, these developments should be tackled with governance approaches tailored to the region's particular requirements and environmental conditions.

Recent scientific assessments describing the regional Arctic landscapes and Arctic ecosystems[1] cover a level of detail that can be summarized only briefly within the limited scope of this chapter. Instead, the chapter's main focus is on setting the scene for the following discussions on governance by providing an overview of aspects to be taken into consideration when considering specific measures and programs in the Arctic. This includes a presentation of the Arctic environment that highlights circumpolar environmental features and the threats they are facing today, focusing on the marine environment, with its links to coastal areas and the mainland where relevant. The impacts on indigenous and other local communities are elaborated on in detail in Chap. 4.

This chapter first gives a brief overview of relevant environmental aspects, with a focus on marine habitats, and highlights distinct geographical particularities within the Arctic region where appropriate (Sect. 2.2). It then turns towards the environmental challenges and threats the region face (Sect. 2.3). Finally, the conclusion (Sect. 2.4) stresses the most critical aspects presented.

2.2 The Arctic Environment

The marine and terrestrial area covered by the term 'Arctic' has been defined in several different ways. The Arctic Monitoring and Assessment Programme (AMAP) working group of the Arctic Council has conducted elaborate studies in the Arctic region and listed existing definitions of the Arctic using geographical, geophysical, and political definitions (AMAP 1998; see also Chap. 1, Sect. 1.2). AMAP's definition of the Arctic is based on the terrestrial and marine areas north of the Arctic Circle (at 66°32′N), with certain exceptions, namely the widening

---

[1] Including, *inter alia*, AOR 2011, summarizing existing Arctic Council assessments. See also AMAP 2011b and IASC et al. 2011. The most comprehensive overview yet will be provided in the upcoming Arctic Biodiversity Assessment (ABA) scientific report, to be completed in 2013, online at <http://www.caff.is/aba > . Accessed 31 Jan 2013.

of the area to 62°N in Asia and to 60°N in North America, thereby including the marine areas north of the Aleutian islands between Russia and the US, and the Canadian Hudson Bay, as well as parts of the North Atlantic Ocean, including the Labrador Sea.

The Arctic region therefore includes the Arctic Ocean basin, which is surrounded almost entirely by land masses, with straits towards the Atlantic and Pacific Oceans. Arctic coastal land is split within five states' jurisdictions. The largest Arctic Ocean coastal states are Russia and Canada, covering together around 9.5 million km^2 (about 70 %) of the Arctic land mass (AMAP 1998). The other regions, from east to west, are the Euro-Barents region (Norway), Greenland, Northern Canada, and the Bering region, including Alaska and the Pacific coast of Russia.

Biodiversity in the Arctic experiences threats from both inside and outside the region. Bird species, fish stocks, and marine mammals in the Arctic environment are not only linked with external systems by their migration patterns, but Arctic habitats are also affected by global ocean currents and pollution transport into the region (CAFF 2010; see also below Sect. 2.3.2). In return, the Arctic environment plays an important part in the biological, chemical, and physical balance of the planet (CAFF 2010, see also below Sect. 2.3.1).

2.2.1 Marine Environment

Nearly two thirds of the Arctic region is covered by ocean waters. These waters are an important part of the global climate system, due to their influence on deep ocean currents and global circulation of the oceans (ACIA 2005). Winds and precipitation also play an important role in mixing warmer waters from the south with colder Arctic waters; this may be subject to change in a warming climate. Arctic ecosystems depend on these interactions and can be highly vulnerable, although it has not been determined with full certainty how potential climatic changes will affect the Arctic environment or if they will lead to a net warming or cooling (ACIA 2005).

The Arctic Ocean provides for diverse habitats, on the surface as well as in the water column, and in open waters as well as in coastal areas. The last ice age and its glacier formations led to a loss in biodiversity that—in combination with extreme climate variability between seasons, sea ice coverage, little solar radiation, and constantly low water temperatures—has resulted in a unique and varied maritime ecosystem. The Arctic Council's PAME (Protection of the Arctic Marine Environment) working group has identified as many as 17 Large Marine Ecosystems (LMEs), described as large ocean areas sharing fundamental oceanographic characteristics (AOR 2011). From a geological point of view, marine ecosystems in the Arctic are rather young, and Arctic food webs can be characterized in a relatively simple structure: The marine food web in the Arctic is based on primary production of algae that is consumed by zooplankton, which is in turn eaten by fish, which are consumed by seabirds and mammals, including humans (ACIA 2005). Despite its simplicity, the functioning of the food web is critically linked to timing. For instance, algal blooms are sensitive to temperature and sea ice retreat, with implications for the entire food web (ACIA 2005).

Box 2.1 The Arctic Ocean and global oceanic currents

The Arctic marine region includes the Arctic Ocean and the surrounding regional and shelf seas, thereby representing an area of approximately 20 million km^2 (AMAP 1998). The Arctic Ocean consists of two basins— Eurasian and Canadian, divided by the transpolar Lomonosov Ridge. The Arctic Ocean is connected to the Pacific Ocean via the Bering Strait between the US and Russia, and to the Atlantic Ocean via the Nordic Seas (Greenlandic Sea, Icelandic Sea, Norwegian Sea) to the East of Greenland and through Davis Strait and Hudson Strait to the West of Greenland.

Currents from both the Atlantic and Pacific Ocean bring water into Arctic waters-warmer waters from the North Atlantic Current via the Fram Strait and the Barents Sea, and comparatively cooler water via the Bering Strait. Due to the narrow and shallow access of the Pacific through the Bering Strait, most of the water in the Arctic ocean originates from the Atlantic, in a ratio of about 80:20 (AMAP 1998).

The Arctic Ocean's vertical water structure is also influenced by these different influxes: Its cold surface waters are divided into a Polar Mixed Layer with low salinity (down to 30–50 m of depth) and a water column of increasing temperature and rising salinity (halocline, down to 200 m of depth) which differs for the incoming Pacific and Atlantic waters. The halocline generally insulates the upper layers from the warmth stored in Atlantic waters in intermediate depths (200–900 m) and thereby also influences the sea ice cover. When the density of the incoming water masses increases (by temperature cooling and rising salinity), water sinks and flows back via the East Greenland Current and the Canadian Straits, in a ratio of about 75:25 (AMAP 1998).

On the ocean surface, sea ice is a dominant feature in the Arctic marine area (CAFF 2001). Almost half of the Arctic Ocean is covered by a permanent ice cap, which grows and shrinks seasonally with maximum cover in March and minimum cover in September (on changes and threats, see Sect. 2.3.1.1). Sea ice determines physical properties, such as exchange of heat between the atmosphere and ocean and light availability, and provides a unique habitat for Arctic species (ACIA 2005). Sympagic organisms live on or immediately below the sea ice and include primary and secondary species dependent on sea ice, with thicker sea ice supporting more complex sympagic communities. They support pelagic ecosystems in the water column in the open ocean, as well as benthic ecosystems on the ocean floor (Molenaar et al. 2008).

Ice algae develop during spring and throughout summer as light becomes available in the polar region. Polar cod, which provide a key link between zooplankton and marine mammals, live in both sea ice and pelagic environments. Nesting seabirds, such as ivory gulls, feed on polar cod and other small fish and zooplankton at the ice edge. Marine mammals, such as the polar bear, walrus, ringed and bearded seals among others, and also whales, such as narwhales, belugas, and bowhead whales, depend on the sea ice for food and survival (CAFF 2010).[2] In the water column and on the seabed, fish, crustaceans and—again—marine mammals also find their habitats, including Atlantic cod, haddock, Alaska pollock, Pacific cod, and the Arctic spider crab (ACIA 2005).

A variety of ocean depth levels adds regional diversification. In the Arctic region, extended continental shelves, particularly along the Russian coastline (see also Chap. 1 for the ongoing claims by circumpolar states) result in rather shallow coastal waters. Across the North Pole, on the other hand, deep sea plateaus are cut by oceanic rims, providing for an entirely different environment. Biodiversity is clustered in areas of higher productivity with warmer waters, especially in the Barents and Chukchi Seas and the Bering Shelf, which host migratory seabirds, marine mammals, and some of the most important fisheries in the world (ACIA 2005).

Within coastal seas, pelagic and benthic organisms together provide for a wealth of ecosystems. Coastal regions also provide shelter, food, and breeding grounds for birds and mammals alike. Some species, such as the common eider, are dependent on benthic organisms in shallow waters. As this seabird also breeds along a vast range of Arctic coastlines, including among others the Barents region, Iceland, both of Greenland's coasts, wide stretches of the Hudson Bay, the Canadian and US Beaufort sea's coasts, and the Bering strait region with the Aleutian islands—it has been used as an indicator of the health of marine of marine environments (CAFF 2010).

2.2.2 Land-Based Impacts on the Marine Environment

Arctic landscapes cover a wide range of topography from bare rock, mountains and glaciers to swamps, meadows, and lowland plains (CAFF 2010). Wetlands cover almost 70 % of land masses in the Arctic region, contribute significantly to freshwater cycles and the exchange of atmospheric gases, including climate forcers, and provide a habitat for many bird species on their migratory routes (CAFF 2010).

Freshwater ecosystems form an important part of Arctic geography and are directly linked with the marine, saltwater ecosystems of the Arctic Ocean. They also span a range of diverse environments. Even in areas with rather low precipitation, freshwater ecosystems can be found, for instance in Arctic lakes, and include

---

[2] See also indicator Number 10 of CAFF 2010 for more details on each of the mentioned species of "Arctic sea-ice ecosystems".

one of the largest freshwater reserves in the world, the Greenlandic ice shield (see below, Sect. 2.3.1.2, for climate impacts on glaciers).

The Arctic region includes some of the largest rivers on the planet, leading an estimated 4,200 km^3 of freshwater together with about 221 million tons of sediment into the Arctic Ocean (AMAP 1998). Inflow from rivers into the Arctic Ocean represents about 2 % of the overall inflow, which is comparatively more than in other oceans (AMAP 1998).

Inputs by several large river systems, such as the Lena, Ob, and Yenisei in the Russian Arctic also provide pathways for pollutants to enter the Arctic Ocean (see Sect. 2.3.2).

Box 2.2 Biodiversity links between Arctic marine and land environments

Biodiversity thrives in the delta regions of these rivers, as they offer a range of various specific habitats. It is important to note that the marine environment is not entirely distinct from the terrestrial environment, as marine mammals, seabirds, and humans are dependent on both for their survival (ACIA 2005). Several species, including Atlantic salmon and some populations of polar bears (CAFF 2001), are specialized in migration between land/freshwater and oceanic habitats, for breeding and hunting grounds respectively, thereby linking the Arctic marine environment to land-based impacts, such as pollution.

2.3 Specific Threats

The Arctic environment suffers from a range of developments and human activities with increasingly adverse impacts. From a perspective directed towards managing these activities, it is beneficial to identify their respective origins and to classify them accordingly.

- From a global perspective (external), there are two broad areas in particular that exhibit Arctic-specific outcomes: climate change and chemical pollution.
- Focusing on the Arctic region itself (internal), there are five areas and activities that show prospects of rising impacts on the Arctic (marine) environment: natural resources, shipping, fisheries, tourism, and military activities.

2.3.1 Climate Change

The most overarching development with severe impacts on the Arctic environment is global climate change. The direct effects of climate change, as well as secondary effects from the increased use of Arctic marine resources, will significantly impact marine systems. Of the numerous risks to the Arctic environment that climate change presents, four particular environmental changes are highlighted here:

- Reduction of sea ice extent, thickness, and distribution,
- Melting of glaciers,
- Thawing of permafrost soil, and
- Ocean acidification

Warming in the Arctic is linked to global increases of greenhouse gas emissions from anthropogenic sources, which increased by 70 % between 1970 and 2004. Atmospheric carbon dioxide (CO_2) concentrations showed an increase of 35 % since the industrial revolution, while atmospheric methane concentrations more than doubled over that time (IPCC 2007a). Hence, anthropogenic sources account for greenhouse gas levels that by 2005 already exceeded the natural range of the past 650,000 years (IPCC 2007b).

As stated in the 2007 IPCC (Intergovernmental Panel on Climate Change) Fourth Assessment Report (AR4), climate change impacts in the polar regions over the next 100 years "will exceed the impacts for many other regions and will produce feedbacks that will have globally significant consequences". However, precise estimates and detailed understanding of the nature and extent of these impacts are still difficult to predict (IPCC 2007b). Models predict general warming in the Arctic with temperature increases ranging from about 3 °C to 6 °C by 2080, even using scenarios in which greenhouse gas emissions are projected to be lower than they have been for the past ten years (AMAP 2011a).

Box 2.3 Climate change and biodiversity

Species which are specially adapted to the harsh conditions of the Arctic region may suddenly find themselves competing with invasive species, where newly warmed Arctic waters create more widely habitable ecosystems. Changes in migration times and routes of birds and ocean mammals may occur, due both to warming and anthropogenic interference. Invasive parasites and pests can threaten both plant and animal populations (ACIA 2005). Such secondary effects of climate change will add stress on Arctic marine biodiversity.

Sea Ice Reduction

September sea ice extent has been declining during the period of 1979–2010 by an average of 11.5 % per decade (NSIDC 2010). Reduced Arctic sea ice extent, especially during the summer months, will rapidly alter the quality of the entire sea ice ecosystem and is expected to impact the entire Arctic marine food web. Sea ice is an important habitat for many Arctic species, including marine mammals such as polar bears, ringed seals, bowhead whales, belugas, and narwhals (AMAP 2011a; see also Sect. 2.2.1).

Ice-dependent species—both land and sea species—are expected to follow the ice edge as it melts and moves further north; however, the abundance of these species is expected to decline due to the rapid shifts in marine conditions (AMAP 2011a). Walruses are also directly threatened by sea ice loss, as the ice provides additional breeding grounds which are reduced and crowded spaces on the coast provide neither a place to raise young, nor sufficient food sources (Reimnitz et al. 1994).

However, it is important to note that some species, especially commercial fish (e.g., cod and herring in the North Atlantic and walleye pollock in the Bering Sea), are expected to benefit from increases of open water leading to increased productivity (Molenaar et al. 2008).

Seasonally occurring changes in the ice coverage of the Arctic region have becoming increasingly extreme in recent years as compared to available data in the earlier 20th century. The past decade is the warmest on record for global surface air temperature with some Arctic regions growing warmer at an even faster pace (IASC et al. 2011).

Shortly after the IPCC stated in its 2007 report that the Arctic could become ice-free in a business-as-usual scenario around the year 2100, Arctic sea ice extent fell far below these modelled estimates. The summer of 2012 brought the latest record of minimal sea ice extent (and thickness) in the Arctic after an already all-time low in recorded history in 2007: On 26 August 2012, the 5-day running average for Arctic sea ice extent was measured with 4.1 million km^2, almost 1.7 % (or 70,000 km^2) less than in September 2007, with the monthly average of 4.72 million km^2 about 38 % (or 2.94 million km^2) less compared to the 1979–2000 average (NSIDC 2012). The extent remained below the 2007 minimum for a total of 40 days (Fig. 2.1).

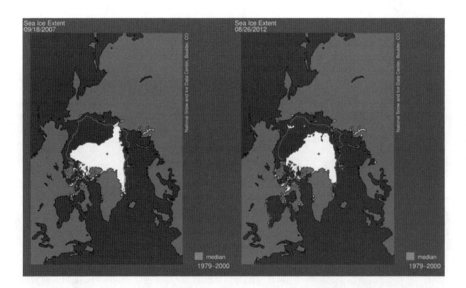

Fig. 2.1 Arctic sea ice extent minima 2007 and 2012 (NSIDC 2012)

Box 2.3 Local warming—the albedo effect

One of the feedback loops identified in relation to Arctic sea ice is the so called albedo effect. It is rooted in the principle that darker surfaces absorb more energy than brighter ones. As sea ice melts, reduction in albedo will likely create a positive feedback effect leading to further global warming (IPCC 2007b): With the opening up of areas in the Arctic Ocean through loss of sea ice cover, the amount of solar radiation absorbed rises, heating up the ocean even more. The higher surface temperatures contribute to further ice melt and slow the growth of new sea ice.

Another albedo-related threat is black carbon. Formed by the incomplete combustion of fossil fuels, biofuels, and biomass—mostly originating outside of the Arctic region—black carbon is a particulate matter with an extent less than 2.5 μm (micrometer) that is emitted into the atmosphere. Its consistency allows it to absorb light, darken surfaces, and thereby increase radiative forcing up to 0.9 W/m^2, second only to CO_2 with an estimated 1.66 W/m^2 (IPCC 2007b; SLCF Task Force 2011). Together with other pollutants that similarly have a powerful, but short-lived climate warming influence, such as tropospheric ozone and the powerful greenhouse gas methane, black carbon is defined as a 'short-lived climate pollutant' (SLCP).

In addition to effects from ice melt and increasing heat absorption, the above-average warming of the atmosphere over the Arctic region can be traced back to other particularities. In contrast to the tropics, for instance, a greater proportion of radiant energy warms the atmosphere above the Arctic, and the Arctic atmospheric layer is shallower (ACIA 2005).

Shrinking sea ice extent also coincides with the loss of ice volume and the loss of multi-year ice (NSIDC 2012). Thin ice melts more rapidly, indicating that the rate of sea ice melt is likely to continue to increase as sea ice continues to become thinner and thicker multi-year ice vanishes (AMAP 2011a). In addition to altering sea ice ecosystems, this ongoing ice loss has the potential to bring dramatic changes to coastal areas. As closed ice cover leads to a degree of protection of coastal lines, these areas will become vulnerable to increasing erosion by waves and storms from open water. A recent study found an average erosion rate of 0.5 m per year for over 60,000 km of sampled coast lines along the Arctic Ocean (Lantuit et al. 2012), with most segments from the Laptev Sea and the Eastern Siberian Sea, followed by the US and Canadian Beaufort Sea. While 89.2 % of the sampled segments (regardless of length) showed erosion rates between 0 and 2 m per year, about 3 % of the length coast lines were found to have a coastal erosion of over 3 m per year.(Lantuit et al. 2012).

Melting Glaciers and Rising Sea Levels

Another impact of climate change observed in recent years is the increasing melt of glaciers, particularly in Greenland. In contrast to sea ice melt, the melt of glacier-bound ice masses contributes to rising sea levels and impacts marine environments by altering the salinity of Arctic waters.

Over 80 % of Greenland's land mass is covered by enormous glaciers with ice sheets as thick as 2–3 km (AMAP 2011a). Average annual ice net mass loss was recently estimated for the years 2005 to 2006 to amount to around 200 (±50) billion tons (AMAP 2011a). In the summer of 2012, satellite imagery showed that more than 97 % of Greenlandic glacier surfaces had begun melting, the largest extent of surface melting observed in three decades of satellite observations.[3] Continued warming will lead to even further melt, although it is as yet unclear to what extent and how fast this will occur. Estimates suggests that total melt of the Greenlandic ice sheet would lead to a potential global sea level rise of 7.5 m (AMAP 2011a).

Rising sea levels can have severe impacts on coastlines and their inhabitants. For land-based species, such as sea birds and marine mammals that rely on bordering marine environments, coastal abrasion by increased erosive forces can lead to the loss of unique breeding spots and feeding grounds (IASC et al. 2011).

There is also concern that significant freshening could impact the thermohaline circulation of the world's oceans, which is a major driver of global weather patterns (IPCC 2007c). With regard to more immediate threats, the freshening of surface layers of the Arctic Ocean could lead to significant changes in the delivery and cycling of nutrients of surface waters, thereby influencing the amount and type of primary production (AMAP 2011a). Sensitive species, such as Greenland Halibut, react to changing salinity conditions in the water by moving their habitat to shelf areas (AMAP 2011a).

Greenhouse Gas Release by Melting Permafrost

In addition to the effects of global warming on oceanic conditions and marine environments, a major feedback loop could be triggered by climate change in land and marine environments. Apart from large areas of land being covered by glaciers (primarily in Greenland and Canada), the upper layer of most land masses of the circumpolar Arctic consists of permafrost—in existing climate conditions, this permanently frozen ground[4] can go as deep as 1,000 m. Permafrost soils provide a generally stable surface that is being used by land animals and birds (as breeding grounds) alike, and have enabled local communities to build necessary infrastructure and housing. It is not limited to upper layers of land but also exists in seabed formations, with the largest hydrate formation under the surface deemed to be on the East Siberian Shelf (AMAP 2011a).

The total coastline affected by permafrost in the northern hemisphere amounts to 407,680 km, thereby representing about 34 % of the world's coastline (Lantuit

---

[3] For images, see NASA Earth Observatory 2012a. Also, in July 2012, a piece of an estimated 120 ± 5 km^2 broke off the Petermann glacier in North-West Greenland that connects the Greenland ice shield with the Arctic Ocean (NASA Earth Observatory 2012b).

[4] The Permafrost Subcommittee of the National Research Council of Canada has established a definition that includes "ground (soil or rock and included ice and organic material) that remains at or below 0°C for at least two consecutive years" (National Research Council of Canada 1988).

et al. 2012). About two thirds of Arctic coastlines consist of permafrost grounds (IASC et al. 2011). These are particularly vulnerable to coastal erosion which is of growing concern due to the decreasing ice coverage of Arctic waters, as highlighted above. In addition, climate change has already brought temperatures at the top of the permafrost layer up by approximately 3 °C since 1980, decreasing the maximum area of frozen ground in the Northern Hemisphere by 7 % since 1900 (IPCC 2007a). The remaining total area in the Northern hemisphere covered by permafrost is estimated to amount to 18.8 million km^2 (Schuur and Abbott 2011).

These increasing temperatures on Arctic land masses have already had severe impacts on living conditions for fauna and flora: Warmer temperatures generally resulted in more greening of land and a shift of habitats of flora towards northern environments, including shrubs and other plants. Reduced permafrost also results in more Arctic wetlands, which release carbon and methane previously contained in the frozen soil into the atmosphere. As regional studies found, the effect of greening Arctic landscapes with greater photosynthetic activity and carbon uptake can offset these releases temporarily, but not completely (AMAP 2011a). Thawing permafrost can also cause an initial expansion of surrounding lakes and groundwater, followed by drainage and disappearance of lakes, which has been detected in Alaska and Siberia (IPCC 2007a).

With regard to the human dimension, thawing permafrost with its consequences for the consistency of the grounds can ultimately lead to the destabilization of building sites and have significant impacts on the coastal lines along the circumpolar shores as well as for indigenous and other local communities (see Chap. 4 for further detail). Permafrost has also sometimes been used as a natural insulation for landfills and containment holding facilities; thawing could now lead to a contamination of ground water and large clean-up costs (IPCC 2007c).

However, the impact of global warming on these frozen grounds goes even beyond the necessity of climate adaptation for communities, as it also releases previously bound carbon compounds, such as carbon dioxide and methane into the air. To stress the importance and possible impact of this reaction: Due to the vast amount of permafrost coverage (see numbers above), it is estimated that worldwide about 1,700 gigatons of carbon are bound in permafrost deposits (Schuur and Abbott 2011)—a number put into perspective by the world's largest emitter's (China's) numbers from 2010, about 7.285 gigatons of CO_2 emitted from fuel combustion (IEA 2012). While current annual emissions from the ice complex along the Siberian coastline, for instance, are estimated at 44 ($\pm$10) megatons of carbon (Vonk et al. 2012), the ever faster cycle of permafrost thawing and subsequent release of stored greenhouse gases is feared to potentially lead towards a climate change 'tipping point', greatly accelerating planetary warming.

Ocean Acidification

The most direct impact of climate change on the Arctic marine environment is caused by acidification. Ocean acidification results from a gas exchange between the oceans and atmosphere, whereby CO_2 dissolves in the water and decreases its pH level. Generally, oceans serve as a sink for CO_2, but increased anthropogenic atmospheric CO_2 concentrations have already led to a higher rate and scope of ocean acidification. This is of particular importance to the Arctic Ocean, since CO_2 is more soluble in cold water. Thus, due to its low water temperature, the Arctic Ocean faces more rapid rates of increasing acidification by absorbing more carbon dioxide than other oceans do (Robbins 2012).

Ocean acidification can lead to a reduction in the diversity and abundance of calcareous organisms, an important marine food source, and thereby affect the rest of the Arctic food chain (CAFF 2010). A combination of the aforementioned effects of climate change further increases the acidification process: The melting of sea ice exposes greater areas of the Arctic Ocean to the atmosphere, and freshwater entering the Arctic Ocean from melting glaciers increases the ocean's potential for CO_2 dissolution, while decreasing its buffering ability.

2.3.2 Chemicals and Air Pollution

Another global threat to the Arctic environment stems from a complex system of interdependencies within air and ocean currents as well as meteorological particularities. The Arctic region has proven to be a sink for pollutants from around the globe, due to its atmospheric conditions: Low air, ground, and water temperatures have a severe impact on the reactivity of chemicals. Once transported into the region—be it via pathways in the air or the ocean, or by riverine discharge—chemicals remain largely in place. The breakdown of chemicals in the Arctic region is slowed down by low temperatures and limited solar radiant. This poses a threat for animals and human beings alike. Accumulation of chemicals over time and rising through the food chain can increase to toxic levels, threatening large predators such as polar bears as well as local communities living on a subsistence lifestyle.

Several kinds of contaminants arrive in the Arctic environment from around the globe. For instance, persistent organic pollutants (POPs) and heavy metals (e.g., mercury) are mainly produced in warmer climates, volatilise, and then spread to the Arctic region through wind, water, and migratory species.

Box 2.4 Transport of contaminants into the Arctic

Air transport is a fast pathway for volatile contaminants and for contaminants that attach to particles. Patterns tend to favour transport of air masses from polluted regions in Europe and Asia during winter months.

Ocean currents continuously exchange water masses from the Arctic Ocean with Atlantic and Pacific waters. Declining ice coverage is likely to

---

cause the Arctic Ocean to emit trapped contaminants back to the atmosphere and increases chemical exchange with the air.

Riverine inputs can also contribute significantly to the flow of pollutants into the Arctic Ocean. Russian rivers lead into Arctic regional seas, including 500,000 t of oil/oily substances every year. The Russian rivers Lena and Ob have carried higher amounts of mercury into the marine environment than atmospheric fluxes (Fisher et al. 2012).

Some non-volatile POPs, including brominated flame retardants (BFRs) for instance, are transported on other particles and thus rely on their transport processes to reach the Arctic. Once in the region, POPs then bio-accumulate in the Arctic marine food web, including humans, and can be stored in layers of ice and permafrost. The latter can lead to so-called secondary emissions, as further melting or thawing could release the POPs now locked in sea ice directly into the food chain (ACIA 2005). For some compounds, such as PCBs and mercury, the levels in some groups of people and wildlife populations are high enough to cause concern about health effects (AMAP 2011b).

Impacts on the environment from contaminants can be severe, as POPs and heavy metals include a number of anthropogenic and natural substances that are toxic to humans and animals under certain circumstances. Populations and ecosystems often experience the impact of several stressors at the same time, which can increase their vulnerability towards them (AMAP 2009).

Climate change is likely to affect both sources and pathways of POPs and mercury through changes in wind patterns or ocean currents and precipitation. Permafrost and glacier melt may also result in higher re-emissions of mercury and other contaminants. However, it is difficult to predict whether long-term climate change will lead to generally increased or decreased loads, as there are processes working in both directions. In terms of affecting long-term levels of and impacts from contaminants in the Arctic, anthropogenic emissions of greenhouse gases may become as important as emissions of the contaminants themselves (AMAP 2004).

Other forms of transboundary air pollution contribute to Arctic haze, a reddish-brown fog in the lower atmosphere at high northern latitudes. It is caused by a mixture of sulphate, black carbon, nitrogen oxides (NOx), sulphur dioxide (SO_2) and other contaminants. These aerosol particles provide a transport pathway for pollution into the Arctic and can also contribute to climate change (ACIA 2005). Black carbon, for example, reduces the albedo of Arctic snow and ice and accelerates warming (see also Sect. 2.3.1 on black carbon).

Industry in and around the Arctic also contributes to acidification and contamination, especially locally. This includes severely contaminated areas with major forest damage around the copper-nickel smelters on the Kola Peninsula and at Norilsk in Siberia (EEA 2007). Highly acidified soils are not able to support plant life.

Box 2.5 Origins of contaminants

Many POPs have been produced for technical applications (e.g., PCBs, BFRs, organic pesticides, PFOs) or are created when the technical products break down (e.g., DDE from DDT). Others are by-products in production of technical products or in various combustion processes (e.g., dioxins and furans).

Mercury and other heavy metals are released via mining, metal processing, or through the respective products. Mercury is also mobilized through coal combustion and also occurs naturally. Re-emissions account for about one third of emissions to the atmosphere and are hard to distinguish from natural sources (e.g., mercury released in forest fires).

2.3.3 Fisheries

The increasing relevance of fisheries activities for the Arctic environment is an indirect impact that can be attributed to climate change. It can be broken down into two aspects:

- Following declining sea ice coverage, Arctic waters provide larger areas for fishing vessels that can also be accessed during longer periods of the season.
- Increasing water temperatures shift inherent temperature gradients for certain species, allowing some to move their habitats further north,[5] but forcing other species to migrate into colder regions or threaten them to become extinct.

These shifts over time add urgency to existing problems in the Arctic marine environment, such as overfishing. Albeit providing only a mere 2.6 % of global fish catches (Rudloff 2012), Arctic fish stocks provide for a substantial part of the European Union (EU)'s supply: For instance, for Norwegian and Icelandic fisheries, the EU was by far the major export destination with 80 and 60 % of fish catches respectively (Rudloff 2012).

The negative impacts of overfishing are numerous. Overfishing can reduce the size of the stock not only temporarily, but can also distort its age structure, for instance by reducing the number of adult fish to an extent which threatens the longer term viability of a stock. As a result of continuously unsustainable quotas or non-compliance with existing quotas by illegal, unregulated, and unreported (IUU) fisheries, stocks of North American cod and Alaska pollock in the Central Bering Sea faced depletion already in the early 1990s (Burnett et al. 2008). More recent data shows that more than half of the Northeast Atlantic regional stocks of cod, haddock, whiting, and saithe are threatened with collapse (UNEP 2005).

---

[5] See, for instance, how the increase in mackerel stocks' abundance rapidly increased Icelandic catches of this species from 2005 on (European Commission 2012). Following the increase of Icelandic quotas, the EU is considering sanctions on Icelandic fishing boats (so called "Mackerel Wars"). See recently Davies 2013.

Apart from impacts to respective fisheries industries, collapsing fish stocks may also have enormous consequences for the Arctic marine environment, the species and ecosystems of which are delicately balanced. Food chains in Arctic ecosystems are usually very simple. Hence, a disruption of a single link in the food chain—for instance, by over-exploitation of stocks—could severely affect the rest of the system. The aforementioned impacts on water temperatures could, however, also lead to beneficial results: Moderate temperature increases are likely to benefit some commercial fish stocks that are currently threatened as well as increase habitat for some species such as cod and herring (IPCC 2007c).

The overall net effect on Arctic fish stocks and commercial fisheries is still uncertain. For one thing, practical approaches for the integrated management of fisheries and the adaptation of management structures will play a role as the effects of climate change continue to emerge (ACIA 2005). Possible conflicts with regional fisheries in the Arctic could arise from the mere physical interference with or from environmental contaminants caused by economic activities, such as transportation and the exploration and exploitation of natural (offshore) resources.

2.3.4 Shipping

The risks from increased shipping in the Arctic region may be considered as indirect effects of climate change. Traditionally, shipping in harsh Arctic conditions and weather (including limited visibility and navigation, sea ice coverage, and a lack of natural light), has been limited to supply vessels for regional settlements, as well as research and fishing vessels in some areas.

Recent developments, however, show several factors that already lead to increased shipping activities in the circumpolar region:

- Sea ice loss (making routes open longer/shipping safer) makes transports economically more viable
- Exploration and development of (offshore) natural resources entails an increase in traffic to support build up and continuing supply
- Fish stocks migrating north (see above) requires more fishing vessels in the Arctic

The economic implications of these developments are of interest to all Arctic coastal states (AMSA 2009). Reductions in sea ice already allow increased shipping within the Arctic marine area, opening up potential routes and widening the time frame available for shipping along them. The historically inaccessible Northwest Passage, for instance, was for the first time in history navigable in 2007 (ESA 2007), and the seasonally accessible Northern Sea Route, was open for five months in 2011 (Corell et al. 2012). After 34 vessels along the Northern Sea Route in 2011, 2012 saw the highest number of 46 vessels—with over 1.25 million tons of cargo transported a more than 50 % increase compared to the previous year (Barents Observer 2012).

With ongoing exploration and exploitation activities in the sector of offshore hydrocarbon resources (see also Sect. 2.3.5), the supply of drilling wells and their maintenance goes hand-in-hand with increased shipping activities by drilling vessels and support barges.

Finally, increasing water temperatures lead to a shift in natural barriers for migratory fish stocks (see Sect. 2.3.3). Following these stocks, fishing vessels will also need to use Arctic waters more frequently and/or for extended periods of time—due to increasing accessibility.

The expected future impacts of increased shipping activities on the Arctic marine environment can be narrowed down to three major influences:

- Operational spills and discharges,
- Accidental spills and discharges, and
- Impacts on marine mammals (including noise)

Increased use of Arctic sea routes is most likely to have positive impacts on shipping, but is yet limited by high operational costs (e.g., due to the need for icebreakers) and other hurdles mentioned above. However, the shipbuilding industry in participating countries could benefit from an increase in demand. On the other hand, the overall increase in traffic holds significant risks for the environment, not only during general operations, but also in the case of accidents or other emergencies (Brigham 2011).

Emergencies can quickly occur in the extreme environments of most Arctic regions throughout the year. The weather and oceanic conditions in the Arctic are generally harsh and difficult to account for. Depending on the season, few hours of daylight may be available—if any—and in areas of broken sea ice coverage larger pieces of floating ice can pose a threat to ships with low or no polar class level constructions. In addition to darkness and climatic conditions, the distances from coastal installations not only make search and rescue operations much more difficult to conduct, but also make clean-up challenging in the case of a spill (AMSA 2009).

Similar conditions, however, can result in different viabilities for shipping routes in Arctic waters: For instance, the scattered Canadian Arctic Archipelago covers a distance of about 2,400 km and leads to several possible "Northwest Passages", the use of which is heavily dependent not only on weather and ice conditions, but also on the respective ship's draft.[6] The Russian regional shelf seas— from the Russian border with the US to Norwegian territory—the Chukchi Sea, the East Siberian Sea, the Laptev Sea, the Kara Sea, and the Barents Sea are characterized in their respective coastal areas by rather shallow water depths, being in between 58 m (East Siberian Sea) and 578 m (Laptev Sea at its northern limit)

---

[6] PAME (2009).

only (AMSA 2009), which influences biodiversity as well as use for shipping activities along the Northern Sea Route.

Recent findings suggest that in the near future (up to 2020), only the Northern Sea Route is expected to become a viable trans-Arctic route, influencing the Bering Strait Region as well. However, an increase in supply traffic is also expected in Canadian Northern Communities (AMSA 2009).

Shipping along these routes has impacts on Arctic marine, air, and coastal environments. It is expected to negatively impact migratory marine mammals that use these routes, as well as increase the risk of oil spills (ACIA 2005). Shipping also contributes to the degradation of air quality from the release of carbon monoxide, nitric oxide, and other chemical substances from ships' combustion engines (Granier et al. 2006). During summer months, surface ozone concentrations in the Arctic could be enhanced by two or three times in the next decades as a result of ship operations through the northern passages (Granier et al. 2006).

2.3.5 Oil and Gas Extraction

Parts of the Arctic environment also face threats from future exploratory drilling for and exploitation of oil and gas resources. According to a much cited US Geological Survey study, approximately 13 % of the world's undiscovered oil resources, 30 % of the world's undiscovered gas resources, and 20 % of undiscovered natural gas liquids are estimated to be located in Arctic region, about 84 % of which are located offshore (Gautier et al. 2008). Direct environmental impacts in the Arctic from energy extraction, including drilling, infrastructure development, and possible accidents, pose a number of threats to ecosystems and communities. Oil pollution poses a particular threat to the fragile Arctic marine environment, which recovers slowly due to low temperatures. Natural recovery from spills is slower due to shorter growing seasons and slower growth rates.

Generally, risks can be grouped into:

- Operational risks, such as discharges and emissions from drilling platforms and transport vessels alike, and
- Accidental risks, such as oil spills

Although many remote onshore sites in Arctic regions are—just like offshore sites—dependent on sea routes for supply and transportation, (exploratory and production) offshore drilling itself can have a much bigger impact on marine habitats. The biggest threats from offshore hydrocarbon exploitation to the Arctic marine area are related primarily to accidental risks such as oil spills.

Oil spills can occur during oil extraction, storage, or transportation from subsea exploration or production and poorly maintained infrastructure in sub-sea pipelines. So far, there have been no major oil spills in the Arctic. However, should this happen—especially during winter months—rescue and clean-up actions in case of accidents are difficult to impossible, due to harsh climate conditions, usually isolated drilling locations, and a lack of effective removal methods in remote icy areas.

Although some climatic conditions might assist with clean up—for instance, ice contains oil, making it easier to prevent further spreading and make the removal more effective –limited experience to date with clean-up measures under Arctic conditions has lead to a shortage of best practices.[7] There is also concern that if a spill from an uncontrolled well in an ice-free area occurs late in the Arctic summer, ice conditions could change so quickly as to prevent drilling a relief well until the following year (Schmidt 2012).

Operational risks are also of concern. Possible discharges of produced waters, drilling liquids, or chemicals from drilling and extracting facilities and building new infrastructure to support operations would have all environmental impacts. Oil and gas flaring releases black carbon emissions, which can reduce albedo and thus increase the rate of regional warming within the Arctic (see Sect. 2.3.1). Air pollutants, such as NOx, SO_2, VOCs, CO_2, methane, and particulate matter (PM), are released into the atmosphere by fuel combustion for onsite power generation, well testing, gas flaring, and other operational leaks. These substances contribute to Arctic haze and have the potential to fasten ice melt (see above Sect. 2.3.2).

In addition, noise from oil and gas activities can interfere with marine animals and temporarily displace them from their habitats. Seismic exploration has affected the migration patterns of bowhead whales and reduces the accessibility of indigenous hunters to their game (CAFF 2001). It may also cause polar bears to abandon their dens and thereby increase cub mortality (CAFF 2001). The effects of drilling activity, pipelines, and subsurface installations on marine communities and seafloors vary and depend on the communities present and their level of sensitivity to disturbances. The geological composition of the sea floor appears to recover from exploratory drilling within a year in some cases (Corrêa et al. 2009).

2.3.6 Tourism

Despite all aforementioned changes in and impacts on the Arctic environment, the circumpolar region is still overall regarded as pristine. As a consequence, and following increased interest in 'adventure tourism', the popularity of Arctic tourism has increased greatly over the past two decades. Marine tourism in the Arctic is highly diversified and *inter alia* driven by tourists looking for sightseeing and observing wildlife species in their natural habitats (AMSA 2009). Concerns over the impacts of climate change and the perception that the Arctic environment's

---

[7] See, for instance, the EPPR working group's presentation at the SAO meeting in Haparanda (Bjerkemo 2012).

landscapes and wildlife are endangered are further driving demand for tourism services in these areas.

Tourism activities have increased in land-based hotels as well as on ship cruises. The number of nights spent at hotels in Greenland increased from about 180,000 in 2002 to more than 235,000 in 2008. Svalbard—an Arctic archipelago off Norway's coast—saw numbers rising from around 30,000 in 1995 to 89,000 in 2008 (Emmerson and Lahn 2012). Cruise passengers landing onshore increased in Svalbard from about 37,000 in 1996 to nearly 70,000 in 2003, while the number of cruise ship landing sites nearly tripled between 1996 and 2010 (Evenset and Christensen 2011).

Tourism activities have the potential to impact both land and marine environments in the Arctic region. For example, tourists can cause significant disturbances for nesting or breeding birds and haul-out sites of marine mammals such as ringed seals or walruses. Pathways of tourist groups to 'points of interest' (viewing points on colonies of birds, for instance) can leave Arctic flora trampled down; in addition, litter on the visited sites remains long after the temporary disturbances are gone.

Marine-based tourism accounts for the largest segment of the Arctic tourism industry in terms of numbers of persons, geographic range, and types of recreational activities (AMSA 2009). Due to limited housing capabilities in the high North, as well as the focus on ships being the main means of transport in the Arctic area, the vast majority of tourists visit Arctic regions on cruise vessels. Cruise ship tourism mostly takes place in areas around Greenland, Iceland, Norway (including Svalbard), and Alaska. The impacts on the marine environment mostly pertain to the aforementioned section on shipping. Particular risks arise where vessels used for tourist operations in Arctic waters do not meet needs for protection against floating ice, equipment for confinement of oil, or waste storage capacities.

2.3.7 Nuclear and Radioactive Waste (Including Military Use)

In general, the Arctic is considered a region of "particular vulnerability" to radioactive contamination (AMAP 2010). Despite radioactive particles from nuclear explosions having decreased since the end of atmospheric nuclear testing in 1963, there has been concern that without a nuclear-weapons-free zone agreement, the Arctic could be threatened by nuclear dumping and the expansion of nuclear activities in the Barents Sea region (ADHR 2004).

A large portion of the dumping, from waste and reactors, can be attributed to the Russian Federation—partially inherited from the former Soviet Union—while both the former Soviet Union and the US are largely responsible for pollution from nuclear testing, with France, China, and the United Kingdom also contributing (Bøhmer et al. 2001).

The specific environmental impacts and risks of military and nuclear waste in the Arctic marine environment depend upon the type of waste and containment.

Contamination from radioactive materials can persist for long periods in soils and plants, and may be revealed in higher concentrations further up the food chain. Radioactive contamination poses particular threats to marine ecology and fisheries with risks increasing where waste settles, rather than diluting and dispersing (Nuclear Threat Initiative 2010). Arctic indigenous peoples are also at risk from exposure from radioactive contaminants. Climate change and its impacts on temperature, permafrost, erosion, precipitation, weather events, sea ice, and oceanic circulation could alter radioactive uptake and distribution in the Arctic (AMAP 2010).

Still, studies have demonstrated that no significant amounts of radioactive materials have migrated from dumping sites and that releases from solid radioactive waste have been small and localized (AMAP 2010). Long-term monitoring has demonstrated that radioactivity is declining, but also stressed the need for significant hazardous operations in relation to the management of spent nuclear fuel (AMAP 2010).

In addition, Russian plans for expanding the use of mobile floating nuclear power plants in the Barents region are ongoing—despite safety concerns from environmental groups (Nikitin and Andreyev 2011). This will ensure ongoing discussions about the use of radioactive materials in, and their impacts on the Arctic environment. For instance, an assessment of whether the current international legal framework and safety standards are applicable and appropriate for transportable nuclear power plants (with particular attention given to floating reactors) was recently conducted by the IAEA (IAEA 2012).

2.4 Conclusion

The Arctic environment is indeed unique and at a point in time that requires international attention and continued efforts to address the many challenges it faces.

The continuing loss of sea ice coverage proves to be not only an imminent threat to species' habitats, but may also trigger climate feedback loops that will hasten the change in the region even further. Climatic changes in the Arctic are linked to global changes such as sea level rise and oceanic circulation, with potentially severely adverse affects. Warming water temperatures could shift species' distribution and confront highly specialized Arctic species with competition from invasive species. Pollutants that enter the region via air, ocean, or river pathways have a disproportionate impact on the Arctic and are now being released from decades of deposits in ice and permafrost. Increasing human activities in fisheries, shipping, and the exploration and exploitation of hydrocarbons add even more pressure to Arctic marine ecosystems. Growing numbers of tourists visit the Arctic to see one of the last pristine environments.

This chapter has sought to provide an overview of answers to the questions: What exactly is at risk in the Arctic environment, why is it at risk, and why does it matter? Once familiar with the urgency that lies in protecting the Arctic marine environment, a logical follow up question comes to mind: What do we do about it?

In the following chapters, specific threats will not only be further elaborated upon, but opportunities, instruments, and approaches will be identified and analyzed to begin answering this question. Particular attention is needed to describe both the influence of global and regional environmental developments and possible gaps in the legal and policy framework for the marine Arctic.

While complex questions are often used as excuses to postpone solutions, the Arctic region's complexity and uniqueness should not be invoked at its expense: There is an urgency and need to act. The diversity of habitats and need to adapt to ongoing changes demand quick responses in governance, however, the existing Arctic patchwork of differing national interests, environmental threats, regulatory approaches, and international fora, implies that there are no simple answers.

References

ACIA (2005) Arctic Climate Impact Assessment. Cambridge University Press, New York

ADHR (2004) Arctic Human Development Report. Stefansson Arctic Institute, Akureyri

AMAP (1998) AMAP Assessment Report: Arctic Pollution Issues. Arctic Monitoring and Assessment Programme, Oslo

AMAP (2004) Assessment 2002: Persistent Organic Pollutants in the Arctic. Arctic Monitoring and Assessment Programme, Oslo

AMAP (2009) AMAP Assessment 2009: Persistent Organic Pollutants in the Arctic. Arctic Monitoring and Assessment Programme, Oslo

AMAP (2010) Assessment 2009: Radioactivity in the Arctic. Arctic Monitoring and Assessment Programme, Oslo

AMAP (2011a) Snow, Water, Ice and Permafrost in the Arctic (SWIPA): Climate Change and the Cryosphere. Arctic Monitoring and Assessment Programme, Oslo

AMAP (2011b) AMAP Assessment 2011: Mercury in the Arctic. Arctic Monitoring and Assessment Programme, Oslo

AMSA (2009) Arctic Marine Shipping Assessment 2009 Report. Arctic Council, Apr 2009

AOR (2011) Phase I report (2009–2011) of the Arctic Ocean review (AOR) project. <www.pame. is>. Accessed 1 Feb 2013

Barents Observer (2012, 23 Nov) "46 Vessels Through Northern Sea Route." Barents Observer. 23 Nov. <http://barentsobserver.com/en/arctic/2012/11/46-vessels-through-northern-sea-route-23-11>. Accessed 20 Feb 2013

Bjerkemo O (2012) Recommended Practices in the Prevention of Marine Oil Pollution in the Arctic (RP3)". Presented at the Arctic Council SAO Meeting, 13 Nov, Haparanda. <http://www.arctic-council.org/index.php/en/about/documents/category/404-presentations?download=1494:4-1-recommended-practices-in-the-prevention-of-marine-oil-pollution-rp3-eppr>

Bøhmer N, Nikitin A, Kudrik I, Nilsen T, McGovern MH, Zolotkov A (2001) The Arctic Nuclear Challenge, vol. 3—2001. Bellona Report. The Bellona Foundation.

Brigham L (2011) Russia Opens Its Maritime Arctic. In: Proceedings of U.S. Naval Institute (May), pp 50–54

Burnett M, Dronova N, Esmark M, Nelson S, Rønning A, Spiridonov V (2008) Illegal Fishing in Arctic Waters. WWF International Arctic Programme

CAFF (2001) Arctic Flora and Fauna—Status and Conservation. Conservation of Arctic Flora and Fauna Working Group, Helsinki. <http://library.arcticportal.org/1295/1/AFF%2DStatus%2Dand%2DTrends.pdf>

CAFF (2010) Arctic Biodiversity Trends 2010—Selected indicators of change. CAFF International Secretariat, Akureyri. <http://www.arcticbiodiversity.is/images/stories/report/pdf/Arctic_Biodiversity_Trends_Report_2010.pdf>

Cavalieri S, McGlynn E, Stoessel S, Stuke F, Bruckner M, Polzin C, Koivurova T, Sellheim N, Stepien A, Hossain K, Duyck S, Nilsson A (2010) Arctic Footprint Policy Assessment Final Report. Ecologic Institute, Berlin. 21 Dec 2008

Corell R, Barry T, Earner J, Hislop L, Kullerud L, Melillo J, Nellemann C, Reiersen L, Samsetz J, Pearce F (2012) UNEP Yearbook Arctic 2012. The View from the Top [Draft], UNEP

Corrêa ICS, Toldo EE, Toledo FAL (2009) Impacts on Seafloor Geology of Drilling Disturbance in Shallow Waters. Deep Sea Res Part II: Top Studies Oceanogr 56(1-2):4–11

Davies C (2013) Fishermen Back Sanctions Against Iceland over Mackerel Catch. The Guardian. 6 Jan <http://www.guardian.co.uk/environment/2013/jan/06/fishermen-sanctions-iceland-mackerel-catch>. Accessed 20 Feb 2013

EEA (2007) Air Pollution in Europe 1990–20042/2007. European Environment Agency, Copenhagen

Emmerson C, Lahn G (2012) Arctic Opening: Opportunity and Risk in the High North. Chatham House, Lloyd's. <http://www.lloyds.com/news-and-insight/risk-insight/reports/arctic-report-2012>. Accessed 6 Feb 2013

ESA (2007) Satellites witness lowest Arctic ice coverage in history. European Space Agency. 14 Sept. <http://www.esa.int/Our_Activities/Observing_the_Earth/Envisat/Satellites_witness_lowest_Arctic_ice_coverage_in_history>

European Commission (2012) Icelandic Fisheries: A Review. European Commission, Copenhagen. <http://www.europarl.europa.eu/committees/en/pech/studiesdownload.html?lan guageDocument=EN&file=73431>

Evenset A, Christensen GN (2011) Environmental impacts of expedition cruise traffic around Svalbard. AECO, Oslo. <http://www.aeco.no/documents/Finalreport.pdf>

Fisher JA, Daniel JJ, Anne LS, Helen MA, Alexandra S, Elsie MS (2012) Riverine Source of Arctic Ocean Mercury Inferred from Atmospheric Observations. Nature Geoscience 5(7):499–504. doi:10.1038/ngeo1478

Gautier DL, Bird K, Charpentier R, Houseknecht D, Klett T, Pitman J, Moore T, Schenk C, Tennyso M, Wandrey C (2008) Circum-Arctic Resource Appraisal: Estimates of Undiscovered Oil and Gas North of the Arctic Circle. U.S. Geological Survey (USGS). <http://pubs.usgs.gov/fs/2008/3049/>

Granier C, Niemeier U, Jungclaus J, Emmons L, Hess P, Lamarque J, Walters, S, Brasseur G (2006) Ozone Pollution from Future Ship Traffic in the Arctic Northern Passages. Geophysical Research Letters 33(13)

IAEA (2012) Nuclear Safety Review for the Year 2012, GC (56)/INF/2. IAEA. <http://www.iaea.org/About/Policy/GC/GC56/GC56InfDocuments/English/gc56inf-2_en.pdf>

IASC, LOICZ, AMAP, IPA (2011) State of the Arctic Coast 2010—Scientific Review and Outlook. International Arctic Science Committee, Land-Ocean Interactions in the Coastal Zone Project, Arctic Monitoring and Assessment Programme, and International Permafrost Association. Helmholtz-Zentrum, Geesthacht, Germany. < http://arcticcoasts.org/>

IEA (2012) CO_2 Emissions from Fuel Combustion—Highlights, 2012 edn. <http://www.iea.org/publications/freepublications/publication/CO2emissionfromfuelcombustionHIGHLIGHTS.pdf>

IPCC (2007a) Climate Change 2007: Working Group I: The Physical Science Basis. Contribution of Working Group I to the Fourth Assessment Report of the Intergovernmental Panel on Climate Change. In: Solomon S, Qin M, Manning M, Chen Z, Marquis M, Averyt KB, Tignor M, Miller HL (eds) Cambridge University Press, Cambridge

IPCC (2007b) Climate Change 2007: Synthesis Report. Contribution of Working Groups I, II and III to the Fourth Assessment Report of the Intergovernmental Panel on Climate Change. Core Writing Team. In: Pachauri RK, Reisinger A (eds) IPCC, Geneva

IPCC (2007c) Climate change 2007: Impacts, Adaptation and Vulnerability. Contribution of Working Group II to the Fourth Assessment Report of the Intergovernmental Panel on Climate Change. In: Parry ML, Canziani OF, Palutikof JP, van der Linden PJ, Hanson CE (eds) Cambridge University Press, Cambridge

Lantuit H, Overduin PP, Couture N, Wetterich S, Aré F, Atkinson D, Brown J (2012) The Arctic Coastal Dynamics Database: A New Classification Scheme and Statistics on Arctic Permafrost Coastlines. Estuaries Coasts 35:383–400

Molenaar EJ, Corell R, Koivurova T, Cavalieri S (2008) Arctic Transform—Introduction to the background papers. Arctic TRANSFORM

NASA Earth Observatory (2012a) Satellites Observe Widespread Melting Event on Greenland. NASA Earth Observatory. <http://earthobservatory.nasa.gov/NaturalHazards/view.php?id=78607>

NASA Earth Observatory (2012b, 18 July) "More Ice Breaks off of Petermann Glacier." NASA Earth Observatory. July 18. <http://earthobservatory.nasa.gov/IOTD/view.php?id=78556>

National Research Council of Canada (1988) Technical Memoranmdum No. 142

Nikitin AK, Andreyev L (2011) Floating Nuclear Power Plants. The Bellona Foundation. <http://www.bellona.org/filearchive/fil_fnpp-en.pdf>

NOAA (2012) NOAA Arctic Theme Page. <http://www.arctic.noaa.gov/index.shtml>

NSIDC (2010, 4 Oct) Arctic Sea Ice News & Analysis—Weather and feedbacks lead to third-lowest extent. National Snow and Ice Data Center. <http://nsidc.org/arcticseaicen ews/2010/10/weather-and-feedbacks-lead-to-third-lowest-extent/>

NSIDC (2012, 27 Aug) Arctic sea ice extent breaks 2007 record low. National Snow and Ice Data Center (NSIDC). <http://nsidc.org/arcticseaicenews/2012/08/arctic-sea-ice-breaks-2007-record-extent/>

Nuclear Threat Initiative (2010) Russia: Spent Fuel and Radioactive Waste. <http://www.nti.org/db/nisprofs/russia/naval/waste/wasteovr.htm>

PAME (2009) Names 5–6 different routes along the Canadian Arctic Archipelago

Reimnitz E, Dethleff D, Nürnberg D (1994) Contrasts in Arctic Shelf Sea-ice Regimes and Some Implications: Beaufort Sea and Laptev Sea. Marine Geol 119:215–225

Robbins L (2012) Studying Ocean Acidification in the Arctic Ocean. United States Geological Survey. <http://pubs.usgs.gov/fs/2012/3058/pdf/fs20123058.pdf>

Rudloff B (2012) The EU as fishing actor in the Arctic. Stocktaking of institutional involvement and existing conflicts. 2010/02. Working Paper, FG 2. SWP, Berlin. <http://www.swp-berlin.org/fileadmin/contents/products/arbeitspapiere/Rff_WP_2010_02_ks.pdf>

Schmidt CW (2012) Offshore Exploration in the Arctic: Can Shell's Oil-Spill Response Plans Keep Up? Environ Health Perspect 120(5):194–199

Schuur EAG, Abbott B (2011) Climate Change: High Risk of Permafrost Thaw. Nature 480:32–33

SLCF Task Force (2011) An Assessment of Emissions and Mitigation Options for Black Carbon for the Arctic Council. Arctic Council, May 2011. <http://www.arctic-council.org/index.php/en/about/documents/category/7-working-groups-scientific-reportsassessments?download=13 5:technical-report-of-the-arctic-council-task-force-on-short-lived-climate-forcers>

UNEP (2005) Ecosystems and Human Well-being: Current State and Trends, Vol. 1. The Millenium Ecosystem Assessment Series. UNEP, Washington, DC. <http://www.unep.org/maweb/en/Condition.aspx>

Vonk JE, Sánchez-García L, van Dongen BE, Alling V, Kosmach D, Charkin A, Semiletov IP et al (2012) Activation of Old Carbon by Erosion of Coastal and Subsea Permafrost in Arctic Siberia. Nature 489:137–140

Chapter 3
Environmental Governance in the Marine Arctic

Susanah Stoessel, Elizabeth Tedsen, Sandra Cavalieri, and Arne Riedel

Abstract This chapter presents an overview of the existing institutional and legal framework at the international and regional levels relevant to environmental governance in the marine Arctic. Examples of both formal and informal environmental governance in the Arctic marine area are presented to demonstrate how different approaches can be used to address particular issues. The chapter highlights the complex interplay of approaches applicable at local, regional, and international scales rather than identifying gaps at the sectoral level. It describes the landscape of governance approaches in place in the Arctic to stimulate discussion regarding the future of environmental governance in the marine Arctic. Many existing institutions and governance mechanisms were developed under political and environmental circumstances that were vastly different from today's reality. As the Arctic is undergoing drastic changes that will spark unprecedented activity in the region, a rethinking of existing structures is required to appropriately address emerging issues.

Based on De Roo C, Cavalieri S, Wasserman M, Knoblauch D, Baush C, Best A (2008) Background Paper: Environmental Governance in the Marine Arctic. Arctic TRANSFORM

S. Stoessel · S. Cavalieri · A. Riedel
Ecologic Institute, Berlin, Germany
e-mail: susanah.stoessel@ecologic.eu

S. Cavalieri
e-mail: sandra.cavalieri@ecologic.eu

A. Riedel
e-mail: arne.riedel@ecologic.eu

E. Tedsen (✉)
Ecologic Institute, 1630 Connecticut Avenue NW, Suite 300, Washington, DC 20009, USA
e-mail: elizabeth.tedsen@ecologic.eu

E. Tedsen et al. (eds.), *Arctic Marine Governance*,
DOI: 10.1007/978-3-642-38595-7_3, © Springer-Verlag Berlin Heidelberg 2014

3.1 Introduction

Beyond the biophysical environment, the Arctic marine area is also geopolitically unique. The Arctic Ocean is an almost enclosed ocean basin surrounded by land, including both nation-state territories as well as international waters. Most of the ocean is within the sovereign territory of the five Arctic coastal states—Canada, Denmark/Greenland, Norway, Russia, and the United States (US)—or is part of their Exclusive Economic Zones (EEZs). Under the United Nations (UN) Law of the Sea Convention (LOS Convention 1982), the centre of the Arctic Ocean (in addition to a few other pockets around its perimeter) is defined as the 'high seas'—*de facto* international water that attracts growing interest from a wide range of nations beyond the Arctic borders. European countries, especially the United Kingdom, France, and Germany, as well as China, Japan, and South Korea are interested in science, energy, and transportation in the Arctic. This growing external interest increases the complexity of policy making in the region (Nordregio 2007).

Melting sea ice will not only result in changes to flora and fauna, but will allow unprecedented access for activities such as shipping, fisheries, and exploitation of offshore hydrocarbon resources, bringing additional threats to species and ecosystems, causing significant impacts for local people and their ways of life, and creating new governance challenges. The way in which actors approach these new challenges is an opportunity to learn from past experience and set precedents for effective regional and international governance regimes (Chapin and Hamilton 2009; Arctic Governance Project 2010). Coherent governance structures at the local, regional, and international levels are especially important in the Arctic as the region both strongly impacts and is impacted by global systems.

This chapter presents an overview of the existing institutional and legal framework at the international and regional levels relevant to environmental governance in the marine Arctic. Examples of both formal and informal environmental governance in the Arctic marine area are presented to demonstrate how different approaches can be used to address particular issues. The chapter highlights the complex interplay of approaches applicable at local, regional, and international scales rather than identifying gaps at the sectoral level.

Sector-based regulation is a necessary component of environmental protection and can provide effective management solutions, but does not always adequately address integrated issues and complex environments. Governance frameworks for marine issues have historically been sectoral in nature, resulting in fragmentation, gaps, and inefficiencies (Cavalieri et al. 2011). In many cases, existing institutional structures are ill-suited for the management of shared and connected resources (Millennium Ecosystem Assessment 2005). Certain sectoral issues such as fisheries, shipping, and offshore hydrocarbon extraction are covered in detail in other chapters of this book.

As seen in the literature and recent policy developments, environmental governance increasingly aims to follow a place-based, ecosystem-based approach (Young et al. 2007; see also Box 3.1 below). However, the practical steps needed to implement the principles of ecosystem-based management will undoubtedly vary based on the specific activities and ecosystems to be addressed. As such, it is important to recognise the value of tailoring and of combining multiple approaches to achieve effective governance for the Arctic marine area.

Effective policies for Arctic marine governance are but one part of a broader Arctic policy framework for environmental governance. Although this book focuses on the marine environment specifically, it should be stressed that terrestrial and marine ecosystems are highly interconnected and that governance approaches should take these interactions and interdependencies into account, especially in the coastal zone.

3.2 Environmental Governance

Governance is the continuous process through which society's goals are defined and met (Jachtenfuchs and Kohler-Koch 2004; Juda and Hennessey 2001). It incorporates a wide range of actors, institutions, and mechanisms that "serve to alter and influence human behaviour in particular directions" to meet societal goals and interests (Juda 1999). Governance encompasses laws and regulations, government policies and institutions, market forces, private sector activities, community actions, and more (Rogers and Hall 2003).

In the environmental context, governance refers to the processes, decision making, and mechanisms by which actors and institutions influence environmental outcomes (Lemos and Agrawal 2006). It encompasses a range of political, social, economic, and administrative systems that are in place to develop and manage natural resources and the delivery of ecosystem services at different levels of society (Rogers and Hall 2003). Policy choices made within these systems define and evaluate opportunities, rules, and acceptable behaviour that result in the management and use of natural resources (Juda and Hennessey 2001).

Governance in the Arctic region was summarized by the 2010 Arctic Governance Project:

> Governance systems emerge to address a variety of societal needs, ranging from the production of public goods (e.g., maintaining healthy populations of living resources subject to human harvesting), to avoidance of public bads (e.g., preventing dangerous climate change or the degradation of large marine ecosystems), internalization of externalities (e.g., curbing the spread of contaminants across borders, avoiding the environmental impacts of oil spills), and protection of human rights (e.g., strengthening the right to self-determination of indigenous peoples) (Arctic Governance Project 2010).

Due to the inherent complexity of natural resource use and the particular difficulty of natural resource management where visible boundaries are lacking between marine ecosystems and vast areas of international waters, governance of the Arctic marine environment involves a myriad of approaches for managing human impacts on the environment. Approaches range from targeting a single species, sector, or issue (e.g., oil pollution) to broader cross-cutting strategies. Depending on the context, various actors are involved from the local to the international level, with participation from stakeholders holding diverse perspectives. Also depending on the context, management occurs through legally binding hard law or non-legally binding soft law measures, with varying levels of enforcement. It is important to note at the outset that legally binding approaches are not necessarily preferable to non-binding approaches.

As described later in this chapter, these various approaches to management in the marine Arctic are often interrelated, overlapping, and at times conflicting. National regulations often differ among countries, causing confusion within sectors. There are also differences in the primary objectives of government agencies and international conventions, resulting in conflicting mandates. Management in the marine Arctic will be further challenged by the impacts of global climate change, as access to and distribution of Arctic resources changes. It is impossible to create an exhaustive list of existing approaches, and often multiple approaches are combined in a single management example. However, it is important to recognise that a spectrum of approaches exists, and that a combination of them could provide the foundation for improved, flexible frameworks for good governance in the marine Arctic.

Box 3.1 Ecosystem-based management (EBM)

Martha McConnell and Dorothée Herr (IUCN Polar Programme)

Maintaining the sustainability of the Arctic's natural and economic resources with the region's cultural integrity requires a flexible, adaptive, and interdisciplinary management approach such as ecosystem-based management.

Ecosystem-based management (EBM) has the potential to provide an organizing framework for decision making in the Arctic by balancing potentially competing priorities and interests while supporting ecosystem resilience to maintain ecological functions and services. Such an approach considers the resilience of an ecosystem primary in order to ensure sustainable economic and social systems. EBM recognizes that humans and their activities are an integral part of ecosystems as a whole. More specifically, EBM assesses and monitors the effects of multiple stressors affecting the same ecosystem in order to facilitate efficient and adaptive science-based decisions. Through multi-stakeholder engagement, the process also helps Arctic inhabitants adapt to a changing ecological and socioeconomic landscape. The following definition of EBM is modified from Laughlin and Speer's (2011) use of the term:

Ecosystem-based management is the comprehensive, integrated management of human activities based on best available scientific knowledge about ecosystems and their dynamics, in order to identify and take action on influences which are critical to the health of ecosystems, thereby achieving sustainable use of ecosystem goods and services and maintenance of ecosystem integrity.

Ecosystem-based management:

- Manages geographic areas that are defined by ecological criteria;
- Balances and integrates the conservation and sustainable use of ecosystems, considering the inter-relationship among living and non-living components;

- Addresses the cumulative impacts of multiple human activities rather than individual sectors;
- Incorporates scientific, traditional, and local knowledge, and is an inclusive process that encourages participation from public and private sectors, indigenous peoples, and other stakeholders;
- Recognizes that a dynamic environment requires a flexible and adaptive management approach.

Arctic states are supporting marine EBM as a goal at the global, regional, and national levels. All eight Arctic states have agreed to one or more of the following global agreements and resolutions that call for EBM or ecosystem approaches: the 2002 World Summit on Sustainable Development Plan of Implementation; the UN Fish Stocks Agreement (1995), the Convention on Biological Diversity (1992); as well as numerous UN resolutions on sustainable fisheries and oceans issues. A report by the Norwegian Polar Institute describes the important progress made by Canada, Denmark/Greenland, Finland, Iceland, Norway, Russia, and the US in implementing ecosystem approaches to oceans management (Hoel 2009). A recent challenge has been to identify how this progress can be enhanced within the Arctic Council.

Through the Arctic Council network, EBM is a guiding principle informing the work of the CAFF (Conservation of Arctic Flora and Fauna) working group, and is reflected in both the Arctic Marine Strategic Plan (Arctic Council 2004) and the approach taken by the Circumpolar Biodiversity Monitoring Programme (CBMP). Other relevant Arctic Council projects include PAME's (Protection of the Arctic Marine Environment working group) work on Large Marine Ecosystems (LMEs), as well as the Best Practices in Ecosystem-based Oceans Management in the Arctic (BePOMAr) document that was endorsed at the 2009 Arctic Council Ministerial Meeting (Hoel 2009).

In May 2011, the Arctic Council Ministers called for the establishment of an expert group on EBM for the Arctic environment. Composed of government experts from Arctic states and representatives from the Arctic Council's permanent participants and working groups, the Arctic EBM expert group was tasked with fostering a common understanding of EBM and its principles across the Arctic Council and providing guidelines or recommendations for advancing EBM in the coastal, marine, and terrestrial ecosystems of the Arctic. The findings and recommendations of the expert group will be considered by the Senior Arctic Officials before the end of the Swedish Chairmanship in May 2013.

Similar to the Arctic states, the EU has also promoted an EBM approach through, for example, its Integrated Maritime Policy (European Commission 2007) and its Marine Strategy Framework Directive (MSFD 2008), and the EU's 2012 Arctic Joint Communication (European Commission and High Representative 2012) notes that the EU will engage on EBM in the Arctic.

Successful environmental governance entails recognition of the interconnected nature of resources. Flexibility in approaches to governance is needed for both complex resource management issues as well as for changing human and natural conditions, such as those brought about by climate change. The broad range of mechanisms encompassed by governance principles may allow for solutions for governance where 'one size doesn't fit all' (Cleaver et al. 2005).

3.3 Legal and Policy Framework

Management of human activities and impacts in marine Arctic ecosystems is undertaken through a complex array of international treaties and programmes, bilateral agreements, national and sub-national laws, and non-governmental and governmental initiatives. Existing Arctic marine governance involves both hard and soft law mechanisms, state and non-state actors, as well as innovative initiatives[1] that incorporate a variety of stakeholders, including indigenous peoples (Young 2002). The following presents a summary overview of the global and regional instruments and bodies involved in governance of the marine Arctic environment.[2]

3.3.1 Global Agreements and Institutions

Global treaties relevant to marine Arctic ecosystems are numerous and address issues ranging from the establishment of protected areas and species protection to reducing greenhouse gas emissions and pollution prevention.

The **UN Convention on the Law of the Sea** (LOS Convention 1982) provides the basic framework for jurisdiction of and resource control in marine areas. The LOS Convention addresses issues related to the protection of the marine environment, with respect to marine pollution, land-based pollution, dumping, and fisheries. The Convention confirms and designates coastal state authority to create and enforce laws to control marine pollution within their national territories and EEZs, designating minimum standards for dumping regulations. The only direct reference to the Arctic is in Article 234, which establishes the right of coastal states to legislate for the "prevention, reduction and control of marine pollution from vessels in ice-covered areas" in their EEZ.

The LOS Convention, despite providing the basic legal framework for the law of the sea, does not claim to cover all aspects of ocean governance, and refers to

---

[1] For example, co-management schemes have been cited in particular as being innovative governance mechanisms.

[2] See Annex for a comprehensive overview of global treaties, conventions, and agreements relevant to Arctic marine ecosystems.

other international instruments and bodies that have competence in this area. It is important to note that the challenge of managing pollution in a transboundary context involves states (within their national territories and EEZ) as well as international space (outside national jurisdiction) in the Arctic Ocean.

Other global conventions dealing specifically with marine pollution include the **International Convention for the Prevention of Pollution from Ships (MARPOL 73/78)** which is the primary international convention for preventing pollution of the marine environment by ships and addresses oil, chemicals, harmful substances in packaged form, sewage, garbage, and air pollution. MARPOL 73/78 (1973/1978) also designates 'special areas' and 'particularly sensitive sea areas' (PSSAs) that are potentially especially vulnerable to vessel-based pollution and therefore require more stringent protection measures through, e.g., routing measures and discharge and equipment requirements. Antarctica was designated a special area in a 1990 amendment, but the Arctic has no areas with this designation. Inadequate compliance with MARPOL 73/78 standards has been cited as a shortcoming due to treaty enforcement through a vessel's flag state, and whereby ships choose to register with states having low enforcement (Rothwell 2000, 2012).

Other agreements relevant to marine pollution include the **Convention on the Prevention of Marine Pollution by Dumping of Wastes and Other Matter** (London Convention 1972) and London Protocol (1996) which regulate dumping in marine areas. The London Convention and London Protocol are particularly relevant to protection of the Arctic marine environment, as there have been problems with dumping of wastes (including radioactive waste) in the Arctic Ocean.[3]

Land-, sea-, and air-based pollution remain major concerns for the fragile ecosystems of the marine Arctic since pollution and contaminants, particularly from sources in lower latitudes, accumulate in the Arctic and adversely affect its inhabitants and marine life. It was therefore a major success when the Stockholm Convention on Persistent Organic Pollutants (2001) was adopted after considerable advocacy efforts by Arctic indigenous peoples' organizations and the Arctic Council. The Convention recognised the negative effect that persistent organic pollutants (POPs) have on humans and the environment, initially banning 12 toxic pollutants. In addition, each party is required to develop a national implementation plan for the reduction of POPs (UNEP/GRID-Arendal 2006). Although all Arctic states have signed the Convention, the US has not yet ratified it and Denmark has entered a territorial exclusion with respect to Greenland.

The **Convention on Long-range Transboundary Air Pollution** (LRTAP Convention 1979) also focuses on the reduction and prevention of air pollution, with specific requirements laid out in several protocols, each concerning different pollutants. Arctic-relevant pollutants are, for instance, covered by the POP

---

[3] Russia has been cited in particular as utilising the Arctic for dumping radioactive waste (see Chap. 2).

Protocol (1998) and the Gothenburg Protocol on Acidification, Eutrophication, and Ground-level Ozone (1999) (amended in 2012 to include limits for black carbon). Protocols for other pollutants which are covered by the Gothenburg Protocol will be considered as terminated once all of their respective parties have ratified, accepted, approved, or acceded to the Gothenburg Protocol (Gothenburg Protocol 1999, art. 18). This includes two protocols on sulphates (Helsinki Protocol 1985; Oslo Protocol 1994), on nitrogen emissions (Sofia Protocol 1988) and on volatile organic compounds (VOCs; Geneva Protocol 1991).

The basic instrument with regard to international action on the issue of climate change is the **UN Framework Convention on Climate Change** (UNFCCC 1992) and its Kyoto Protocol (1997). Signatories to the latter include the EU and its Member States who, along with other so-called Annex I Parties, signed up to binding, specific, quantified emissions reductions listed in Annex B of the Protocol for the timeframe of 2008–2012. During the Conference of the Parties (COP) in December 2012, parties agreed to a second commitment phase for the Kyoto Protocol from 2013 to 2020.[4] Canada has withdrawn from the Protocol while others seem to lack interest in binding reduction targets. As of January 2013, the participation of the Russian Federation—along with Belarus, Ukraine, and Kazakhstan –in the second commitment period is unclear (Murray 2013). The US is a party to the UNFCCC, but has not ratified the Kyoto Protocol. A new "outcome with legal force" is to be concluded under the UNFCCC by 2015 and is supposed to take effect by 2020.[5]

While neither of these instruments contain Arctic-specific rules or requirements (due to the global distribution of greenhouse gases and global participation within the convention), the Arctic is one of the regions affected the most by climate change: the decrease of annual sea ice, the melting of glaciers, and the thawing of permafrost soils already shows severe impacts on the Arctic environment (see also Chap. 1).

As a more recent initiative in this context, the **Climate and Clean Air Coalition** (CCAC) targets short-lived climate pollutants (SLCPs)—covering greenhouse gases under the Kyoto Protocol such as methane and HFCs (hydrofluorocarbons), but also black carbon and other SLCPs. The actions of the CCAC complement efforts underway within the UNFCCC, aiming to leverage high-level political will among its partners to effectively scale-up existing activities to reduce these emissions (e.g., reduce methane and black carbon emissions from oil and natural gas production). The CCAC was launched by six countries and the UN

---

[4] Outcome of the work of the Ad Hoc Working Group on Further Commitments for Annex I Parties under the Kyoto Protocol at its sixteenth session, UNFCCC Decision 1/CMP.7, 11 Dec 2011, UN Doc. FCCC/KP/CMP/2011/10/Add.1.

[5] Establishment of an Ad Hoc Working Group on the Durban Platform for Enhanced Action, UNFCCC Decision 1/CP.17, 11 Dec 2011, UN Doc. FCCC/CP/2011/9/Add.1.

Environment Programme in February 2012 and is a voluntary partnership with an international scope. As of its first anniversary, the Coalition is comprised of more than 50 partners, including 27 countries, the European Commission, 22 non-governmental organizations, and six intergovernmental organizations. While the initiative is still in its early stages, a key focus of the CCAC's outreach activities has been the Arctic region, as black carbon is a contributing factor to the rapidly melting polar ice cap. Three Arctic states—Canada, Sweden, and the US—were among the founding partners of the CCAC, with Denmark, Finland, and Norway joining soon afterwards. The CCAC and Arctic Council have not yet formally engaged in cooperative activities, however, through efforts of Arctic Council Member States who are also CCAC Partners, there is informal cooperation to advance the shared objectives.

Land-based sources of marine pollution still lack regulation both in the Arctic and around the world. Efforts of the UN with the **Global Programme for Action for the Protection of the Marine Environment from Land-based Activities**,[6] adopted in 1995, and the Arctic Council's **Regional Programme for Action for the Protection of the Arctic Marine Environment from Land-based Activities**,[7] a regional effort adopted in 1998, are possible first steps toward legally binding measures, but remain ultimately non-binding in nature.

Fisheries management on a global and non-species-specific level is regulated through the UN Fish Stocks Agreement (1995) and complemented by the soft law **FAO Code of Conduct for Responsible Fisheries** (FAO 1995). All Arctic states are parties to the Fish Stocks Agreement (for more, see Chap. 5).

Although wildlife management and protection take place largely at the national level, there are a number of relevant international instruments geared toward this purpose. Species-specific initiatives include the **International Convention on the Regulation of Whaling** (ICRW 1946) and the **International Agreement for the Conservation of Polar Bears** (Polar Bear Agreement 1973). Frameworks like the **Convention on the Conservation of Migratory Species of Wild Animals** (CMS 1979) aim at broader protection of wildlife. Although the CMS Convention has no particular focus on the Arctic, there are numerous migratory species that inhabit the Arctic for parts of the year (UNEP/GRID-Arendal 2006). Only four of the eight Arctic countries[8] are parties, however, limiting its efficacy. The CITES Convention (1973) (see Box 3.2) regulates the trade of currently endangered species and those that could be threatened if no action is taken. It requires a permit for the export and import of listed species and certain species, if designated, may not be traded at all.

---

[6] See <http://www.gpa.unep.org./>

[7] See <http://www.pame.is/index.php/regional-program-of-action>

[8] Denmark, Finland, Norway, and Sweden.

Box 3.2 Polar bear management and climate change challenges

After mounting evidence in the 1960s that the harvesting of polar bears was endangering populations, the five polar bear range states (Canada, Denmark/Greenland, Norway, Russia, and the US) entered into the Polar Bear Agreement in 1973. The agreement established research coordination and conservation of polar bears through the preservation of polar bear habitat. Furthermore, it prohibited the 'taking'[9] of polar bears except for scientific and indigenous subsistence purposes. The agreement represented a historic international cooperation during the Cold War era. The Polar Bear Agreement outlines framework goals and the contracting parties have an obligation to advance these in accordance with national law. Implementation of management frameworks for polar bear stocks uses varying approaches (e.g., prohibitions on takings, quota systems, action plans).

Until recently, polar bear stocks were generally considered to be stable in the Arctic, and the Polar Bear Agreement and national frameworks—in conjunction with bilateral cooperation[10]—deemed effective. However, recent climate change-induced environmental effects have altered this outlook. In 2005, the International Union for Conservation of Nature (IUCN) Polar Bear Specialist Group (PBSG) unanimously recommended that the species be classified as 'vulnerable' according to IUCN Red List criteria (see IUCN 2013) and in 2009, the five range states began development of a coordinated, range-wide action plan to guide polar bear conservation efforts.[11] This joint circumpolar action plan covers the entire circumpolar range of polar bears.

In 2009, the US recommended that the polar bear be transferred from listing under Appendix II of the CITES Convention, which limits international trade of polar bears and their parts, to Appendix I, shifting trade restrictions from the requirement of a CITES permit for import or export to a prohibition of international commercial trade in polar bear specimens. The proposal was rejected at the CITES COP in 2010, with the opposition led by Canada, the only country that allows polar bear harvesting for export purposes (CITES 2010; Cummings and Siegel 2012; Telecky and Smith 2012). In October 2012, to the dismay of several indigenous groups and

---

[9] 'Taking' is defined by the Agreement as hunting, killing, and capturing [art. 1(2)].

[10] See Polar Bear Management Agreement for the Southern Beaufort Sea; Memorandum of Understanding between Environment Canada and the United States Department of the Interior for the Conservation and Management of Shared Polar Bear Populations (2008); Agreement on the Conservation and Management of the Alaska-Chukotka Polar Bear Population; Memorandum of Understanding for the conservation of the Baffin Bay and Kane Basin sub-populations (Meeting of the parties to the 1973 Agreement 2011).

[11] Meeting of the parties to the 1973 Agreement 2011.

some environmental organizations (e.g., WWF 2008), the US again submitted a proposal to uplist polar bears, however, this was rejected by the Parties in March 2013. Those in opposition emphasized that the current Appendix II listing is appropriate given that the main driver threatening polar bears is climate change and not international trade. During the March 2013 vote, the EU offered a compromise position to regulate trade with export quotas and a tagging system. The compromise was rejected and all 27 EU Member States abstained from voting because they failed to agree on a common position.

The evolving management of the polar bear demonstrates the complexity of options surrounding a single species and its habitat. It highlights that Arctic nations are influenced by political decisions taken outside the Arctic and at the same time, suggests decisions taken outside the Arctic could impact governing decisions for the region as a whole.

The **Convention on Biological Diversity** (CBD 1993) marks a departure from issue-specific agreements, concentrating on the conservation and sustainable use of biological diversity as well as the fair use of its resources. For the *components* of biological diversity, the CBD applies only in areas within the national jurisdiction of each party. However, with respect to *processes and activities* impacting biological diversity, the CBD applies regardless of where their effects occur, provided that they are carried out under a party's jurisdiction or control. This would include processes and activities carried out in Arctic waters including the high seas. The CBD includes a strong emphasis on the establishment of protected areas, which are utilised to conserve certain species or areas that have unique biodiversity or hold special importance.

Although no specific programme dealing with the Arctic environment exists within the CBD and the treaty text does not address marine and coastal biodiversity, the 'Jakarta Mandate on Marine and Coastal Biological Diversity' (1995) is relevant for the protection of Arctic marine species as it focuses on priority areas for action regarding marine and coastal biodiversity. Additionally, the CBD Secretariat signed a Memorandum of Cooperation with the Secretariat of the Arctic Council's working group on **Conservation of Arctic Flora and Fauna** (CAFF) in April 2009. This cooperation is intended to build and share knowledge, create awareness, and enhance capacity for implementation of the CBD in the Arctic region. The CBD helps place Arctic biodiversity within a global framework, while CAFF helps inform the CBD on the status and trends of biodiversity in the region.

Other global instruments that provide for the designation of protected areas are the **Convention for the Protection of the World Cultural and Natural Heritage** (UNESCO 1972) and the Ramsar Convention on Wetlands (1971), along with many others, including the above mentioned ICRW. One of the largest protected Ramsar wetlands in the world is Queen Maud Gulf in Nunavut, Canada, and numerous other Ramsar, as well as Natural Heritage, sites exist throughout Greenland, Scandinavia, Siberia, and on Svalbard and Iceland.

In 2013, a new global treaty for reducing mercury emissions was negotiated. The **Minamata Convention on Mercury** provides controls and reductions for products, processes, and industries where mercury is used or released, using a combination of legally binding and voluntary approaches, including trade measures, national strategies, and technologies (UNEP 2013). Over 140 states took place in the negotiations, which were launched at the 2009 session of the UNEP Governing Council. The Minamata Convention on Mercury will be open for signature at a Ministerial meeting in Japan in October 2013. The Inuit Circumpolar Council (ICC) has noted that it is pleased with the efforts to reduce mercury levels, an issue of particular concern for Arctic peoples (Alaska Dispatch 2013) (Table 3.1).

3.3.2 Regional and Sub-Regional Regimes

The marine Arctic is governed by the global instruments described above, as well as regional and sub-regional initiatives, numerous bilateral agreements, and national legislation. Of particular importance is the **Arctic Environmental Protection Strategy** (AEPS 1991) by the eight Arctic countries, a regional initiative stemming from broad recognition of the need for more international cooperation in the Arctic. The **Arctic Council** was created in 1996 to strengthen the AEPS as an intergovernmental forum for discussion and policy making for the Arctic environment, as well as monitoring trends in the Arctic environment. The Arctic Council and its six working groups have effectively brought together actors and stakeholders in the Arctic to address environmental issues. The Arctic Council's key strengths lie in its engagement with indigenous peoples and its ability to produce pan-Arctic scientific assessments that raise the visibility of Arctic environmental issues. The Council has produced many important scientific assessments, such as the Arctic Climate Impact Assessment (ACIA 2005), Arctic Biodiversity Assessment, and Arctic Marine Shipping Assessment (AMSA 2009), which have brought attention to important regional issues and promoted policymaking. PAME is currently leading the Arctic Ocean Review (AOR) project and assessing the status, trends, and recommendations for Arctic marine governance. The Council also serves as the most important forum for discussion between and within countries on many overarching issues (both sectoral and cross-sectoral).

The Arctic Council is a non-regulatory body. However, it is beginning to show tendencies towards becoming a decision making body: While political and academic spheres debated about whether it should be more policy-oriented or eventually have regulatory capacities, in 2011 the Member States of the Arctic Council concluded a search and rescue agreement (Arctic SAR Agreement) under its auspices; however, although the agreement will be binding once the states have implemented it into their respective national legal regimes, the Arctic Council itself remains without regulatory powers. Taking this model approach forward, its member states put the development of a binding agreement on the agenda concerning preparedness and response to oil pollution, to be concluded at the Council's May 2013 Ministerial Meeting. The Arctic Council's role in the governance of the area could be described as a decision-shaping body evolving into a negotiating forum (Munk-Gordon 2012).

Table 3.1 Participation of Arctic countries in global treaties

	Marine	Pollution			Environment				Biodiversity
	UNCLOS	MARPOL[1]	London convention	POPs	CMS	Heritage	Ramsar	Fish stocks	CBD
	1982	1973/1978	1972	2001	1979	1972	1971	1995	1992
Canada	Yes	Yes[b]	Yes[a]	Yes	No	Yes	Yes	Yes	Yes
Denmark	Yes	Yes[b]	Yes[a]	Yes[2]	Yes	Yes	Yes	Yes	Yes
Finland	Yes	Yes[b]	No	Yes	Yes	Yes	Yes	Yes	Yes
Iceland	Yes	Yes	Yes[a]	Yes	No	Yes	Yes	Yes	Yes
Norway	Yes	Yes[b]	Yes[a]	Yes	Yes	Yes	Yes	Yes	Yes
Russia	Yes	Yes[b]	Yes	Yes	No	Yes	Yes	Yes	Yes
Sweden	Yes	Yes[b]	Yes[a]	Yes	Yes	Yes	Yes	Yes	Yes
USA	No	Yes	Yes	No	No	Yes	Yes	Yes	No
Total Arctic countries	7	8	7	7	4	8	8	8	7
(EU)	(Yes)	(No)	(No)	(Yes)	(Yes)	(No)	(No)	(Yes)	(Yes)
Total parties globally	164	152	86	178	117	190	163	80	193

Source Adapted from UNEP/GRID-Arendal (2006)

As of 3 December 2012

[1]Annexes I and II

[2]This ratification was with territorial exclusion of Greenland

[a]Indicates that this country has also ratified the 1996 Protocol of the London convention

[b]Indicates that this country has ratified all of the Annexes to MARPOL

Other state-linked, regional groups with more focused interests or limited mandates include the Nordic Council of Ministers, the Northern Dimension, the Conference of Arctic Parliamentarians, International Arctic Science Committee, the Barents Euro-Arctic Council, and many other regionally based organizations.

Also at the regional level, a Canadian initiative resulted in the drafting of the **Guidelines for Ships Operating in Arctic Ice-covered Waters** (Arctic Shipping Guidelines 2002) in the International Maritime Organization (IMO), that were extended in their applicability to Antarctic waters in 2009, now called the **Guidelines for Ships Operating in Polar Waters** (Polar Shipping Guidelines 2009). These guidelines outline non-binding principles and standards for construction, equipment, and safety procedures for ships in polar regions. Albeit their non-binding status, these guidelines have advanced several national initiatives in Arctic countries with regard to shipping safety (Vidas 2000). Since 2010, a mandatory 'Polar Code' has been under development through the IMO and is planned to contain aspects of shipping safety as well as environmental requirements, for instance with regard to discharges and emissions, but also covering liabilities and insurance requirements. A final decision on the Polar Code is not to be expected before 2014 implementation via existing IMO agreements could defer its entry into force years later.

Box 3.3: Comparing environmental governance in the Arctic and Antarctic

A number of studies have broached the topic of using the Antarctic governance framework as a model for a future regime in the Arctic (see Lennon 2008). The Arctic and Antarctic environments are both high-latitude, circumpolar regions with extreme environmental conditions. Both regions have large deposits of natural resources such as coal, natural gas, and offshore oil reserves. Furthermore, because most of the Central Arctic Ocean is outside of national jurisdiction, arguably, the threat of claiming territories in the Arctic is similar to the situation in the Antarctic before the Antarctic Treaty (1959) was established.

Major differences between the two regions exist, however. The Arctic is primarily oceanic, whereas the Antarctic primarily consists of an ice-covered land mass. The Arctic is characterized by the presence of indigenous populations, whereas the Antarctic has virtually no permanent residents (Millennium Ecosystem Assessment 2005). Arctic nations have active territorial and marine claims in the Arctic, whereas territorial claims are on hold in the Antarctic, as stated in the Antarctic Treaty, signed 50 years ago.

The Antarctic environment is governed primarily through the legally binding **Antarctic Treaty** and its accompanying Protocol on Environmental Protection to the Antarctic Treaty (1991), together with two separate conventions on the Conservation of Antarctic Seals (1972) and the Conservation of Antarctic Marine Living Resources (1980) and some 200 recommendations adopted at the Antarctic Treaty Consultative Meetings, altogether referred to as the Antarctic Treaty System (ATS). These constitute a regional cooperative effort in the sense that the treaty system exclusively addresses one region (Vidas 2000). Emphasising that Antarctica should be used exclusively for peaceful purposes, the Antarctic Treaty promotes scientific research and international cooperation and the 1991 Protocol essentially designates the region as a nature reserve, prohibiting claims to mineral deposits, and regulating waste management and marine pollution (Antarctic Treaty 1959, arts. 1–3). The **Convention on the Conservation of Antarctic Marine Living Resources**, one of the agreements belonging to the ATS, utilises a precautionary and ecosystem-based approach to regulate in particular krill fishing and all marine resources excluding seals and whales, which are governed by other instruments (Rayfuse 2008). Despite the differences between the Antarctic and the Arctic, some aspects of the ATS, particularly its focus on security and peace, could serve as a model for a future Arctic regime.

Aside from comprehensive Arctic regimes, sub-regional, multilateral cooperation has also been cited as an important component of Arctic marine governance (Vidas 2000). Intergovernmental initiatives addressing a portion of the Arctic include the **Barents Euro-Arctic Region (BEAR)** with its associated council and the **Norwegian/Russian Commission on Environmental Protection**.

In the Barents region, bilateral agreements have often preceded multilateral agreements. In 1975, the Soviet-Norwegian Fishing Commission was established and fisheries agreements in the Barents Sea were agreed between Norway and the former Soviet Union. Since that time, Norway and Russia have worked together to manage fish stocks through the Joint **Norwegian-Russian Fisheries Commission** (Stokke 2001; UNEP 2004). In 1988, the **Joint Norwegian-Russian Commission on Environmental Protection** was established as a bilateral, intergovernmental commission focused on environmental protection through control of economic activities (e.g., petroleum-related operations and oil refuelling from ship to ship). Scientific cooperation supports Norwegian-Russian cooperation on fisheries and environmental protection, such as through the Knipovich Polar Research Institute of Marine Fisheries and Oceanography and Bergen-based Institute of Marine Research, which work together to provide scientific data used to help sustainably harvest fish stocks in the Barents Sea, and through the Research Council of Norway and the Russian Foundation for Basic Research, fostering collaboration to strengthen knowledge-based, integrated management of the Barents Sea (Mila 2012). With stocks in the Barents Sea at an all-time high, Norwegian-Russian cooperation has shown that concerted action against illegal, unreported, and unregulated (IUU) fishing can indeed be successful (Fisheries.no 2011).

Other sub-regional legally binding approaches include the six-nation **Convention on the Conservation and Management of Pollock Resources in** the **Central Bering Sea** (CBS 1994) and the **Convention on the Protection of the Marine Environment of the North-East Atlantic**[12] (OSPAR Convention 1992). A specific aim of the OSPAR Convention (Annex V) is to apply an "integrated ecosystem approach', although fisheries management and maritime transport are outside of its mandate.[13] OSPAR has also promoted the establishment of a network of marine protected areas, four of which, as of 2007, are in the Arctic[14] (OSPAR Commission 2007). Together with the North East Atlantic Fisheries Commission (NEAFC), OSPAR is currently working to describe 10 areas in the northeast Atlantic that have been jointly identified as meeting the CBD scientific criteria for ecologically or biologically significant areas (EBSAs), in preparation for their possible inclusion in the CBD Repository of EBSAs.

The OSPAR Commission has signed five MOUs (memorandum of understanding) to enhance collaboration and coordination with the International Council for the Exploration of the Sea (ICES)—for scientific information; the European Environment Agency (EEA)—for compatibility in data collection and assistance with information dissemination; the UN Economic Commission for Europe (UNECE)—to provide data and analysis of airborne pollutants from regional monitoring centres across the OSPAR area; the International Seabed Authority (ISA)—to help ensure appropriate coordination of measures in order to reconcile the development of mineral resources with protection of the marine environment; and NEAFC—to promote mutual cooperation towards the conservation and sustainable use of marine biological diversity. In addition, the OSPAR Commission and the International Atomic Energy Agency (IAEA) have established Practical Arrangements on sharing data on concentrations of radioactive substances in the OSPAR Maritime Area.

The **North Atlantic Marine Mammal Commission (NAMMCO)** has been recognized for its ecosystem-based approach to marine mammal protection, involving state and non-state actors and including indigenous populations (Young 2002). NAMMCO is a sub-regional cooperation and co-management framework for whales, seals, and walruses among Norway, Iceland, Greenland, and the Faroe Islands (AHDR 2004).

Norway, Iceland, Greenland, and the Faroe Islands manage beluga whales through the sub-regional NAMMCO, while the US and Canada rely on **co-management agreements** between indigenous communities and federal agencies (e.g., Alaska Beluga Whale Committee and the Indigenous People's Council

---

[12] The OSPAR Maritime Area covers the northeast Atlantic and therefore includes but is not limited to part of the marine Arctic area referred to in this paper; OSPAR Region 1 covers part of the marine Arctic.

[13] Fisheries management lies within the mandate of the relevant Regional Fisheries Management Organisations (RFMOs) and EU and domestic authorities. For initiatives related to shipping, OSPAR relies primarily on the IMO, an observer to the Convention with which it has an Agreement of Cooperation.

[14] The four marine protected areas in the Arctic were nominated by Norway.

Table 3.2 Examples of informal approaches focused on environmental governance

Type of informal approach	Description
Government initiatives involving non-governmental groups	Co-management of resources and wildlife: empowering local communities by allowing independent choice of practices and ways to achieve government goals
Cooperation between researchers and local communities	Involving researchers and local communities in providing vital information for projects and bridging the gap between science and local knowledge
Cooperation between NGOs and local communities	Involving NGOs as donors and project managers and local communities in implementing project objectives
Initiatives governed by local communities	Involving local communities who have sought alliances for a common cause within their community and inside or outside their region
Cooperation initiatives between researchers or research institutes	Involving researchers and institutes with a common research interest in the Arctic

for Marine Mammals in the US and Nunavut Wildlife Management Board in Canada). These signed agreements are non-binding, but outline agreed principles of management and methods of communication and collaboration. Although some argue that these agreements do not sufficiently transfer power to indigenous peoples, co-management has been widely applauded as an effective tool to increase user participation (ACIA 2005) and has resulted in increased knowledge about species' health and distribution for hunters and scientists (Fernandez-Gimenez et al. 2006).

3.3.3 Informal Approaches and Initiatives

In addition to these legally binding and non-legally binding approaches to environmental governance at the international and regional levels, there are a large number of informal approaches and initiatives that play important roles. Common characteristics of such informal approaches are: a lesser degree of institutionalisation, cooperation emerging on an ad-hoc basis, less complex decision making processes, and less formal cooperation structures, such as verbal agreements.

Roughly five different types of informal approaches to environmental governance in the Arctic region can be distinguished, as shown in Table 3.2.

3.4 Analysis of Governance Shortcomings

Identifying general shortcomings in Arctic environmental governance can help to identify areas for (transatlantic) policy action. In order to address future challenges and governance needs in the Arctic marine area, the literature suggests different ways forward, as well as areas requiring further examination. The following section discusses certain weaknesses—and strengths—of the existing legal and policy framework in light of current conditions and foreseeable changes.

Environmental governance in the marine Arctic is characterised by a patchwork of rules and institutions that reflects the mix of national jurisdictions and international space in the region, as well as its historical realities. As shown, there is currently no overarching governance body specifically mandated to adopt and enforce legally binding rules for the marine Arctic; rather, governance mechanisms come from a variety of sources. The primary Arctic institution—the Arctic Council—does not have the mandate to develop or impose legally binding obligations on its participants, although, as mentioned, the Council has recently begun taking a stronger role in policy formulation by helping to negotiate binding treaties. This means that capacity for legally binding governance measures currently lies primarily in the hands of the Arctic states themselves and through any bilateral and multilateral initiatives they might undertake.

According to the Arctic Governance Project, a strengthening of the complex of institutions and arrangements is needed "to form an interlocking suite of governance systems for the Arctic in which the idea of stewardship is central and the whole is greater than the sum of the parts" (Arctic Governance Project 2010). Absent a more integrated approach, current governance systems are not set up to manage for resilience in the face of the uncertain but inevitable and rapid changes that are occurring in the region. There is a high risk that a 'business as usual' approach will be the default position regarding both national and international waters. Encouragingly, although there is more urgency related to climate change now and less time to implement responses, the ability to establish adaptive management systems is greater now than it was 20 years ago.

Integrated, cross-sectoral governance is still evolving, for instance, through the work of the Arctic Council. The Arctic Council is promoting cross-sectoral approaches for integrated management of Arctic resources through several programmes such as the Arctic Marine Strategic Plan and the Integrated Oceans Management project and several working groups of the Arctic Council, notably PAME and AMAP, have made specific recommendations on cross-sectoral Arctic governance issues. Given the lack of another pan-Arctic forum, and its leading work to date, the Arctic Council may be the best means to implement an ecosystem-based management approach requiring coordination across sectors and countries. The Council's scientific activities and forum for dialogue have the potential to facilitate greater state action, but modifications to its current mandate may be necessary.

While sector-specific policies are critical for managing fishing, hydrocarbon, and shipping activities, a holistic, EBM approach is necessary to ensure that adequate environmental safeguards are established in the marine Arctic. The development of such an approach is fundamentally international in nature and should be based on a system of international principles, standards, and rules that address the interactions and interdependencies among countries, stakeholders, and institutions in the context of climate change.

3.5 Perspectives on the Way Forward: Policy Pathways[15]

As useful as it is to consider a set of policy options, it is perhaps even more fitting to approach environmental governance in terms of policy pathways. The idea of pathways particularly allows for an adaptive management approach and an evolution of policy over time. For example, a pathway approach could enable a precautionary beginning to environmental protection—e.g., moratoriums on certain activities in specific regions or protected areas and then, where warranted by the evidence, a gradual easing of environmental restrictions. Especially in the face of so much uncertainty, there is a need to adopt a precautionary approach regarding the Arctic environment. Chapin and Hamilton (2009) note:

> A key choice in approaching environmental governance is the extent to which a precautionary approach should be undertaken. Typically, environmental rules are developed only after problems of overexploitation or destruction of the natural environment and its resources have already emerged. Such an approach is also possible in the modern Arctic context, but the relatively pristine nature of the Arctic, its heightened fragility and past failures elsewhere point to a need to take a higher level of precaution.

Thinking in terms of pathways also encourages a plurality of approaches within and among the governing institutions involved, perhaps informed by shared principles and improved through dialogue on emerging best practices. Pathways can emerge, diverge, and merge as needed.

Figure 3.1 illustrates possible scales of various environmental governance activities. Consistent with the idea of policy pathways, it is possible to envision multiple starting points on any particular issue. It is also possible to identify approaches that can be implemented relatively quickly as a means of developing best practices and measures in the near-term.

There are several components of a multi-pronged approach that need to be further developed to improve environmental governance at local, national, bilateral, and international scales. These approaches include those that are currently underway through sector-based approaches, national approaches, and multilateral approaches as well as the potential development of a new set of instruments to improve multilateral cooperation. Chapin et al. (2006) argues that institutions should be identified that are poised to implement policies at appropriate scales (or governance levels) and that different issues should be dealt with at different levels.

---

[15] Many of the ideas in this section came from the paper prepared by the Arctic TRANSFORM Project's Environmental Governance Working Group, co-chaired by Dr. Stuart Chapin and Dr. Neil Hamilton (2009). <http://arctic-transform.org/download/EnvEX.pdf>. Accessed 11 Feb 2013.

Activity	Possible scales		
	(easier / less ambitious) ← → (more difficult / ambitious)		
Scientific research	species-specific – LME research – full Arctic Ocean assessment		
Management level	local/regional – national – bilateral – multilateral		
Marine protection	species/stocks – LMEs and MPAs – reserve networks – ocean		
Legal structures	soft law –national regulation – bilateral agreement – multilateral treaty		

Fig. 3.1 Governance scales

3.5.1 Principles of Environmental Governance[16]

The following set of core governance principles can provide a foundation for effective governance of Arctic marine ecosystems:

- *The principle of fit*: Create arrangements that avoid or minimize spatial and temporal mismatches among biophysical systems, socioeconomic activities, and governance practices. Multi-level governance is an example of this principle. Different system components operate at different scales, and effective regime design implies attention to relevant scales.
- *The principle of multiple use*: Develop integrated approaches that can mediate among different uses of marine resources and establish priorities when such uses are incompatible.
- *The principle of cooperation*: Ensure that all interested stakeholders have a voice in decision making and decisions are made in a transparent fashion at the appropriate level of governance.
- *The principle of adaptive management*: Design and operate governance systems to promote adaptation and social learning as knowledge improves regarding the relevant biophysical systems, human activities, and their interactions.
- *The principle of policy flexibility*: Marine ecosystems in the Arctic are changing rapidly, and ecosystem functions requiring protection will be different in the future than the present. Attention solely to issue-based threats is thus highly unlikely to be effective unless framed within an overarching context. Resilience and learning are significant elements of this principle, and EBM is an example approach.
- *The principle of precaution*: Frameworks for environmental protection should recognise that preserving healthy ecosystems and functioning ecosystem

---

[16] This list of principles was developed by the Arctic TRANSFORM Project's Environmental Governance Working Group, co-chaired by Dr. Stuart Chapin and Dr. Neil Hamilton.

services requires a precautionary approach, especially in conditions as vulnerable and relatively pristine as those found in the Arctic. This would ideally entail putting regulations in place *before* human activities increase.

3.5.2 Conclusion and Questions for Discussion

To conclude, four key questions are posed to serve as a starting point for identifying transatlantic policy options for governance in the Arctic marine area. Alongside consideration of the pathways and principles laid out above, the following questions provide a possible starting point for further reflection and discussion on environmental governance needs and opportunities:

- *Uniqueness*: What are the unique opportunities and threats in the Arctic marine area that could guide the adaptation of governance regimes in light of future changes?
- *Content*: Where are there gaps and overlaps in the current governance structure?
- *Approaches*: What are the advantages and trade-offs of the various possible approaches? (e.g., flexibility versus enforceability)
- *Transatlantic contribution*: How can transatlantic policies contribute to effective environmental governance in the marine Arctic and to adaptation to climate change?

This chapter has highlighted the complexity of Arctic environmental governance from multiple perspectives. It is an attempt to describe the landscape of governance approaches in place in the Arctic and to stimulate discussion regarding the future of environmental governance in the marine Arctic. Many existing institutions and governance mechanisms were developed under political and environmental circumstances that were vastly different from today's reality. As the Arctic is undergoing drastic changes that will spark unprecedented activity in the region, a rethinking of existing structures is required to appropriately address emerging issues.

References

ACIA (2005) Arctic climate impact assessment. Cambridge University Press, New York
AEPS (1991) Arctic Environmental Protection Strategy, 14 Jan 1991, 30 I.L.M. 1624
AHDR (2004) Arctic Human Development Report. Stefansson Arctic Institute, Akureyri
Alaska Dispatch (2013) Mercury treaty adopts legal framework welcomed by Arctic indigenous peoples.<http://www.alaskadispatch.com/article/mercury-treaty-adopts-legal-framework-welcomed-arctic-indigenous-peoples>. Accessed 20 Feb 2013
AMSA (2009) Arctic Marine Shipping Assessment 2009 Report. Arctic Council, Apr 2009
Antarctic Treaty (1959) The Antarctic Treaty, 1 Dec 1959, 402 U.N.T.S. 71. Entered into force 23 June 1961
Arctic Council (2004) Arctic Marine Strategic Plan, 24 Nov 2004. Available at <http://www.pame.is/images/stories/AMSP_files/AMSP-Nov-2004.pdf>. Accessed 5 Feb 2013

Arctic Governance Project (2010) Arctic Governance in an Era of Transformative Change: Critical Questions, Governance Principles, Ways Forward. <http://arcticgovernance.customp ublish.com/the-arctic-governance-project.142454.en.html>. Accessed 8 Feb 2013

Arctic SAR Agreement (2011) Agreement on Cooperation on Aeronautical and Maritime Search and Rescue in the Arctic, 12 May 2011, 50 I.L.M. 1119 (2011). Entered into force on 19 Jan 2013

Arctic Shipping Guidelines (2002) Guidelines for Ships Operating in Arctic Ice-Covered Waters, IMO MSC/Circ. 1056, MEPC/Circ. 399, 23 Dec 2002

Cavalieri S, Reid A, Lang S, McGlynn E, Cicin-Sain B, Balgos M, Orbach M et al (2011) Policy recommendations for improved EU and US cooperation in maritime governance. <http://www.calamar-dialogue.org/sites/default/files/Policy_recommendations_full.pdf>. Accessed 8 Feb 2013

CBS Convention (1994) Convention on the Conservation and Management of Pollock Resources in the Central Bering Sea, 16 June 1994, 34 I.L.M. (1995). Entered into force 8 Dec 1995

Chapin FS et al (2006) Building Resilience and Adaptation to Manage Arctic Change. AMBIO J Human Environ 35(4):198–202

Chapin S, Hamilton N, Arctic TRANSFORM Environmental Governance Working Group (2009) Policy options for Arctic environmental governance. Prepared by the Environmental Governance Working Group. Arctic TRANSFORM. <http://arctic-transform.org/download/EnvEX.pdf>. Accessed 8 Feb 2013

CITES (2010) Consideration of Proposals for Amendment of Appendices I and II. In: 15th Meeting of the Conference of the Parties, Doha, 13–25 Mar 2010

CITES Convention (1973) Convention on International Trade in Endangered Species of Wild Fauna and Flora, 3 Mar 1973, 993 U.N.T.S. 243. Entered into force 1 July 1975

Cleaver F, Franks T, Boesten J, Kiire A (2005) Water governance and poverty: what works for the poor?. Bradford Centre for International Development, University of Bradford, Bradford

CMS Convention (1979) Convention on the Conservation of Migratory Species of Wild Animals, 23 June 1979, 1651 U.N.T.S. 33. Entered into force 1 Nov 1983

Convention on Biological Diversity (1992) Convention on Biological Diversity, 5 June 1992, 1760 U.N.T.S. 79. Entered into force 29 Dec 1991

Convention on the Conservation of Antarctic Seals (1972) Convention on the Conservation of Antarctic Seals, 1 June 1972, 11 I.L.M. 251 (1972). Entered into force 11 Mar 1978

Convention on the Conservation of Antarctic Marine Living Resources (1980) Convention on the Conservation of Antarctic Marine Living Resources, 20 May 1980, 19 I.L.M. 841 (1980). Entered into force 7 Apr 1982

Cummings B, Siegel K (2012) A critical examination of the inconsistent status designations of the polar bear under United States, Canadian, and international legal regimes. In: IPY 2012 Conference. Montreal. <http://132.246.11.198/2012-ipy/Abstracts_On_the_Web/by_theme.html>. Accessed 8 Feb 2013

De Roo C, Cavalieri S, Wasserman M, Knoblauch D, Baush C, Best A (2008) Background Paper: Environmental Governance in the Marine Arctic. Arctic TRANSFORM

European Commission (2007) Communication from the Commission to the European Parliament, The Council, the European Economic and Social Committee and the Committee of the Regions: An Integrated Maritime Policy for the European Union, 10 Oct 2007, COM (2007) 575 final

European Commission and High Representative (2012) Joint communication to the European Parliament and the Council. Developing a European Union Policy towards the Arctic region: progress since 2008 and next steps. 26 June 2012, JOIN (2012) 19 final

FAO (1995) Code of Conduct for Responsible Fisheries. Adopted by the Twenty-eight Session of the FAO conference, 31 Oct 1995

Fernandez-Gimenez M, Huntington H, Frost K (2006) Integration or co-optation? Traditional knowledge and science in the Alaska Beluga Whale Committee. Environ Conserv 33(4):306–315

Fish Stocks Agreement (1995) Agreement for the implementation of the provisions of the United Nations convention on the law of the sea of 10 December 1982 relating to the conservation and management of straddling fish stocks and highly migratory fish stocks, 4 Aug 1995, 2167 U.N.T.S. 3. Entered into force 11 Dec 2001

Fisheries.no (2011) Fisheries no. Norway. <http://www.fisheries.no/resource_management/International_cooperation/Fisheries_collaboration_with_Russia/>. Accessed 16 Nov 2012

Geneva Protocol (1991) Protocol to the 1979 Convention on LRTAP concerning the Control of Emissions from Volatile Organic Compounds or their Transboundary Fluxes, 18 Nov 1991, 2001 U.N.T.S. 187. Entered into force 29 Sept 1997

Gothenburg Protocol (1999) Protocol to the 1979 Convention on Long-range Transboundary Air Pollution to Abate Acidification, eutrophication and Ground-level Ozone, 30 Nov 1999, EB.AIR/1999/1. Entered into force 17 May 2005

Helsinki Procotol (1985) Protocol to the 1979 Convention on Long-Range Transboundary Air pollution on the Reduction of Sulphur Emissions or their Transboundary Fluxes by at least 30 percent, 14 June 1985, 1480 U.N.T.S. 215. Entered into force 2 Sept 1987

Hoel AH (2009) Best practices in ecosystem-based oceans management in the Arctic. Report no. 129 of the Norwegian Polar Institute, Polar Environmental Centre

ICRW (1946) International Convention for the Regulation of Whaling, 2 Dec 1946, 161 U.N.T.S. 72. Entered into force 10 Nov 1948

IUCN (2013) The IUCN Red List of Threatened Species. International Union for the Conservation of Nature. <http://www.iucnredlist.org/details/6335/0>. Accessed 11 Feb 2013

Jachtenfuchs M, Kohler-Koch B (2004) The dynamics of European integration: Why and when EU institutions matter. Palgrave, Basingtoke

Jakarta Mandate (1995) Jakarta Mandate on the Conservation and Sustainable Use of Marine and Coastal Biological Diversity, Nov 1995, COP 2 Decision II/10

Juda L (1999) Considerations in developing a functional approach to the governance of large marine ecosystems. Ocean Dev Int Law 30:89–125

Juda L, Hennessey T (2001) Governance profiles and the management of the uses of large marine ecosystems. Ocean Dev Int Law 32:43–69

Kyoto Protocol to the United Nations Framework Convention on Climate Change (1997) 11 Dec 1997, 2303 U.N.T.S. 148. Entered into force 16 Feb 2005

Laughlin T, Speer L (2011) IUCN/NRDC workshop on ecosystem-based management in the Arctic marine environment. Workshop report. International Union for the Conservation of Nature and Natural Resources Defense Council, Rekjavik

Lemos M, Agrawal A (2006) Environmental governance. Ann Rev Environ 31:297–325

Lennon E (2008) A tale of two poles: A comparative look at the legal regimes in the Arctic and the Antarctic. Sustain Dev Law and Policy 8(3)

London Convention (1972) Convention on the Prevention of Marine Pollution by Dumping of Wastes and Other Matter. 29 Dec 1972, 1046 U.N.T.S. 120. Entered into force 30 Aug 1957

London Protocol (1996) Protocol to the Convention on the Prevention of Marine Pollution by Dumping of Wastes and Other Matter, 7 Nov 1996, 36 I.L.M. 7 (1997). Entered into force 24 Mar 2006

LRTAP Convention (1979) Convention on Long-range Transboundary Air Pollution, 13 Nov 1979, 1302 U.N.T.S. 217. Entered into force 16 Mar 1983

LOS Convention (1982) United Nations Convention on the Law of the Sea, 10 Dec 1982, 1833 U.N.T.S. 396. Entered into force 16 Nov 1994

Marine Strategy Framework Directive (2008) Directive 2008/56/EC of the European Parliament and the Council of 17 June 2009 establishing a framework for community action in the field of marine environmental policy, June 17 2009, 2008 O.J. (L 164)

MARPOL 73/78 (1973/1978) International Convention for the Prevention of Pollution from Ships, 2 Nov 1973, 2 I.L.M. 1319 (1973) as modified by the 1978 Protocol Relating to the International Convention for the Prevention of Pollution from Ships [17 Feb 1978, 17 I.L.M. 546 (1978)] and the 1997 Protocol to Amend the International Convention for the Prevention of Pollution from Ships (26 Sept 1997) and as regularly amended. Entry into force varies for each Annex. At the time of writing Annexes I-VI were all in force

Meeting of the parties to the 1973 Agreement (2011) Meeting of the parties to the 1973 Agreement on the Conservation of Polar Bears. Outcome of Meeting. Iqaluit, Nunavut, Canada. 24–26 Oct 2011 <www.polarbearmeeting.org/content.ap?thisId=500040871>. Accessed 8 Feb 2013

Mila M (2012) The research council strengthens research on the environment and resources of the Barents Sea. The Research Council of Norway. <http://www.forskningsradet.no/prognett-geopolitikknord/Nyheter/The_Research_Council_strengthens_research_on_the_environment_and_resources_of_the_Barents_Sea/1253978601513/p1253961979600>. Accessed 16 Nov 2012

Millenium Ecosystem Assessment (2005) Ecosystems and human well-being: current state and trends. Island Press, Washington

Munk-Gordon (2012) Canada as an Arctic power: Preparing for the Canadian Chairmanship of the Arctic Council (2013–2015). The Munk-Gordon Arctic Security Program. <http://www.gordonfoundation.ca/sites/default/files/publications/CanadaasanArticPower_1.pdf>. Accessed 8 Feb 2013

Murray J (2013) Russian row over Kyoto extension rumbles on. Business Green. <http://www.businessgreen.com/bg/news/2239407/russian-row-over-kyoto-extension-rumbles-on>. Accessed 8 Feb 2013

Nordregio (ed) (2007) People and politics of the Arctic. J Nordregio 4(7)

Oslo Protocol (1994) Protocol to the 1979 Convention on Long-Range Transboundary Air Pollution on Further Reduction of Sulphur Emissions, 14 June 1994, 122 U.N.T.S. 2030. Entered into force 5 Aug 1998

OSPAR Commission (2007) 2006 Report on the status of the OSPAR network of marine protected areas

OSPAR Convention (1992) Convention for the Protection of the Marine Environment of the North-East Atlantic, 22 Sept 1992, 32 I.L.M. 1072 (1993). Entered into force 25 Mar 1998

Polar Bear Agreement (1973) International Agreement on the Conservation of Polar Bears, 15 Nov 1973, 13 I.L.M. 3 (1974). Entered into force 26 May 1976

Polar Shipping Guidelines (2009) Guidelines for ships operating in polar waters. IMO Assembly Resolution A, 1024(26), 2 Dec 2009

POPS Protocol (1998) Protocol to the 1979 Convention on LRTAP on Persistent Organic Pollutants (POPs), 24 June 1998, 2230 U.N.T.S. 79. Entered into force 23 Oct 2003

Protocol on Environmental Protection to the Antarctic Treaty (1991), 4 Oct 1991, 30 I.L.M. 1461 (1991). Entered into force 14 Jan 1998

Ramsar Convention (1971) Convention on Wetlands of International Importance Especially as Waterfowl Habitat, 2 Feb 1971, 996 U.N.T.S. 245. Entered into force 21 Dec 1975

Rayfuse R (2008) Protecting marine biodiversity in polar areas beyond national jurisidiction. Rev Eur Community Int Environ Law 17(1):3–13

Rogers P, Hall AW (2003) Effective water governance. Global Water Partnership Technical Committee (TEC) background papers no. 7

Rothwell D (2000) Global environmental protection instruments and the polar marine environment. In: Vidas D (ed) Protecting the Polar Marine Environment: Law and Policy for Pollution Prevention. Cambridge University Press, Cambridge

Rothwell D (2012) Legal Challenges for Maritime Operations in the Southern Ocean. In: 2012 Comité Maritime International Beijing Conference, 18 Oct 2012

Sofia Protocol (1988) Protocol to the 1979 Convention on Long-Range Transboundary Air Pollution Concerning the Control of Emissions of Nitrogen Oxides or their Transboundary Fluxes, 31 Oct 1988, 28 I.L.M. 212 (1989). Entered into force 14 Feb 1991

Stockholm Convention on Persistent Organic Pollutants (2001) 2256 U.N.T.S. 119. Entered into force 17 May 2004

Stokke OS (2001) Managing fisheries in the Barents Sea loophole: Interplay with the UN fish stocks agreement. Ocean Dev Int Law 32:241–262

Telecky T, Smith Z (2012) The science and policy interface in a United Nations Treaty: Polar bears and CITES. In: IPY 2012 Conference. Montreal. <http://132.246.11.198/2012-ipy/Abstracts_On_the_Web/by_theme.html>. Accessed 16 Nov 2012

UNEP (2004) Matishov G, Golubeva N, Titova G, Sydnes A, Voegele B. (eds) Barents Sea, GIWA Regional Assessment 11. Kalmar. University of Kalmar, Sweden

UNEP (2013) Minamata Convention Agreed by Nations. UNEP News Centre. <http://www.unep.org/newscentre/default.aspx?DocumentID=2702&ArticleID=9373>. Accessed 20 Feb 2013

UNEP/GRID-Arendal (2006) Background report for the seminar on multilateral environmental agreements and their relevance to the Arctic, Arendal, Norway, 21–22 Sept 2006

UNESCO (1972) Convention Concerning the Protection of the World Cultural and Natural Heritage, 16 Nov 1972, 1037 U.N.T.S. 151. Entered into force 17 Dec 1975

United Nations Framework Convention on Climate Change (1992) 1771 U.N.T.S. 107. Entered into force 21 Mar 1994

Vidas D (2000) The polar marine environment in regional cooperation. In: Vidas D (ed) Protecting the Polar Marine Environment: Law and Policy for Pollution Prevention. Cambridge University Press, Cambridge

WWF (2008) A new sea: The need for a regional agreement on management and conservation of the Arctic marine environment. World Wildlife Fund. <http://wwf.panda.org/what_we_do/where_we_work/arctic/publications/?122260/A-New-Sea-The-Need-for-a-Regional-Agreement-on-Management-and-Conservation-of-the-Arctic-Marine-Environment>. Accessed 15 Mar 2008

Young O (2002) Arctic governance: Preparing for the next phase. Presented at the Arctic Parliamentary Conference Tromsø 11-13 Aug 2002. <http://www.arcticparl.org/resource/images/conf5_scpar20021.pdf>. Accessed 29 Apr 2008

Young O, Osherenko G, Ekstrom J, Crowder LB, Ogden J, Wilson JA, Day JC, Douvere F, Ehler CN, McLeod KL, Halpern BS, Peach R (2007) Solving the crisis in ocean governance: Place-based management of marine ecosystems. Environment 49(4):20–32

Chapter 4
Arctic Indigenous Peoples and the Challenge of Climate Change

Adam Stepien, Timo Koivurova, Anna Gremsperger, and Henna Niemi

Abstract This chapter presents climate change impacts on indigenous traditional harvesting, cultures, identities, traditional knowledge, economies, societies, health, and infrastructure in light of overall socioeconomic and political changes in the Arctic. Responses to these stressors can be autonomous (e.g., ad-hoc responses within communities) or planned (i.e., governmental strategies). Responses are evaluated here in light of the predominant scientific and political discourse on vulnerability and adaptive capacity. This dominant vulnerability-adaptation approach has had a major influence on policy developments and research, though requires greater problematization and critical overview. Therefore, notions of intervention, trusteeship, power, and the use of the language of crisis are discussed. As an outcome of these deliberations, further and genuine empowerment is presented as a primary response to climate change impacts and adaptation challenges.

Based on Koivurova T, Tervo H, Stepien A (2008) Background Paper: Indigenous Peoples in the Arctic. Arctic TRANSFORM. This chapter has been restructured and expanded with recent scholarship and in-depth discussion on the concepts of vulnerability and adaptation.

A. Stepien · T. Koivurova · A. Gremsperger · H. Niemi (✉)
Arctic Centre, University of Lapland, Yliopistonkatu 8, 96300 Rovaniemi, Finland
e-mail: htervo@ulapland.fi

A. Stepien
e-mail: adam.stepien@ulapland.fi

T. Koivurova
e-mail: timo.koivurova@ulapland.fi

A. Gremsperger
e-mail: annagremsperger@gmail.com

E. Tedsen et al. (eds.), *Arctic Marine Governance*,
DOI: 10.1007/978-3-642-38595-7_4, © Springer-Verlag Berlin Heidelberg 2014

4.1 Introduction

The aim of this chapter is to present Arctic indigenous peoples in relation to the changing climate and marine environment. This overview is based largely on scientific and policy discussions on vulnerability and adaptation. The objective is not to present climate change impacts to indigenous peoples and adaptation options in detail, but rather to draw a general picture of climate change challenges alongside currently advocated approaches and ideas for adaptation. This allows for recommendations with an emphasis on empowerment as a primary response to holistically-viewed Arctic change. As this book focuses on the Arctic marine environment, issues typical for coastal, mainly Inuit, communities are highlighted.

The Arctic region is home to many culturally diverse indigenous peoples who have been exposed to a wide range of changes and pressures that include economic and cultural globalization, colonial policies, modernization, industrialization, major development projects, modern transport, and communication. These processes have brought about new industries, the emergence of mixed economies, acculturation, and social problems—producing both positive and negative outcomes for indigenous peoples. Climate change poses new and additional challenges for indigenous communities and facilitates creation of a 'total environment of change' (Moerlein and Carothers 2012).

Indigenous peoples have traditionally been adaptive and resilient to change. However, the scope and pace of climate change are seen as being beyond indigenous adaptive capacity—capacity that is already undermined by existing social pressures. The concepts of vulnerability, adaptive capacity, and adaptation must be viewed within the context of existing political and legal systems, which influence the ability of communities to cope with changing environmental conditions. In particular, traditional harvesting in marine areas is under threat, as these practices usually depend on sea ice and the availability of species. Culture, traditional ecological knowledge (TEK), economy, and health are bound with harvesting activities and are expected to undergo serious disturbances under climate change. Other major affected areas include community infrastructure and housing.

This chapter opens with an introduction to Arctic indigenous peoples and their situation in the face of climate change. Next, climate change impacts are presented, followed by a discussion on adaptive capacity and on autonomous and planned adaptation options, including an example from Alaska. This is juxtaposed with a critical discussion on the dominant adaptation discourse and concluded with certain options for moving forward.

The term 'indigenous' is used throughout the paper, although other terms—such as 'natives', 'numerously small peoples', or 'original peoples'—are also applied around the circumpolar North. As the United Nations (UN) Declaration on the Rights of Indigenous Peoples (UNDRIP 2007) uses 'peoples' as a leading term, this is preferred to 'people', 'populations', or 'groups'—although these terms can be used as synonyms without prejudice.

4.2 Arctic Indigenous Peoples

The Arctic region is home to many indigenous peoples with diverse cultural, social, economic, and historical backgrounds, including the Inuit of Russia, Alaska [United States (US)], Canada, and Greenland, Aleut, North American Natives (Athabascans, Gwitch'n, Métis), Sámi people of Fennoscandia, and numerous groups in Russia (*inter alia* Chukchi, Nivkhi, Sámi, Eveny, Evenky, and Nenets). Indigenous peoples constitute 10 % of the total Arctic population of 4 million (Nuttall 2000; IPCC 2007).

The definition of 'indigenous peoples' can be problematic. Indigenous peoples themselves are uncomfortable with definitions and emphasize (for instance in the UN Permanent Forum on Indigenous Issues) self-identification and common-sensical approaches (UNPFII n.d.). Nevertheless, international documents contain certain defining characteristics, as is the case with Special Rapporteur Martinez Cobo's report or the International Labour Organization (ILO) Convention No. 169 Concerning Indigenous and Tribal Peoples (1989) (Thronberry 1994; UNPFII n.d.). Indigenous peoples are defined as non-dominant groups descending from populations inhabiting certain areas before the time of conquest, colonization, or establishment of present state borders. These groups identify themselves as indigenous, have retained some of their own social, economic, cultural, and political institutions, and have maintained special relationships with their land and environment (ILO 1989, art. 1). The historic legacies and current positioning of indigeneity have bearing on climate change vulnerability, adaptation options, and empowerment (AHDR 2004).

Indigenous communities are often viewed from the perspective of *inter alia*, traditional livelihoods, cultures with close connections to the environment, and disadvantaged minority positions. While these characteristics have salience for vulnerability, other equally important factors include the development of mixed economies, technological progress, and colonial legacies. Policies for resettlement or assimilation and large-scale industrial and infrastructure projects have in some cases been imposed without consideration of indigenous social structures and within frameworks of colonial trusteeship and paternalism (Young 1995; Moerlein and Carothers 2012). Discriminatory practices, acculturation, usurpation of ownership and management of natural resources, and a rise in poverty are undeniable and persistent parts of the Arctic's historical and political landscape. Recent developments have aimed at enabling greater empowerment, participation, and the resolution of past injustices towards indigenous groups, though have in some cases resulted in producing new, complex co-management[1] and decision making frameworks that can be overwhelming for indigenous communities (Ford et al. 2010b).

Moreover, the Arctic no longer serves primarily as a harvesting ground for indigenous hunters, but has become a region that is increasingly crowded with other activities such as transport, shipping, resource extraction, commercial

---

[1] Co-management regimes are governance systems where resources, the environment, conservation, or land use are managed via joint (national, local, and indigenous) institutions and with strong participation from regional, local, community, and indigenous actors in policy making.

fishing, tourism, and nature conservation (AHDR 2004). Climate change is an additional factor affecting livelihoods that are already under pressure. Therefore, to fully understand climate impacts on indigenous peoples, these must be viewed within the context of modern indigenous societies and pertinent economic and political factors (Smit and Wandel 2006; Ford et al. 2006; Njåstad et al. 2009).

4.2.1 Traditional Harvesting and Mixed Economies

Understanding Northern mixed economies is a key component of assessing vulnerability and adaptive capacity (ACIA 2005; Pearce et al. 2012). Mixed economies are a combination of formal economies based on cash flows and traditional methods of acquiring food, clothing, and commodities. Harvesting and pastoralism remain important parts of Arctic livelihoods, with significance not only for health and food security, but also for culture and social ties. Today, however, complete reliance on harvesting is no longer possible due to changes in lifestyles, new standards of living, and declines in natural resources (Nuttall 2002; AHDR 2004; Ford et al. 2006; Poppel et al. 2007). Further, traditional subsistence is now subject to a range of regulations and resource management regimes, harvesting quotas, and access limitations (Nuttall 2000; ACIA 2005; Ford et al. 2006).

Financial resources support contemporary lifestyles, as well as modern harvesting in the Arctic, where costly technology such as snowmobiles, rifles, and GPS are now used (ACIA 2005; ICC 2005; IPCC 2007). In mixed economies, money is acquired through wage income (often through government jobs or resource extraction), governmental transfers, leisure hunting tourism, and selling of harvested goods and handicrafts (Nuttall 2000; AHDR 2004; Ford et al. 2010b).

Indigenous peoples' involvement in the formal, global economy creates both opportunities and challenges for adaptation governance. Extraction of hydrocarbons and minerals, tourism, military facilities, commercial fishing, and transport can offer the revenues required to support adaptation actions, but at the same time, make communities more dependent on world markets and undermine the long-term viability of traditional livelihoods. While many indigenous communities endorse industrial development, particularly when having legal control over land and resources, there is at the same time also strong opposition due to anticipated negative effects on environment and indigenous cultures (ACIA 2005; IPCC 2007; Bone et al. 2011).

An example of an external policy claimed to have bearing on indigenous mixed economies is the European Union's (EU) ban on the import and marketing of seal products (European Community 2009; Hossain 2012). The ban is seen to have resulted in adverse impacts to Arctic indigenous livelihoods and, consequently, the future adaptive capacity of Arctic communities. Although the EU's ban included an exemption for products coming from traditional indigenous seal hunting, it nonetheless caused a collapse of the market for seal products and thus limited potential Inuit income from traditional activities (European Parliament 2012). From the Inuit perspective, the EU trade regulation was an additional stressor in the Arctic's 'total environment of change'.

4.2.2 Challenges for Indigenous Societies and Culture

For centuries, newcomers to the Arctic have introduced new lifestyles, technologies, culture, and food, along with new policies and industries, which have increasingly affected all aspects of life in the Arctic (AHDR 2004; ACIA 2005; Pearce et al. 2009b).

Following such modernization and changes, unemployment and poverty have emerged as increasing phenomena in indigenous communities. Social and health problems such as alcohol and drug abuse, domestic violence, depression, and high suicide rates need to be taken into consideration when studying climate change impacts and vulnerability (AHDR 2004; ACIA 2005; Poppel et al. 2007).

For most Arctic peoples, family and communal ties are of great importance (Poppel et al. 2007), although traditional family structures are changing, with the nuclear family replacing the multi-generational model (AHDR 2004; Ford et al. 2006). Important parts of community identity and integrity, such as sharing and reciprocity systems and the position of elders, are in decline.

Problems with transmission of knowledge, culture, and language to younger generations are visible throughout the Arctic. For instance, changes to lifestyles, boarding schools, and educational systems have induced acculturation. They have, however, also created indigenous elites and economic opportunities (AHDR 2004; ACIA 2005; Furgal and Seguin 2006). Differentiated access to modern technologies has introduced additional inequalities (ACIA 2005; Ford et al. 2006; Pearce et al. 2010).

4.2.3 Political and Legal Framework

Political systems and indigenous governance vary throughout the Arctic. Past colonial policies aimed to assimilate indigenous peoples, but modern states' approaches have gradually shifted towards endorsing indigenous knowledge and values, protecting cultures, and introducing more participatory structures of governance (AHDR 2004). Today, trends of globalization, democratization, devolution, and human rights promotion are common in the Arctic region, although to varying degrees (for instance, in the Russian North, some opposing trends occur, especially regarding devolution). In areas where indigenous peoples constitute a majority, models of public government have been used (e.g., in Greenland and Nunavut), while in other areas, dual systems of governance prevail and indigenous and public governments coexist (e.g., Nordic Sámi Parliaments, the representative bodies of the Sámi). Powers have been redistributed to indigenous peoples in arrangements varying from granting legislative (e.g., agreements with Yukon First Nations) or decision making responsibilities, to providing consultation rights. In North America, land claims agreements have been a significant means of redistributing power and securing rights to territories and resources (AHDR 2004; Tennberg 2012).

Increasing recognition of human rights, including collective rights, such as self-determination, group rights, and land rights is visible both internationally and in the Arctic. Universal, European, and American human rights instruments [including the International Covenant on Civil and Political Rights (ICCPR 1966), International Covenant on Economic, Social, and Cultural Rights (1966), the European Convention on Human Rights (1950), and the Convention on Elimination of All Forms of Racial Discrimination (1966)] are generally applicable to indigenous peoples, although the acceptance of instruments differs among Arctic nations. The ICCPR has been ratified by all of the Arctic states and both Article 27 concerning minority rights and Article 1 concerning self-determination are seen by the Human Rights Committee as applicable to indigenous peoples, including their right to culture, traditional ways of life, land, and resources (HRC 1994, 1999/200/2002/2004). In addition, indigenous-specific instruments have been developed: in particular, ILO Convention No. 169 on Indigenous and Tribal Peoples in Independent Countries (ILO 1989), which has been ratified by Denmark and Norway (with Finland and Sweden considering ratification) and the UN Declaration on the Rights of Indigenous Peoples (UNDRIP 2007), which has been endorsed by all of the Arctic states except the Russian Federation. The goal of the currently negotiated draft Nordic Sámi Convention (see Koivurova 2008) is to harmonize and safeguard Sámi rights, including rights to marine resources.

Land and resource rights are critical for indigenous groups facing environmental change and resource extraction pressures (AHDR 2004). ILO Convention 169 (ILO 1989 arts. 14, 15) and UNDRIP give land and resources high priority, addressing recognition of ownership and possession of lands traditionally occupied, the requirement of indigenous free, prior, and informed consent, and states' positive obligation to identify and protect indigenous lands.

National developments in the Arctic have followed international trends. For example, the 2005 Finnmark Act in Norway gives the Sámi Parliament certain influence over management and titling of public lands in Norway's northernmost county (Bull 2008). For the most part, Arctic land and resources are publicly owned, although this picture is diverse and complex. In Alaska, for instance, 14 % of land is owned by Native Corporations, while in Russia, post-Soviet land ownership is often unclear due to complex legislation and legacies of Soviet collectivisation. Perhaps the most multifarious situation is in Canada, where a variety of treaties, land claim agreements (including the Nunavut Land Claim Agreement and Inuvialuit Final Agreement) create a patchwork of co-management and ownership, with situations differing greatly between provinces and territories (AHDR 2004).

4.2.4 Arctic Cooperation

Both the Arctic Council and Barents Euro-Arctic Region work to engage indigenous peoples and perform substantial work on the issues of the climate change and adaptation (Koivurova and Hasanat 2009; Hasanat 2010). Since 1996, indigenous organizations are to be fully consulted in the Arctic Council as permanent participants, a unique status within the international context. By participating in all of the activities of the Arctic Council, including working groups, indigenous organizations have been able to raise their concerns and emphasize the importance of their traditional knowledge, as with the Arctic Climate Impact Assessment (ACIA) and the Arctic Marine Shipping Assessment (ACIA 2005; AMSA 2009). The salience of the permanent participant model is visible in the demands to adopt a similar solution within the Barents Euro-Arctic Region (Koivurova and Heinämäki 2006; Koivurova 2011b).

4.3   Climate Change Impacts, Stressors, and Indigenous Vulnerability

Rapid non-linear environmental changes are expected to push the social-ecological systems of the Arctic beyond tipping points (Pearce et al. 2010; Carmack et al. 2012). Following the ACIA and the International Polar Year 2007–2009, a surge of research on Arctic community vulnerability and adaptation was seen (Ford 2009). The UN Framework Convention on Climate Change (UNFCCC 1992) has introduced vulnerability language by calling for attention to particularly vulnerable regions such as the Arctic (UNFCCC 1992, art. 3.2; Smit and Wandel 2006). Vulnerability approaches have dominated the scientific and political discourse on climate change impacts (Pearce et al. 2009b; Tennberg 2012; Cameron 2012).

Vulnerability can be understood as a function of the "exposure of the community to climatic conditions and adaptive capacity to deal with such conditions" (Smit and Pilifosova 2003) with both human and bio-physical determinants playing a role at multiple spatial and temporal scales (see also Ford 2009). When considering vulnerability, researchers examine who is vulnerable and how, to what stresses, in what way, and what the local capacity to adapt is. Social, economic, political, and ecological features (experienced diversely at a community level) are assessed. Understanding of climate change impacts has been influenced by indigenous observations and TEK, supplementing scientific observations (Smit and Wandel 2006; Pearce et al. 2010).

Climate change directly impacts traditional livelihoods and has indirect effects on indigenous societies. For Arctic indigenous peoples, climate change affects hunting, whaling, and fishing activities as sea ice, weather conditions, and the availability and health of harvested species change – all crucial aspects of successful and relatively safe harvesting. Furthermore, changes or decreases in harvesting activities have implications for indigenous economies, societies, cultures, and health. Not all observed and predicted climatic changes are necessarily negative, but overall impacts are seen as having increasingly adverse effects on indigenous livelihoods (ACIA 2005; Moerlein and Carothers 2012; see Fig. 4.1 for a summary of impacts).

4.3.1   Primary Impacts on Livelihoods, Harvesting, Health, and Infrastructure

Traditional harvesting is still seen as a necessary and important element of Arctic indigenous lifestyles and cultures (Ford et al. 2006, 2010b; Pearce et al. 2010).

Sea ice, on which many traditional harvesting activities depend, is an integral part of life for many Arctic peoples and is tied to some of climate changes' major impacts. The health, location, and abundance of harvested animals are adversely affected by disappearing sea ice, on which species depend, as well as changing

Elements of climate change	Harvesting/ availability of resources	Infrastructure and transport	Economy	Society and culture	Health
Changing ice and snow conditions	-Less access to certain, particularly ice-dependent, species -More polynyas make fishing and duck hunting easier -Changes in harvesting cycles -Difficulties building igloos	- Thinning ice reduces available travel time and increases risks - Challenges for ice-dependent communities to maintain connections with population centres - Better sea access for coastal settlements	- Travelling and harvesting become more expensive - Replacing country food with imported food creates additional expenses	- Ice, an integral part of indigenous cultures, is lost - Risk of loss of hunting cultures - Sharing systems may decline	- Possible rise in accidents during harvesting and travel on ice - Loss of country food, which is often healthier than imported products - Increased risk of health problems such as diabetes, obesity, or cardiovascular diseases
Warming waters	-Sea mammals and fish species shift locations and change in numbers	- In some cases, necessary travelling of longer distances to reach resources increases costs and hazards	- Loss of important food sources (as above)	- Decrease in hunting and fishing success and/or frequency of harvesting activities	- Less country food (as above)
Thawing permafrost, coastal erosion, rising sea level, and decreasing river run-off	-Decreased fish and migratory duck populations	- Damage to houses, roads, power lines, airstrips, water supplies and sewage systems - Community relocation - Damage to permafrost or ice-based food storage - Changes to water supply	- Rising costs of infrastructure maintenance - High costs of settlements relocation - Higher costs of food preservation	- Damage to sacred sites - Estrangement to the environment and changing landscapes - Traditional knowledge no longer applicable	- Possible rise in accidents

Fig. 4.1 Selected impacts of climate change on coastal indigenous communities. Only the most vital of many concurrent factors have been mentioned and taken out of the broader social-economic-political context. Table is based partly on tables available in ACIA 2005

General warming, intensification of extreme weather events, and other impacts	- Mitigation options limited due to impacts to terrestrial species (e.g., reindeer herding affected by snow and river conditions) - Worsened health of harvested animals - Shorter and/or less frequent harvesting trips	- Uncertainty of the weather, more storms - Houses not prepared for warmer average temperatures - Damage to infrastructure and housing	- Rising costs of infrastructure maintenance - Savings on heating - More jobs due to development of agriculture, tourism, shipping, and resource extraction - Increased need for expensive technologies to ensure travel safety	- Traditional knowledge no longer useful - Migration of settlers from the south may cause acceleration of acculturation - Effects on transfer of traditional knowledge and skills to younger generations	- Invasive insect and animal species bringing new diseases - Inadequate housing - UV radiation - Rise in accidents - Less country food (as above)
Autonomous adaptations	- Changing timing and frequency of harvesting - Harvesting new species - Travelling longer distances - Use of modern technologies - Carrying more equipment	- Acquiring information on changing snow/ice conditions - Ad-hoc construction strengthening - Changing modes of transportation	- Supplementing country food with store food - Engaging to a greater extent in wage economy		
Planned adaptations	- Harvester support programmes - More flexible regulatory frameworks responsive to changes in environment	- Need for better information - Spatial planning adjusted to changing environments - Funding for infrastructure renewal - Relocation - Improving search and rescue	- Job creation in the North - Community freezer programmes - Facilitating cash acquisition via traditional harvesting	- Supporting the transfer of traditional knowledge - Cultural programmes	- Better information on health impacts - Health programmes responsive to additional stressors - Better surveillance
General actions	- Need for general empowerment - Better designed decision making processes - Research on adaptation planning and implementation, sensitive to indigenous perceptions - Human rights-based approach - Integration of various policy fields				

Fig. 4.1 Continued

ocean productivity, altered water salinity due to melting ice and river run-off, and the emergence of new diseases and insects (ACIA 2005; ICC 2005; IPCC 2007). These changes additionally result in worsened quality of meat and hides and increased reliance on store food, which can be more costly and have negative dietary implications. Other climate change impacts to traditional harvesting activities range from changes in hunting logistics (e.g., impossibility of igloo construction) to communities being cut off from hunting areas owing to later autumn freeze-up and earlier spring melt (Nuttall 2002; Abate 2007; Poppel et al. 2007).

Hunting is seen as an increasingly dangerous activity as ice undergoes various transformations, such as the occurrence of more polynyas and areas of unusually thin ice. Under climate change, travelling becomes more dangerous, less frequent, and often requires either shorter or much longer distances, depending on local sea ice conditions and availability of harvested species. Hunters increasingly rely on modern technologies. Climate change and changing ice conditions also influence other livelihoods practiced by indigenous coastal communities such as reindeer herding and hunting and fishing (ACIA 2005; Abate 2007). Less commonly, communities' situations may improve; for example, due to newly gained access to large fish stocks (ACIA 2005; Ford et al. 2006).

Impacts on transportation in the Arctic, as noted in relation to harvesting activities, affect connections between communities and population centres (Pearce et al. 2010). Thawing permafrost and coastal erosion cause roads to disintegrate. In the spring and autumn, routes based on ice and snow cover are affected and in the summer, rivers may be un-navigable due to decreases in run-off. On the other hand, summer sea access for many coastal communities may improve (ACIA 2005; ICC 2005; IPCC 2007).

Construction of buildings and infrastructure such as sewage systems, airstrips, power lines, and roads built on permafrost and eroding coastlines are endangered or have already experienced climate change impacts, due not only to thawing permafrost but also to enhanced storm strength, river run-off, and flooding (ACIA 2005; IPCC 2007). Some communities may be confronted with the prospect of relocation (ICC 2005; Ford et al. 2006; Pearce et al. 2010).

For Arctic indigenous peoples, the health impacts from climate change are multiple and complex. Experts predict the introduction and expansion of new diseases and allergies due to the arrival of southern plant, insect, and animal species as well as changes in diet that may contribute to obesity, diabetes, and cardiovascular disease. Increases in sunburns due to UV radiation are expected (Furgal and Seguin 2006; ACIA 2005; ICC 2005; IPCC 2007). Freshwater supply and traditional methods of food preservation are also impacted by climate change, and the probability of contamination due to damage to pipelines and permafrost-based waste containers increases (ACIA 2005; ICC 2005; IPCC 2007). Conversely, health benefits are also possible, such as a reduction in winter mortality and cold-related injuries (IPCC 2007).

4.3.2 Impacts on Northern Economies, Societies, Cultures and Health

The formal, cash economy is an alternative to traditional harvesting and is an essential part of mixed economic systems, though many Arctic regions today experience high levels of unemployment and poverty (ACIA 2005; ICC 2005; IPCC 2007). Climate change may increase economic hardship due to rising living costs (infrastructure maintenance, harvesting costs, transport, increased demand for store food), despite expected savings in heating and insulation (ACIA 2005; IPCC 2007). Infrastructure damage by climate change will adversely impact access to communities, transport costs, trade, development of tourism, and job creation, endangering aspects of the indigenous formal economy (ACIA 2005; IPCC 2007; Pearce et al. 2010).

At the same time, increased agricultural and fisheries productivity and opportunities for resource extraction may create more jobs and promote investment in the region, although many of the jobs offered to indigenous communities are likely to be of low quality due to both disparities in education and training and the character of positions (ACIA 2005; IPCC 2007). Moreover, an influx of southern employees to fill new positions might increase acculturation pressures (AHDR 2004).

Changes to the Arctic landscape and harvesting activities are expected to influence social structures, affect communities' distinct ways of life, and even endanger survival of some groups as distinctive peoples (ACIA 2005; Bravo 2009). Although, as discussed in the previous section, social and economic impacts currently pose a greater threat to indigenous societies than climate change itself, climate change impacts are expected to increase the pressures and speed of certain negative processes (ACIA 2005; ICC 2005).

Social and cultural features perceived as vulnerable in the context of the 'total environment of change' are numerous, and many are especially connected to climate-induced impacts to traditional harvesting, impoverishment, out-migration, and changing landscapes (IPCC 2007; Pearce et al. 2010). Such social and cultural features include social and family ties, traditional sharing systems, and the authority of elders. Historical, sacral, and cultural sites, such as cemeteries, are threatened by the environmental impacts of climate change (AHDR 2004; ACIA 2005; Ford et al. 2006). Changes in landscape and declines in sea ice—a cultural, social, and economic component of indigenous coastal communities—affect the sense of place and belonging (ICC 2005). Limitations to traditional harvesting also have adverse impacts on self-esteem, tied to notions of lack of opportunities. Climatic pressures on culture, in connection with other social pressures, are expected to amount to a level of social crisis and to have effects on mental health. Critical impacts include depression, alcoholism, drug addiction, permanent unemployment syndromes, and rising suicide rates (ACIA 2005; Furgal and Seguin 2006).

Traditional knowledge provides community members with an understanding of their environment and binds them with nature and each other, but is claimed to be less applicable to an environment undergoing new and unprecedented changes (AHDR 2004; Abate 2007). Coupled with a decrease in harvesting activities, this causes alienation to land and disruptions to community relationships and inter-generational transfers of knowledge, values, and culture (ACIA 2005; IPCC 2007; Moerlein and Carothers 2012).

4.4 Adaptive Capacity and Proposed Responses to Climate Change

The hazardous impacts of climate change are seen as already exceeding national and community capacities to cope through use of traditional strategies and behaviours (Smit and Pilifosova 2003). There is a need for better and more feasible adaptation responses and options, both of an autonomous and planned character (Carmack et al. 2012; Smit and Wandel 2006; Hovelsrud and Smit 2010).

4.4.1 The Concepts of Adaptation and Adaptive Capacity

Adaptation can be described as a "process, action, or outcome in a system in order for the system to better cope with, manage or adjust to some changing condition, stress, hazard, risk, or opportunity" (Pearce et al. 2010). Adaptation is not a term specific to climate change, and generally accompanies all processes of human learning and development. It is a multi-stage, problem-solving process in response to situations where normal responses cease to be sufficient, but it is also a conflict-inducing challenge for social organization. Climate adaptation policy can be viewed as transition management of changes and impacts that are inherently uncertain (Smit and Wandel 2006; Tennberg 2012).

Adaptation may be either autonomous or planned (Smit and Wandel 2006). In the former, communities and societies adapt to changing environmental conditions on a continuous basis. Conversely, planned climate change adaptation measures, such as spatial planning or building code adjustments, are still in the early phases of development and implementation. Adaptation is neither a linear nor a homogenous process, with some communities or sectors adapting better or faster than others (Keskitalo 2010; Tennberg 2012).

Discussion on adaptive capacity has been instigated by the International Panel on Climate Change (IPCC), where adaptive capacity is defined as the "ability or potential of a system to respond successfully to climate variability and change, including adjustments in behaviour and in resources and technologies" (IPCC 2007). Concepts closely associated with adaptive capacity include coping ability, management capacity, stability, and flexibility, as well as resilience, conceptualized as the capacity of ecosystems and populations (or socio-ecological systems) to recover from stress or change (Tennberg 2012).

All elements of the vulnerability-adaptation-resilience scientific model are seen as place-specific and dynamic, and are subject to on-going alterations that are dependent on policies, environmental variations, individual and community values, and changing conditions within communities (Smit and Wandel 2006; Tennberg 2012).

Assessments of both vulnerability and adaptive capacity depend on knowledge of past experience with climatic conditions, traditional responses to climate variations, future climate change projections, and understanding of non-climatic factors (Pearce et al. 2009b). Adaptive capacity is dependent on factors such as managerial abilities, infrastructure, institutional and political environments, and access to financial, technological, and information resources. Kinship and social networks, social capital, strength of local institutions, and flexibility are also of importance. Broad community support for adaptation policies, willingness to change, and good linkages to external governance institutions are often necessary for successful adaptation. Socio-environmental systems need to be analysed holistically, on not only on a community level but on all interdependent scales (e.g., regional, national, international) (Smit and Wandel 2006; Carmack et al. 2012). Enhancing adaptive capacity requires addressing pertinent locally identified vulnerabilities and involving local stakeholders (Smit and Pilifosova 2003; Carmack et al. 2012).

4.4.2 Autonomous Adaptations

Research has shown Arctic communities to be highly adaptive and resilient, but also increasingly overwhelmed by environmental changes (AHDR 2004; Ford et al. 2010b; Pearce et al. 2010). Arctic indigenous peoples are accustomed to natural variability and unpredictability. For example, fish stocks and wildlife location and availability or sea ice extent can change significantly from year to year. Communities have developed flexible social structures (i.e., group size, multiple means of livelihood), which allow them to adapt without significant cultural loss (ACIA 2005). Nonetheless, the current and projected rate and scope of climate change in the Arctic presents a significant challenge in comparison to earlier experienced phenomena (ACIA 2005; Pearce et al. 2009b). Moreover, climate change will impact communities that are already affected by social, economic, cultural, and political pressures (ACIA 2005; AHDR 2004; Pearce et al. 2010). For example, inequality of access to traditional and monetary resources weakens social networks, and weaker social network in turn affect transmission of traditional skills and knowledge to younger generations (Moerlein and Carothers 2012; Ford et al. 2006; Bone et al. 2011).

A number of autonomous climate adaptation strategies are already being applied in the Arctic, although most are reactive and individual in nature. Traditional adaptation mechanisms relevant for climate change adaptive capacity include changing the timing and frequency of hunting activities, harvesting new species, and lengthening travelled distances to follow moving animals (Ford et al. 2006; Pearce et al. 2009a, 2010). While these traditional mechanisms already exist, time is still needed for learning new harvesting and livelihood practices in an altered environment. For example, hunters must prepare themselves for lengthier ventures, impossibility of igloo construction, and more frequent storms by taking additional equipment and carefully planning journeys. Moreover, hunters and herders must acquire current information on weather and ice conditions, utilize different modes of transportation, travel routes, hunting areas, and species harvested. Modern technologies (e.g., snowmobiles, GPS, satellite phones, all-terrain vehicles) play a significant role in such autonomous adaptations. Technological adaptations, however, make hunting also more costly (ACIA 2005; ICC 2005; IPCC 2007) (See Fig. 4.1 for the summary of impacts and adaptations).

4.4.3 Adaptation Planning and Governance

Autonomous adaptation measures at the community level need to be considered in a broader governance context and coupled with adaptation planning. Governmental programmes are seen as crucial for adaptive capacity-building and preservation of indigenous hunting cultures. Adaptation planning is hoped to further social and economic well-being by insulating communities from threats and enhancing realisation of opportunities connected with climatic and economic changes (Pearce et al. 2010; Lukovich and McBean 2009; Tennberg 2012). Although the field of adaptation planning appears to be underdeveloped, many Arctic municipalities and regions are already designing climate change adaptation plans (Tennberg 2012; Hovelsrud and Smit 2010).

Adaptation governance does not occur in a political, economic, or administrative vacuum and is closely tied to existing political and administrative structures and practices (Tennberg 2012). In addition to climate change impacts, many adaptation policy options simultaneously address pressing economic, social, cultural, and health concerns pertaining to northern communities and may bring near-term positive benefits even before the occurrence of major climate change impacts (Ford et al. 2010b; Pearce et al. 2010). Addressing climate change may be impossible in communities facing serious social problems (ACIA 2005; IPCC 2007; Smit and Wandel 2006). Due to the perceived urgency of climate change impacts, locating adaptation initiatives within existing decision making procedures is often advised; for instance, within established resource management systems, disaster preparedness planning, and sustainable development programmes. Adaptation initiatives should therefore be incremental, modifying rather than replacing current structures (Smit and Pilifosova 2003; Smit and Wandel 2006).

Proposals for culturally appropriate and acceptable adaptation strategies include: creation of forums between government decisionmakers and indigenous leaders; promotion of assessment, data gathering, and information, and integration or mainstreaming[2] of climate change adaptation into policies (e.g., land-use plan-

---

[2] Mainstreaming is a policy concept whereby a certain issue is considered across different policy areas so that each addresses aspects of the issue concerned.

ning, fish and wildlife management, water management). In order to successfully implement adaptation actions, ownership, use, and access to land and resources must be settled—an issue that continues to be a major challenge in many parts of the Arctic.

Adaptation strategies must meaningfully involve indigenous peoples at the outset and secure their participatory rights. Policies should be designed in a cooperative manner and founded on cooperative research to be both legitimate and potentially successful. Without local engagement, policies would be either based on inaccurate goals or meet with resistance from within communities (ACIA 2005; Ford et al. 2010b). This may require authorities to enhance the financial, human, and technical capacity of indigenous organizations and governments in order to enable them to draw up their own visions of adaptation (ACIA 2005).

Options for adaptation actions are numerous (Wenzel 2001; Ford et al. 2010b). For instance, programmes may support TEK transmission and training in use of modern technologies. Financial support for adaptation (e.g., harvester support or community freezer programmes) should be provided alongside enabling more cash acquisition through traditional harvesting via sport hunting, tourism, and commercial sale of hunting outcomes (Ford et al. 2006). Safety threats require: enhanced search and rescue capabilities, forecasting of weather, ice conditions, and hazards, and protection of major infrastructure. Integrated land use planning, moving new construction away from endangered areas, and promoting awareness of climate change are seen by researchers as vital measures to take (ACIA 2005; IPCC 2007; Tennberg 2012). Health issues are identified as particularly urgent. Researchers have called for more surveillance, establishing early warning systems, and culturally specific health assessments (Ford et al. 2010b).

Adaptation planning must be a dynamic process based on the on-going production of knowledge. Governments, institutions, researchers, and communities must perpetually monitor, evaluate, and redesign adaptation strategies as underlying conditions change. Consequently, legal and regulatory arrangements should also be incremental and flexible in nature (ACIA 2005). For instance, reassessment of hunting and fishing regulations (e.g., spatial limitations or quotas) as well as conservation laws may prove necessary as the conditions upon which they are premised are radically altered (Dahl et al. 2001; ACIA 2005). In more heavily bureaucratized Arctic governance environments, adaptive capacity depends on institutional flexibility. Despite the progress of co-management regimes, there is still a need for more flexible, multi-level governance schemes with the ability to serve as arenas of joint decision making, conflict resolution, and information sharing. Evolution of such flexible systems requires co-production of knowledge, exchange of best practices, local and decentralized decision making, and acknowledgment of traditional rights. However, governance changes are not expected to be smooth, as so far the prior reconstruction of governance in the North has been a difficult undertaking underpinned with inter-group conflicts (Dahl et al. 2001; Ford et al. 2010b).

In order to enhance adaptive capacity, education, and exchange of information on vulnerability, behavioural adaptation options are advocated with rising focus on indigenous youth and trans-generational transmission of culture and knowledge (Magga et al. 2011; Carmack et al. 2012). Diverse, audience-specific approaches and materials integrating indigenous history and culture are needed for education, communication, capacity-building, outreach, and training (Njåstad et al. 2009; Pearce et al. 2010; ACIA 2005).

Both the viability and transmission of TEK is increasingly seen by indigenous politicians and researchers as central to adaptation and resilience (Magga et al. 2011; Ford et al. 2010b). Integration of TEK and community observations with scientific knowledge and policy making is seen as crucial for production of applicable research and effective policies. Involving local people at all stages of research projects and taking into account their observations is intended to make research more holistic, participatory, and relevant for communities (Pearce et al. 2009a; Berkes and Jolly 2001; Bravo 2009).

Research has played and is to play a major, if not primary, role in the debate on Arctic climate change (Cameron 2012; Bravo 2009). Assessment of impacts in regard to long-term scenarios, locally-focused analysis of specific adaptation options, and comparative studies are encouraged (ACIA 2005; Smit and Wandel 2006; IPCC 2007) as well as studies on already implemented measures and adaptations processes (Tennberg 2012; Ford et al. 2010b).

Calls have been voiced for a comprehensive Arctic strategy dedicated to human security and human rights (Ford et al. 2010b; Lukovich and McBean 2009). There is a movement towards inclusion of human well-being, adaptation, and vulnerability in Arctic security discourse, or, alternatively, framing these issues within the language of human rights. For example, food security, when expressed as a right to food, produces a duty on the side of public authorities. From an adaptation standpoint, governments, in light of their human rights obligations, would need to tackle inequalities and provide resources for adaptation (Adger 2010; Ford et al. 2010a).

4.4.4 Barriers to Adaptation

Both autonomous adaptation and governmental adaptation planning have limitations. Major constraints include sociopolitical values, institutional capacity, and lack of economic, technological, or information resources (Ford et al. 2010a, 2010b). Local governments often lack the resources necessary to conduct adaptation activities (e.g., in the Russian Federation; see Tennberg 2012), and such deficiencies are even more visible in indigenous communities. Adaptive capacity depends on financial resources (e.g., ability to purchase equipment and engineering responses to infrastructural damages) and the ability to use modern technologies. Social values, too, which are often neglected in planning, are crucial and include the use of social networks for sharing risks, readiness to adjust behaviour, and transmission of traditional knowledge (Ford et al. 2006; Pearce et al. 2009b, 2010). Developing adaptation responses can also result in conflicts between diverse political, economic, legal, and conservation interests. Engagement in adaptation governance and planning also depends on different perception of risks by public authorities and communities (Tennberg 2012; Keskitalo 2010; ACIA 2005).

Indigenous people have become more dependent on the outside world, which increases their adaptive capacity, but, paradoxically, also amplifies their vulnerability to the impacts of climate change. A settled way of life may improve educational opportunities and increase material standards of living, but at the same time, people may become dependent on mechanized transportation and fossil fuels.

Today, some past adaptation options, such as reconstruction or community migration are either unavailable, limited, or very costly. Modern, permanent settlements have elaborate infrastructure, raising the costs of both maintenance and resettlement. For example, engineering solutions to climate change impacts on infrastructure and constructions in Canada's Northwest Territories were estimated at 200-420 million CND and relocation of one community of Inupiaq was predicted to cost 50 million CND (Ford et al. 2010b; Pearce et al. 2010). However, even where financial resources are available, relocation is often seen as an unacceptable option, particularly due to past indigenous experiences of forced and semi-forced population movements such as in the 1950s–1960s in Canada and in the Stalinist Soviet Union. Relocation also results in social problems or deepening of cultural loss (ACIA 2005; IPCC 2007; AHDR 2004).

Education is a priority issue in the Arctic if autonomous and planned adaptations are to be successfully implemented (Lukovich and McBean 2009). Weakening of TEK and its transmission to younger generations has already been observed and not only increases vulnerability by affecting autonomous and planned adaptations, but also adversely impacts indigenous cultural development (ACIA 2005; Ford et al. 2006; Tennberg 2012). Young people are not always interested in traditional livelihoods and there is a rising knowledge gap between older and younger generations. New methods of knowledge transfer, such as cultural programmes and trainings, and documentation of knowledge are seen by some as a remedy for this (Bone et al. 2011; Moerlein and Carothers 2012). Young people have also been viewed as not fully prepared to assume the responsibilities of older generation in indigenous politics and resource co-management (Ford et al. 2010b; Pearce et al. 2010).

Box 4.1 Alaskan experiences with adaptation governance

Over the past 50 years, Alaska has been warming at more than twice the rate of the rest of the US (USGCRP 2009). The US Government Accountability Office has reported that most of Alaska's 200 native villages have been affected by flooding and erosion, likely in relation to climate change, and relocation options have been considered for several communities (GAO 2004).

In general, experiences with adaptation governance around the circumpolar North are limited, and that is also the case for Alaska. In 2010, the Alaskan state government prepared a report on climate change impacts and possible adaptive action (Adaptation Advisory Group 2010), although the document has not been fully endorsed by state authorities. The report analyzes climatic changes and expected impacts (with conclusions similar to those presented earlier in the chapter) focusing particularly on fisheries, infrastructure, and traditional livelihoods. It contains adaptation options designed to reduce the vulnerability of Alaska's natural and human systems to climatic changes and examines potential economic opportunities (e.g., due to a longer summer tourism season or increased navigation).

Having future adaptation options as a point of reference, the strategy operationalized the notion of ecosystem services, seeking to sustain ecosystem services in order to meet essential needs for food, water, renewable resources, community stability and safety, and cultural well-being. Five areas for adaptation covering broad spectrum of governance were identified: fish and wildlife, fisheries management, wildland fire, freshwater management, and invasive and eruptive species (Adaptation Advisory Group 2010).

Actual implementation of adaptation policies in Alaska remains somewhat limited, although action has been initiated at the state, national, and local levels. Some funding, primarily from the US federal government, has been provided for further research on future impacts and for community adaptation. The US Department of Interior established the Alaska Climate Science Center (part of the US Geological Survey), which finances research on, *inter alia*, impacts on communities and models for future climate change-related phenomena such as coastal erosion (AK-CSC 2012). In 2011, the Alaska Native Fund was introduced to support, among other areas, the protection of cultures and communities from future climate change impacts. Community capacity-building or trainings are among the supported actions (MFPP 2012). Other initiatives include the university-based Marine Advisory Programme (MAP) funded by the State of Alaska and US Department of Commerce, which seeks to provide Alaskan coastal communities with advice on adaptation to the changing environment. MAP proposes, for example, utilizing new species in hunting grounds, engaging elders to teach youth adaptability, avoiding construction on frozen subsurfaces, rerouting transportation routes to avoid vulnerable permafrost areas, and identifying more secure sources of water (MAP website). Some Alaskan municipalities have also taken action to implement adaptation planning projects (see Feifel 2010 on the case of Homer).

4.5 Criticism Towards Vulnerability and Adaptation Approaches

Vulnerability and adaptation discourses are predominant in research and policy making regarding climate change impacts on Arctic indigenous communities. Participatory research and governance, co-management, and adaptive capacity-building actions are highly promoted. There is, however, also criticism to such approaches—to notions and narratives they produce and to the resulting exclusion of certain issues from the scientific and political discourse.

4.5.1 Crisis Narrative and Resilience Language

Bravo (2009) sees the climate change debate in the Arctic as dominated by resilience/instability/crisis narratives. These narratives seek to make abstract climate science tangible and morally relevant by utilizing the plight of vulnerable indigenous peoples, and therefore are charged with moral overtones and emotional rhetoric. Indigenous peoples, however, often reject being presented as populations on the edge of extinction and cultural collapse. Vulnerability discourses are seen as solidifying the victimization of indigenous communities (Lindroth 2011; Niezen 2003). Conversely, living from the land means an on-going process of negotiating one's position in a changing environment (Forbes and Stammler 2009), and adaptation itself is a crucial part of life, not a catastrophe. Still, developing an alternative climate change narrative free of crisis dimension appears to be a difficult task (Bravo 2009).

Consequently, the language applied in adaptation discourse is viewed as disturbing by some researchers. Ideas coming from natural science are applied to complex human societies and relationships in a highly politicized Arctic milieu. For example, adaptation and resilience debate is criticized for the introduction of reactive and path-dependent thinking and the notion of permanent instability—where the aim is to govern something as indeterminate and uncertain as climate change (Tennberg 2012; Carmack et al. 2012). Moreover, there is a paradox in the fact that at the same time that adaptive capacity of indigenous communities has decreased due to modernization, dependence on public transfers, and so-called 'technology-induced environmental distancing'[3] (Bone et al. 2011), the same modern technology and dependence on financial resources, wage economies, and state support are among proposed adaptations (Cameron 2012).

These applied terms, by taking a singular focus, can also reduce the visibility of other related problems. Adaptation discourse often averts issues of colonial legacies and injustices, not only making adaptation strategies potentially ineffective, but also possibly strengthening processes leading to dispossession, inequality, and further political marginalization (Cameron 2012; Bravo 2009).

Cameron (2012) sees a danger of confining our understanding of indigeneity in the context of climate adaptation to what is *local* and *traditional* (see also Tennberg 2012). As a result, in climate change debates indigenous groups are often excluded from discussions on issues located outside the *traditional* and *local* (although these notions are of major importance), such as hydrocarbon and mineral extraction, shipping, sovereignty, or militarization of the region (e.g., ICC

---

[3] 'Technology-induced environmental distancing' means that the more technology is applied in daily lives, the less the individual and community understand and interact with surrounding environment.

2009). The role of permanent participants in the Arctic Council can be seen here as a positive example, a step outside of the local/traditional limit and endorsing indigenous peoples as equal and capable actors and not only victims of change (Koivurova and Heinämäki 2006). Such a participatory and empowering approach should be more pronounced in governance structures.

4.5.2 Adaptation Governance as Intervention

Many adaptation policies are presented as new forms of state intervention—usually of a neoliberal, i.e., market-based and technical character. These mirror past well-intentioned, but currently criticized, interventions into people's lives and relationships with the environment. Adaptation ideas are often limited to technical approaches and economic and market logic, including cost-benefit, cost-effectiveness, and efficiency calculations. Economic concerns, such as high relocation and infrastructure costs, harvesting expenses, unemployment, and low wages, have indeed dominated the debate (Moerlein and Carothers 2012).

To make management of climate change impacts possible, risks need to be construed as manageable and governable. Communities should be encouraged to take responsibility and urgently adjust to changes (Tennberg 2012; Slezkine 1996; Cameron 2012). Due to notions of crisis and urgency, the *will to improve* indigenous lives (i.e., secure their well-being) may take a form of trusteeship, resembling colonial way of thinking (Cameron 2012; Escobar 1995; Li 2007).

The consequence of presenting environmental change and adaptation as purely technical problems is that it depoliticizes climate change and makes it subject to expert direction, often confined to existing systems of governance and excluding related issues that lie outside of purely technical discussion (such as resource extraction, commercial fisheries, or colonial legacies) (Cameron 2012; Bravo 2009). Governmental routines, including participatory procedures, may be seen as ways in which indigenous resistance to interventions is de-politicized, managed, and limited to bureaucratic forms (Tennberg 2012). External experts claim stakeholder and stewardship status in regard to resources, governance, and the environment of the region (Bravo 2009; Nuttall 2002).

4.5.3 Using Traditional Knowledge

There are also critical voices regarding the use of TEK in predominant research and policy making. There are problems with translation, representation, and de-contextualisation of TEK as something produced in interactions between communities and researchers where information based on TEK is taken out of holistic spiritual systems and local practices and used as data for western science (Cameron 2012). Often, terms and ideas used in research are not in line with the actual perceptions in communities and are rather abstract, exogenous, managerial, or scientific concepts like *TEK*, *reindeer management*, or *climate change* (Forbes and Stammler 2009; Moerlein and Carothers 2012; Njåstad et al. 2009). Moreover, climate change perceptions are shaped not exclusively by community observations, but also by interpretations and ideas coming from outside. Views of natural hazards and their impacts and likelihood are socially constructed through research, politics, and media, and therefore, economic, social, and political conditions have bearing on perception of and responses to such impacts (Ford et al. 2006; Bravo 2009).

Criticism of vulnerability-adaptation approaches, at least within this chapter, is not intended to dismiss discourse on the need for adaptation to a changing environment. Even if narratives on cultural collapse often appear exaggerated, indigenous communities indeed depend on traditionally used resources, with broader implications for society and culture (Bravo 2009). Vulnerability-adaptation research and recommendations form the only relatively coherent and practically applicable proposition so far for responding to climate change. However, there is a need to identify and problematize the shortcomings of the vulnerability-adaptation-resilience approach—as it has been directing policy making—and to supplement it by a broader view of human communities and cultures than that offered by human ecology studies (Cameron 2012). Consequently, as the debate on community adaptation cannot be confined to technical terms, community participation and empowerment of indigenous peoples need to play central roles.

4.6 Empowerment as a Primary Response

An overview of climate change impacts, vulnerability, and adaptive capacity studies, autonomous and planned adaptations and barriers, as well as criticism to current climate change discourses, all lead to the conclusion that further and more genuine empowerment need to be a primary response to climate change and to interconnected socioeconomic-political stressors, even before planned adaptation actions are taken. This may offer Arctic governance a safe passage between the need for active adaptation policies and the danger of new state interventionism and paternalistic policies. Without genuine empowerment, truly inclusive decision making and meaningful participation, adaptation, and adaptive capacity-raising efforts may prove ill-conceived, ineffective, costly, or turn into a continuation of a colonial trusteeship project. The voices critical of the adaptation and vulnerability discussion also highlight that research, even if it incorporates traditional knowledge, and science-based policy making cannot offer an ultimate response to climate change challenges. In order to be effective and legitimate, these need to be combined with the participation of local actors. Furthermore, meaningful participation in decision making has to be based on capacity-building.

Any deliberations on policy options for adaptation for indigenous peoples need to begin with the needs, perspectives, and perceptions of indigenous peoples and communities. Advocated holistic, community-based, and participatory research focused on application of traditional knowledge (Cochran et al. 2009) cannot take place without empowerment of peoples and communities. Empowerment, to be meaningful, should refer also to challenging issue areas such as self-determination, sovereignty, lands, and resources, and not only to hunting practices and cultural activities (Cameron 2012).

Empowerment allows communities to participate to a greater extent in research activities and have greater control on how their traditional knowledge is produced and used. In decision making, empowerment also helps guarantee effective participation. Communities that are more autonomous in terms of financial and human resources can be more active in the design and implementation of strategies and actions. Resolved land rights issues establish clearer frameworks for discussion and decisions on how policies are implemented. They allow communities to plan better and provide those engaged in harvesting with more certainty of their access rights and flexibility when the adjustments in light of changing environment are needed. Efficient co-management systems may allow for faster response to changing conditions, for example, regarding limit-setting on the take of fish and wildlife (see e.g., MAP website).

Strong, capable community leadership can, in an informed manner, enter into negotiations with authorities, especially in cases where major actions such as infrastructure reconstruction and relocation are to be taken. That is of primary importance if the history of improving Arctic indigenous lives by force, with decisions made in southern capitals (as was the case with relocations occurring in the past, for instance in Canada and the former Soviet Union) is not to be repeated in the case of adaptation policies. The need for participation will rise with the increase in climate change impacts and development of adaptation policies.

Climate change could, in fact, perversely enhance indigenous empowerment and participation in governance. Arctic indigenous activists attempt to utilize the discussion on climate change to strengthen advocacy of their rights and to empower their peoples and organizations (Bravo 2009; ICC 2005, 2009). Adaptation governance is seen as an opportunity for facilitating social learning, public participation, and strengthening deliberative democracy (Tennberg 2012).

There is clearly a trend and a normative pressure towards indigenous empowerment and participation in the Arctic, which is expected to also have positive effects on communities' adaptive capacities. Available short-term options include strengthening co-management systems, adjusting participation mechanisms to indigenous capacities and traditions, providing communities with feedback on the outcomes of their participatory engagements, utilizing indigenous rights frameworks, and applying a human rights-based approach.

4.6.1 Co-management, Participatory Capacities, and Clear Outcomes of Participatory Engagement

The vital elements of this empowerment trend are self-government, self-determination, strengthening local governments, and co-management arrangements, including management and use of land and resources. Newly developed institutions can provide communities with the possibility to transmit their concerns and enhance learning and self-organization processes (Berkes and Jolly 2001; ACIA 2005).

Management of marine natural resources is a particularly sensitive issue. Indigenous territorial use and perception, where waters and ice are an extension of traditional territory, should be taken into consideration in the restructuring of governance systems. For example, owing to the Nunavut Land Claims Agreement, local communities have the possibility to influence offshore developments (NLCA 1993: para.11.1.2; Royal Commission on Aboriginal Peoples 1996).

Nevertheless, what is understood by participation and empowerment as well as the practical implementation of these principles remains vague (Pearce et al. 2010). First, it is often not clear what changes follow indigenous contributions, as is the case in the Canadian Arctic or with consultations between Sámi Parliaments and Nordic governments. Secondly, there can be too much consultation, with indigenous institutions overburdened with complex procedures and processing hundreds of documents, without a clear picture of the consultation outcome. Balancing between providing indigenous groups with the genuine possibility to be fully consulted in an informed manner and the capacities of these actors, needs to be taken into account when designing decision making procedures (Henriksen 2008, 2010; Huntington et al. 2012).

Devolution is not a straightforward or clearly empowering process. Neither does bringing power closer to the communities and putting it in the hands of indigenous politicians automatically result in the improvement of social, economic, and political conditions. Experiences of Nunavut with mismanagement, corruption, inefficient spending, and questionable policies show that empowerment is a complex and long-term process and not a clear-cut solution (Loukacheva and Garfield 2009).

4.6.2 Indigenous Rights

The application of indigenous rights, which has evolved within human rights mechanisms over recent decades (Niezen 2003), is seen as having empowerment potential. Indigenous rights extend beyond the boundaries of traditional and local livelihoods and provide for indigenous involvement in issues of resource extraction or shipping (Hansen and Bankes 2008; Lukovich and McBean 2009; Cameron 2012). In the 2009 Arctic TRANSFORM project, the Indigenous Peoples Working Group argued for treating indigenous peoples not as stakeholders, but rather as rights-holders and incorporating rights language into Arctic governance, for example by establishing an Indigenous Rights Review Working Group under the Arctic Council, which would analyze legal and institutional adaptation barriers from the human rights perspective (Cochran et al. 2009).

Indeed, climate change impacts are increasingly presented as an issue of human rights (Humphreys 2010) with an obligation of states, the international community, and industries (seen as duty-bearers) to assist those affected. The aforementioned Arctic TRANSFORM Working Group also proposed establishing an Arctic Trust Fund, through which those benefiting from exploitation of Arctic resources—particularly extractive industries—would counterbalance risks these activities create (Cochran et al. 2009).

Indigenous peoples around the circumpolar North have attempted to utilize human rights avenues, including the Human Rights Committee or the European Court of Human Rights (Koivurova 2011a). A prominent example is the unsuccessful, but influential within climate change debates, Inuit Circumpolar Council's (ICC) petition to the Inter-American Commission on Human Rights, where US "acts and omissions" in climate policy were argued to have violated Inuit's human rights (ICC 2005; Koivurova 2008; Abate 2007). In addition to developments in litigation, the Human Rights Council recognized climate change as "an immediate and far-reaching threat to people and communities around the world and [which] has implications for the full enjoyment of human rights" (HRC 2008).

4.7 Conclusion: A Holistic Response

Empowering Arctic communities in the current social, economic, environmental, and political landscape of the region is challenging. Available literature and governance practices continue to offer few practical propositions or concrete best practices. Ideally, empowerment, co-management, and participatory decision making should be tailor-made to communities' individual capacities and perpetually re-evaluated and attuned to changing conditions. Participatory efforts need to have clear relevance in decision making processes or risk turning into purely formal acts resulting in disappointment rather than empowerment.

Adaptation interconnected with empowerment needs to be a holistic endeavour. Increasing adaptive capacity is not only foreseen to address climate change impacts, but may potentially help respond to other present-day challenges. Addressing climate change impacts cannot become the only goal of implemented policies as, for instance, poverty alleviation or protection of culture could be equally important (Ford 2009). Rather, engagement should take place alongside planning and engineering solutions and at multiple scales, including the national, regional, and community levels and addressing diverse challenges, with values pertinent to each society playing a crucial role in choosing adaptation strategies (Hovelsrud and Smit 2010).

Financial constraints are one of the primary barriers to implementing adaptation measures. The high costs of adaptation will continue to be an ongoing problem in policy making, but if the desired goal is protection of invaluable indigenous cultures and preserving settlements in the Arctic, policymakers need to accept that certain costs need to be borne.

Adaptation to climate change and concern for particular vulnerability of indigenous peoples are nowadays increasingly presented as issues of human rights and human security (Government of Canada 2009; Ford et al. 2010b; Lukovich and McBean 2009). These approaches have advocacy potential to trigger greater action on the side of duty-bearers and security-providers such as states, public authorities, private companies, or NGOs. At the same time, however, the language of securitization and perception of indigenous peoples exclusively as victims can be seen as introducing old notions of paternalism and trusteeship, and therefore must be applied with caution (Bravo 2009; Cameron 2012; Pearce et al. 2012).

Adaptation policies are already being developed and implemented (Tennberg 2012) and they need to be monitored, evaluated, and reflected upon on an ongoing basis. Therefore, additional research on actual adaptation planning and implementation is needed.

The use of a common definition of adaptation in policy making needs to be critically questioned. The meaning of adaptation—a complex term in itself—within diverse cultural settings has to be better understood by decisionmakers (Cochran et al. 2009). Different concepts of climate change impacts and adaptation may exist even in neighbouring communities, especially as climate change in the Arctic is a heterogeneous phenomenon (Pearce et al. 2010).

Research and sharing experience on current adaptation planning is of major significance broadly, as well as between the US and the EU (including EU Member States). Partners from both sides of the Atlantic could learn much from one another regarding indigenous governance and adaptation. Alaskan policies could prove particularly useful for EU policymakers as they represent decades of experience with various indigenous governance arrangements (e.g., North Slope Borough set-up and land claims agreements). Greater cooperation and dialogue may contribute to sensitizing various actors—particularly on the side of the EU, which has been perceived as a regional outsider—to the specificity of economic, social, cultural, and political developments in the indigenous North. On the other hand, the US could utilize the EU's experiences with its progressive climate change policies, including adaptation actions at multiple levels.[4]

The Arctic is seen as a harbinger of change, but also a potential test-bed for developing creative responses, adaptive policies, resilient communities, capacity for management, responsible governance, and community-based sustainable (i.e., not hypothesis-testing) research. It is believed that in the Arctic, such tools for managing and coping with emerging global issues may best be developed (Carmack et al. 2012; Berkes and Jolly 2001). However, Arctic communities cannot be seen as locations of 'adaptive experiments' for the benefit of global governance and their own needs should always come to the fore (Carmack et al. 2012; Tennberg 2012).

References

Abate RS (2007) Climate change, The United States, and the impacts of Arctic melting: A case study in the need for enforceable international environmental human rights. Symposium: Climate Change Liability and the Allocation of Risk. Stanford J Int Law 43A:3–76

---

[4] See e.g, the EU Commission's 2009 White Paper on climate change adaptation (European Commission 2009), as well as Mettiäinen (2012) on an example of regional adaptation strategy supported by the EU funding.

ACIA (2005) Arctic Climate Impact Assessment. Cambridge University Press, New York

Adaptation Advisory Group (2010) Alaska's climate change strategy: Addressing climate change in Alaska. Final Report Submitted by the Adaptation Advisory Group to the Alaska Climate Change Sub-Cabinet (Final Draft Report). Alaska Department of Environmental Conservation. <http://www.climatechange.alaska.gov/aag/aag.htm>. Accessed 20 Aug 2012

Adger WN (2010) Climate change, human well-being and insecurity. New Polit Econ 15(2):275–292

AHDR (2004) Arctic Human Development Report. Stefansson Arctic Institute, Akureyri

AK-CSC (Alaska Climate Science Centre) (2012) Annual Action Plan FY2012. <http://csc.alaska.edu/AK%20CSC%20annual%20action%20plan%202012.pdf>. Accessed 7 Nov 2012

AMSA (2009) Arctic Marine Shipping Assessment 2009 Report. Arctic Council, US 2009

Berkes F, Jolly D (2001) Adapting to climate change: Social-ecological resilience in a Canadian western Arctic community. Conserv Ecol 5(2):18–32

Bone C, Alessa L, Altaweel M, Kliskey A, Lammers R (2011) Assessing the impacts of local knowledge and technology on climate change vulnerability in remote communities. Int J Environ Res Public Health 8(3):733–761

Bravo MT (2009) Voices from the sea ice: the reception of climate impact narratives. J Hist Geogr 35:256–278

Bull KS (2008) Historisk fremstilling av retten til fiske i havet utenfor Finnmark. Official Norwegian Reports, NOU 2008:5

Cameron ES (2012) Securing Indigenous politics: A critique of the vulnerability and adaptation approach to the human dimensions of climate change in the Canadian Arctic. Global Environ Change 22:103–114

Carmack E, McLaughlin F, Whiteman G, Homer-Dixon T (2012) Detecting and coping with disruptive shocks in Arctic marine systems: a resilience approach to place and people. Ambio 41:56–65

Cochran P, Nuttal M, Arctic TRANSFORM Indigenous Peoples Working Group (2009) Policy Options for Arctic Environmental Governance. Prepared by the Indigenous Peoples Working Group. Arctic TRANSFORM. <http://arctic-transform.org/download/IndPeEX.pdf>

Dahl J, Hicks J, Jull P (eds) (2001) Nunavut. Inuit regain control of their lands and their lives. IWGIA, Copenhagen

Escobar A (1995) Encountering development: The making and unmaking of the third world. Princeton University Press, Princeton

European Commission (2009) Adopting to climate change: Towards a European framework for action (White Paper), 1 Apr 2009, COM (2009) 147 final

European Community (2009) Regulation No. 1007/2009 of the European Parliament and of the Council of 16 Sept 2009 on trade in seal products, 31 Oct 2009, 2009 O.J. (L 286/36)

European Convention on Human Rights (1950) European Convention for the Protection of Human Rights and Fundamental Freedoms, 4 Nov 1950, 213 U.N.T.S. 221. Entered into force 3 Sept 1953

European Parliament (2012) The impact of the EU seal ban on the Inuit population in Greenland. Seminar chaired by Pat the Cope Gallagher, Member of the European Parliament, Brussels

Feifel K (2010) Homer, Alaska Climate Action Plan. Case study on a project of the City of Homer and ICLEI. Product of EcoAdapt's State of Adaptation Program. <http://www.cakex.org/case-studies/711>. Accessed 12 Nov 2012

Forbes BC, Stammler F (2009) Arctic climate change discourse: the contrasting politics of research agendas in the West and Russia. Polar Res 28:28–42

Ford JD (2009) Dangerous climate change and the importance of adaptation for the Arctic's Inuit population. Environ Res Lett 4

Ford JD, Smit B, Wandel J (2006) Vulnerability to climate change in the Arctic: A case study from Arctic Bay, Canada. Global Environ Change 16:145–160

Ford JD, Berrang-Ford L, King M, Furgal C (2010a) Vulnerability of Aboriginal health systems in Canada to climate change. Global Environ Change 20:668–680

Ford JD, Pearce T, Duerden F, Furgal C, Smit B (2010b) Climate change policy responses for Canada's Inuit population: The importance of and opportunities for adaptation. Global Environ Change 20:177–191

Furgal CH, Seguin J (2006) Climate change, health and vulnerability in Canadian northern aboriginal communities. Environ Health Perspect 114(12). doi:10.1289/ehp.8433

GAO (2004)Testimony before the committee on Appropriations, U.S. Senate: ALASKA NATIVE VILLAGES—Villages Affected by Flooding and Erosion Have Difficulty Qualifying for Federal Assistance. Government Accountability Office. <http://www.gpo.gov/fdsys/pkg/GAOREPORTS-GAO-04-895T/pdf/GAOREPORTS-GAO-04-895T.pdf>. Accessed 7 Nov 2012

Government of Canada (2009) Canada's Northern Strategy: Our north, our heritage, our future. Ottawa: Minister of Indian Affairs and Northern Development and Federal Interlocutor for Métis and Non-Status Indians

Hansen KF, Bankes N (2008) Human rights and Indigenous Peoples in the Arctic: What are the implications for oil and gas industry. In: Mikkelsen A, Langelle O (eds) Arctic Oil and Gas: Sustainability at Risk?. Routledge, New York, pp 291–316

Hasanat W (2010) Cooperation in the Barents Euro-Arctic region in the light of international law. Yearb Polar Law 2:279–309

Henriksen J (ed) (2008) Sámi self-determination: Scope and implementation. Galdu Čala. J Indigenous Peoples Rights 2

Henriksen J (ed) (2010) Sámi self-determination: Autonomy and economy—The authority and autonomi of Samediggi in the health and social services sector. Galdu Čala. J Indigenous Peoples Rights 2

Hossain K (2012) The EU ban on the import of seal products and the WTO regulations: neglected human rights of the Arctic indigenous peoples? Polar Record. doi:<http://dx.doi.org/10.1017/S0032247412000174>

Hovelsrud GK, Smit B (eds) (2010) Community adaptation and vulnerability in Arctic regions. Springer, Berlin

HRC (1994) General Comment No. 23 (50th Session, 1994) by the Human Rights Committee, UN Doc. HRI/GEN/1/Rev 3

HRC (1999/2000/2002/2004) Human Rights Committee. Concluding Observations (multiple): Canada (UN doc. CCPR/C/79/Add.105 (1999)); Mexico (UN Doc. CCPR/C/79/Add.109 (1999)); Norway (UN Doc. CCPR/c/79/Add.112 (1999)); Australia (UN Doc. CCPR/CO/69/AUS (2000)); Denmark (UN Doc. CCPR/CO/70/DNK (2000)); Sweden (UN Doc. CCPR/CO/74/SWE (2002)). Finland (UN Doc. CCPR/CO/82/FIN (2004))

HRC (2008) Human Rights Council. Resolution 7/23 on human rights and climate change. <http://ap.ohchr.org/documents/E/HRC/resolutions/A_HRC_RES_7_23.pdf>. Accessed 28 Aug 2012

Humphreys S (2010) Human rights and climate change. Cambridge University Press, Cambridge

Huntington HP, Lynge A, Stotts J, Hartsig A, Porta L, Debicki CH (2012) Less ice, more talk: The benefits and burdens for Arctic communities of consultations concerning development activities. Carbon Climate Law Rev 1:33–46

ICC (2005) Petition to the Inter American Commission on human rights, violations resulting from global warming caused by the United States, 7 Dec 2005 by Sheila Watt-Cloutier with support of Inuit Circumpolar Council. <http://www.inuitcircumpolar.com/files/uploads/icc-files/FINALPetitionICC.pdf>. Accessed 20 Aug 2012

ICC (2009) A Circumpolar Inuit Declaration on Sovereignty in the Arctic. Inuit Circumpolar Council. Apr 2009

ICCPR (1966) International Covenant on Civil and Political Rights, 16 Dec 1966, 999 U.N.T.S. 17. Entered into force 23 Mar 1976

ILO Convention No. 169 (1989) Convention concerning Indigenous and Tribal Peoples in Independent Countries, 27 June 1989, 28 I.L.M. 1382 (1989). Entered into force 5 Sept 1991

International Covenant on Economic, Social, and Cultural Rights (1966) 16 Dec 1966, 993 U.N.T.S. 3. Entered into force 3 Jan 1976

International Convention on the Elimination of All Forms of Racial Discrimination (1966) 7 Mar 1966, 660 U.N.T.S. 195. Entered into force 4 Jan 1969

IPCC (2007) Climate change 2007: Impacts, Adaptation and Vulnerability. Contribution of working group II to the fourth assessment report of the intergovernmental panel on climate change.In: Parry ML, Canziani OF, Palutikof JP, van der Linden PJ, Hanson CE (eds) Cambridge University Press, Cambridge

Keskitalo ECH (ed) (2010) Developing adaptation policy and practice in Europe: Multi-level governance of climate change. Springer, Berlin

Koivurova T (2008) The draft Nordic Sámi Convention: Nations working together. Int Commun Law Rev 10:279–293

Koivurova T (2011a) Jurisprudence of the European court of human rights regarding Indigenous Peoples: Retrospect and prospects. Int J Minori Group Rights 18:1–37

Koivurova T (2011b) The status and role of Indigenous Peoples in Arctic international governance. Polar Law Yearb 3:169–192

Koivurova T, Hasanat W (2009) Climate policy of the Arctic Council. In: Koivurova T, Heinämäki L (2006) The participation of Indigenous Peoples in international norm-making in the Arctic. Polar Record 42(221):101–109. In: Koivurova T, Keskitalo C, Bankes N (eds) Climate Governance in the Arctic. Springer, Berlin

Koivurova T, Heinämäki L (2006) The participation of Indigenous Peoples in international norm-making in the Arctic. Polar Rec 42(221):101–109

Li TM (2007) The will to improve: Governmentality, development, and the practice of politics. Duke University Press, Durham

Lindroth M (2011) Paradoxes of power: Indigenous Peoples in the Permanent Forum. Cooperation Conflict 46(4):542–561

Loukacheva N, Garfield MD (2009) Sustainable human rights and governance: The quest of an Arctic entity in transition. Polar Law Yearbook 1:283–305

Lukovich JV, McBean GA (2009) Addressing human security in the Arctic in the context of climate change through science and technology. Mitig Adapt Strat Glob Change 14:697–710

Magga OH, Mathiesen SD, Corell RW, Oskal A, Benestad R, Bongo MP, Burgess P, Degteva A, Etylen V (2011) Reindeer herding, traditional knowledge and adaptation to climate change and loss of grazing land. EALAT project executive summary. Arctic Council and International Centre for Reindeer Husbandry. Fagtrykk Ide AS, Alta

MAP (n.d.) Marine Advisory Programme. <http://seagrant.uaf.edu/map>. Accessed 8 Nov 2012

Mettiäinen I (2012) Knowledge to action question in regional climate change policy formation. In: Adam S, Mettiäinen I, Soppela P (eds) Science-Policy Interface – Societal Impacts of Arctic Research. Abstract book. The 10th Annual Seminar of the ARKTIS Arctic Doctoral Programme and the Final Seminar of the project "Knowledge in Environmental Planning and Decision-making" (2010-2012). 15-16 Mar 2012, Rovaniemi, Finland. Arctic Centre. University of Lapland. University of Lapland Printing Centre, Rovaniemi. pp 67

MFPP (2012) Funding opportunity—Alaska native fund could speed climate adaptation training. Model Forest Policy Programme. <http://www.mfpp.org/?p=2326>. Accessed 8 Nov 2012

Moerlein KJ, Carothers C (2012) Total environment of change: Impacts of climate change and social transitions on subsistence fisheries in Northwest Alaska. Ecol Soc 17(10)

Niezen R (2003) The Origins of Indigenism. University of California Press, Berkeley

Njåstad B, Kelman I, Rosenberg S (2009) Vulnerability and adaptation to climate change in the Arctic (VACCA). Brief Report Series No. 12. Norwegian Polar Institute and Arctic Council Sustainable Development Working Group. Tromso: Norwegian Polar Institute

NLCA (1993) Nunavut Land Claims Agreement, extract. <http://npc.nunavut.ca/eng/npc/article11.html>. Accessed 5 May 2008

Nuttall M (2002) Protecting the Arctic, Indigenous Peoples and cultural survival. Routledge, London

Nuttall M (2000) Indigenous Peoples, self-determination and the Arctic environment. In: Nuttall M, Callaghan TV (eds) The Arctic—Environment, people, policy. Harwood Academic, Amsterdam

Pearce TD, Ford JD, Laidler GJ, Smit B, Duerden F, Allarut M, Andrachuk M, Baryluk S, Dialla A, Elee P, Goose A, Ikummaq T, Joamie E, Kataoyak F, Loring E, Meakin S, Nickels S, Shappa K, Shirley J, Wandel J (2009a) Community collaboration and climate change research in the Canadian Arctic. Polar Res 28:10–27

Pearce T, Smit B, Duerden F, Ford JD, Goose A, Kataoyak F (2009b) Inuit vulnerability and adaptive capacity to climate change in Ulukhaktok, Northwest Territories, Canada. Polar Rec 46:157–177

Pearce T, Ford JD, Duerden F, Smit B, Andrachuk M, Berrang-Ford L, Smith T (2010) Advancing adaptation planning for climate change in the Inuvialuit Settlement Region (ISR): a review and critique. Reg Environ Change 11:1–17

Pearce T, Ford JD, Caron A, Kudlak BP (2012) Climate change adaptation planning in remote, resource-dependent communities: An Arctic example. Reg Environ Change. doi:10.1007/s10113-012-0297-2

Poppel B, Jack K, Gérard D, Larissa A (2007) Survey of living conditions in the Arctic (SLiCA): Results. Institute of Social and Economic Research, University of Alaska Anchorage, Anchorage

Royal Commission on Aboriginal Peoples (1996) Restructuring the relationship. Canada Communication Group Publishing, Part two. Report

Slezkine Y (1996) Arctic mirrors: Russia and small Peoples of the North. Cornell University Press, Ithaca

Smit B, Pilifosova O (2003) From adaptation to adaptive capacity and vulnerability reduction. In Smith JB, Klein RJT, Huq S (eds) Climate change, adaptive capacity and development. Imperial College Press, London

Smit B, Wandel J (2006) Adaptation, adaptive capacity and vulnerability. Global Environ Change 16:282–292

Tennberg M (ed) (2012) Governing the Uncertain. Adaptation and Climate in Russia and Finland. Springer, Dordrecht

Thronberry P (1994) On some implications of the UN Declaration on Minorities for Indigenous Peoples. In: Gayim E, Myntti K (eds) Indigenous and Tribal Peoples Rights—1993 and after. University of Lapland, Rovaniemi

UNDRIP (2007) United Nations Declaration on the Rights of Indigenous Peoples. General Assembly Resolution. New York, 13 Sept 2007. A/RES/61/295

United Nations Framework Convention on Climate Change (1992) 9 May 1992, 1771 U.N.T.S. 107. Entered into force 21 Mar 1994

UNPFII (United Nations Permanent Forum on Indigenous Issues) (n.d.) Who are indigenous peoples? Factsheet. United Nations

USGCRP (2009) U.S. Climate Impacts Report—Regional Climate Change Impacts. United States Global Change Research Program. <http://globalchange.gov/images/cir/pdf/alaska.pdf>. Accessed 7 Nov 2012

Wenzel G (2001) Inuit subsistence and hunter support in Nunavut. In: Dahl et al (eds) Nunavut. Inuit regain control of their lands and their lands. International Work Group for Indigenous Affairs (IWGIA), Copenhagen

Young E (1995) Third world in the first: Development and Indigenous Peoples. Routledge, London

Part II
Impacts and Activities in the Marine Arctic

Chapter 5
Status and Reform of International Arctic Fisheries Law

Erik J. Molenaar

Abstract Marine capture fisheries are among the maritime uses that are expected to expand and intensify in the marine Arctic. Fishing could intensify in existing fishing areas and expand into areas where marine capture fisheries have never taken place. This chapter assesses the adequacy of the current international legal and policy framework for Arctic fisheries conservation and management in light of the current and expected impacts of global climate change on the marine Arctic. It provides an overview of the international legal and policy framework as well as national regulation and policy, identifies the main gaps therein, and suggests options for addressing them. These options include increased efforts in the sphere of research and data gathering, national regulation aimed at avoiding unregulated fishing, fisheries arrangements between Arctic Ocean coastal states, and a new regional fisheries management organization (RFMO) or Arrangement for part of the (Central) Arctic Ocean. Separate attention is devoted to the potential for cooperation between the European Union and the United States in this regard.

Writing this chapter was facilitated by funding from the Netherlands Polar Programme. The author is grateful for comments by Ted McDorman and others on an earlier version of the chapter. This chapter builds on Molenaar EJ, Corell R (2009) Background Paper. Arctic Fisheries. Arctic TRANSFORM.

E. J. Molenaar (✉)
Netherlands Institute for the Law of the Sea (NILOS), Utrecht University, Utrecht,
The Netherland
Faculty of Law, University of Tromsø, Tromsø, Norway
e-mail: E.J.Molenaar@uu.nl

E. Tedsen et al. (eds.), *Arctic Marine Governance*,
DOI: 10.1007/978-3-642-38595-7_5, © Springer-Verlag Berlin Heidelberg 2014

5.1 Introduction

Marine capture fisheries are among the maritime uses that are expected to expand and intensify in the marine Arctic. Fishing could intensify in existing fishing areas and expand into areas where marine capture fisheries have never taken place so far. This chapter assesses the adequacy of the current international legal framework for Arctic fisheries conservation and management in light of the current and expected impacts of global climate change on the marine Arctic. After providing some context and background information on Arctic fish stocks, fisheries, and climate change in Sect. 5.2, an overview of the international legal and policy framework for Arctic fisheries management is presented in Sect. 5.3. Section 5.4 then devotes attention to national regulation and policy. Gaps in the international legal and policy framework and national regulation and options for addressing them are covered succinctly in Sect. 5.5 and the potential for EU–US cooperation is examined in Sect. 5.6. Some conclusions are offered in Sect. 5.7.

This chapter relates to marine capture fisheries that target 'fishery resources', which are defined as fish, molluscs, crustaceans, and (other) sedentary species. Inland fisheries, aquaculture, and harvesting of marine mammals are thus excluded. For the purposes of this chapter, the term regional fisheries management organization (RFMO) includes a so-called 'Arrangement', which is understood to be a bilateral or (sub-)regional cooperative mechanism other than an intergovernmental organization, but otherwise has in principle the same characteristics as an RFMO.[1] Due to its predominantly sectoral perspective, the international component of this chapter will be restricted to instruments and bodies that relate to, or pursue, conservation as well as management of fishery resources. No attention will therefore be paid to those that focus exclusively on conservation of species and habitats by various means, including by the regulation of international trade.

As explained in Chap. 1, there are no generally accepted geographical definitions for the terms 'Arctic' and 'Arctic Ocean' and for the purposes of this book, the term 'Arctic' has an identical meaning as the term 'AMAP area' adopted by the Arctic Monitoring and Assessment Programme (AMAP) of the Arctic Council (AMAP 1997). The waters within the AMAP area are in this chapter referred to as the 'marine Arctic'. The 'Arctic Ocean' is defined here as the marine waters north of the Bering Strait, Greenland, Svalbard, and Franz Josef Land, excluding the Barents Sea. Canada, Denmark/Greenland, Norway, the Russian Federation, and the United States (US) are 'Arctic Ocean coastal states'. These five states and Finland, Iceland, and Sweden are 'Arctic states' by virtue of their membership of the Arctic Council. There are four high seas pockets in the marine Arctic, namely

---

[1] The term 'Arrangement' is derived from the term 'arrangement' as defined in Article 1(1)(d) of the Fish Stocks Agreement (1995). The main differences between an RFMO's constitutive instrument and an Arrangement are that the latter (a) does not establish an international organization, (b) does not have to be legally binding, and (c) can be bilateral.

the so-called 'Banana Hole' in the Norwegian Sea, the so-called 'Loophole' in the Barents Sea, the so-called 'Donut Hole' in the central Bering Sea, and the so-called 'Central Arctic Ocean'.

5.2 Arctic Fish Stocks, Fisheries, and Climate Change

There are a number of potentially significant commercial fish stocks in the marine Arctic. The ranges of distribution of some of these are confined to the North Pacific or the North Atlantic, while others have a circumpolar distribution. Important North Pacific fish stocks include Alaska pollock (*Theragra chalcogramma*), Pacific cod (*Gadus macrocephalus*), snow crab (*Chionoecetes opilio*), and various Pacific salmon species (*Oncorhynchus* sp.). As regards the North Atlantic, important fish stocks include North-East Arctic cod (*Gadus morhua*), haddock (*Melanogrammus aeglefinus*), Norwegian spring-spawning (Atlanto-scandian (AS)) herring (*Clupea harengus*), Atlantic salmon (*Salmo salar*), and red king crab (*Paralithodes camtschaticus*). Significant circumpolar fish stocks include capelin (*Mallotus villosus*), Greenland halibut (*Reinhardtius hippoglossoides*), and northern shrimp (*Pandalus borealis*). Polar cod (*Boreogadus saida*) and Arctic char (*Salvelinus alpinus*) also have circumpolar distribution, but the former is only marginally targeted by commercial fisheries and the latter is predominantly fished for subsistence purposes (ACIA 2005; AOR 2011; Anchorage Science Meeting Report 2011; Zeller et al. 2011).[2]

While Arctic marine ecosystems have always been highly dynamic and variable (AOR 2011), both qualitatively and quantitatively, the impacts of climate change on the marine Arctic—e.g., increasing water temperature, reduced sea ice coverage and thickness, reduced salinity, and increasing acidification (Anchorage Science Meeting Report 2011)—are likely to make these changes more rapid, more profound, and probably also more difficult to predict. Some existing fish stocks may collapse and never recover, others may become more dominant, and new fish species may successfully invade the marine Arctic. The various assessments and projects currently undertaken within the Arctic Council, such as the Arctic Biodiversity Assessment, the Arctic Change Assessment, and the Arctic Resilience Report, are expected to shed more light on this.

While there are large-scale commercial fisheries in the more southerly waters of the marine Arctic—namely the Bering Sea, Barents Sea, Baffin Bay, and along the coast of east and west Greenland—in the Arctic Ocean there are currently mainly small-scale subsistence fisheries and no significant commercial fisheries, and in the Central Arctic Ocean no fisheries at all. It seems more likely that new fishing opportunities will occur within coastal state maritime zones before occurring on the high seas. According to some commentators, in the short term, it is

---

[2] See also information at <arcticportal.org/fishing-portlet>.

unlikely that abundance of fish stocks in the high seas portion of the Arctic Ocean will allow for commercially viable fisheries (Hoel 2011). Others disagree while pointing to Polar cod, which has a circumpolar distribution, both inshore and off-shore, and may be highly abundant in view of the pivotal role it plays at the bottom of Arctic marine ecosystems.[3] Finally, as reduced ice coverage and thickness will also enable other human activities—most importantly shipping and offshore hydrocarbon activities—these activities may compete with fishing in a spatial sense or affect them by pollution—including noise—and other impacts.

The impact of current and future Arctic fisheries on the marine environment and marine biodiversity in the Arctic is not likely to be fundamentally different from fisheries impacts to the marine environment and biodiversity in other parts of the globe. Arctic fisheries could lead to over-exploitation of target species and a variety of impacts on non-target species, for instance on dependent species due to predator–prey relationships, on associated species due to bycatch, and on benthic species and habitats due to bottom fishing techniques. In view of the broad spatial scope of the marine Arctic, such undesirable effects are without doubt already occurring, even though not necessarily on a very serious scale.

5.3 International Legal and Policy Framework for Arctic Fisheries Management

5.3.1 Interests, Rights, Obligations, and Jurisdiction

The international legal and policy framework for fisheries conservation and management seeks to safeguard the different interests of the international community with those of states that have rights, obligations, or jurisdiction in their capacities as flag, coastal, port, or market states or with respect to their natural and legal persons. While the term 'flag state' is commonly defined as the state in which a vessel is registered and/or whose flag it flies (LOS Convention 1982, art. 91(1)), there are no generally accepted definitions for the terms 'coastal state', 'port state', or 'market state'. For the purposes of this chapter, however, the term 'coastal state' refers to the rights, obligations, and jurisdiction of a state within its own maritime zones over foreign vessels.

The term 'port state' refers to the rights, obligations, and jurisdiction of a state over foreign vessels that are voluntarily in one of its ports. In order to avoid an overlap with jurisdiction by coastal states, this chapter regards port state jurisdiction as relating to fishing by foreign vessels beyond the coastal state's maritime zones as well as over violations of conditions for entry into port (Molenaar 2007).

While there is no universally accepted definition for the term 'market state', this chapter uses a definition that was proposed during the negotiation of the SPRFMO

---

[3] For additional information see <www.arcodiv.org>.

Convention (2009), but did not make it to the final text. That definition reads "a State […] which imports, exports, re-exports or has a domestic market for fish or fish products derived from fishing in the Convention Area" (SPRFMO 2008, art. 1(m)).

Both flag and coastal states have, in principle, an interest in the long-term exercise of their entitlements over marine living resources in the various maritime zones. However, as coastal states have exclusive access to marine living resources within areas under their national jurisdiction, their commitment to that objective may often be stronger than that of flag states. A port state will commonly pursue socio-economic interests related to the port and its *hinterland*. States generally have interests, rights, obligations, and jurisdiction in more than one capacity. This commonly leads to a more balanced compromise position, but occasionally also to contradictory positions of the same state within different fora. There is no reason or indication to assume that Arctic states are different in this regard.

The interests of the international community—e.g., sustainable utilization, protection, and preservation of the marine environment and conservation of marine biodiversity—normally overlap with those of states within the various capacities in which they can act, but are usually broader and more general. The interests of some states, however, clearly undermine those of other states and the international community, for instance, by not ensuring that their ships comply with international minimum standards or by allowing foreign vessels in their ports to be non-compliant with international minimum standards. These states, vessels, and ports thereby have a competitive advantage over states, vessels, and ports that *do* comply with international minimum standards. Such 'free riders' clearly benefit from the consensual nature of international law—meaning that a state can only be bound to a rule of international law when it has in one way or another consented to that rule.

5.3.2 Substantive Fisheries Standards

Fisheries conservation and management authorities often make use of the following substantive fisheries standards:

1. Restrictions on catch and effort, for instance, by setting the total allowable catch (TAC) and allocating the TAC by means of national quotas;
2. Designated species for which targeted fishing is prohibited;
3. Minimum size limits for target species;
4. Maximum bycatch limits, for instance, in terms of the number of individuals (e.g., in relation to marine turtles and marine mammals) or as a percentage of the target catch;
5. Gear specifications, for instance, minimum mesh sizes, bycatch mitigation techniques (e.g., turtle excluder devices, bird-scaring lines); and
6. Temporal/seasonal or spatial measures (e.g., closed areas) aimed at avoiding catch of target species (e.g., nursing and spawning areas) or non-target species (e.g., important feedings areas) or avoiding impact on sensitive habitat (e.g., cold water coral reefs).

5.3.3 Global Bodies and Instruments

All the relevant global intergovernmental bodies and instruments discussed in this subsection also apply to the marine Arctic, however defined. The main intergovernmental bodies of relevance to this chapter are the United Nations General Assembly (UNGA) and the United Nations Food and Agriculture Organization (FAO). While the mandate of these bodies would not preclude them from dealing specifically with Arctic fisheries as such, most of the Arctic Ocean coastal states would oppose this (Molenaar 2012a).

The following are the main global legally binding and non-legally binding fisheries instruments:

1. LOS Convention (1982);
2. Fish Stocks Agreement (1995);
3. Compliance Agreement (1993);
4. Port State Measures Agreement (2009);
5. Other FAO fisheries instruments, most importantly, the FAO Code of Conduct for Responsible Fisheries (FAO 1995)—including its Technical Guidelines and international plans of action [IPOAs; e.g., the IPOA-IUU (FAO 2001)], the International Guidelines on Deep-sea Fisheries in the High Seas (FAO 2008), and the International Guidelines on Bycatch Management and Reduction of Discards (FAO 2010); and
6. Certain (parts of) UNGA Resolutions, which have contributed to the phase-out of large-scale pelagic driftnet fishing and imposed innovative restrictions on bottom-fisheries on the high seas.[4] Both initiatives were predominantly aimed at the conservation of non-target species and vulnerable marine ecosystems.

The provisions on marine capture fisheries in the LOS Convention and the Fish Stocks Agreement have a so-called 'framework' character. They contain overall objectives and basic rights and obligations for states, but not the key substantive fisheries standards set out here. Actual fisheries regulation is carried out by states individually or collectively, including through RFMOs (see Sect. 5.3.4).

The key objectives of the LOS Convention are (a) avoidance of overexploitation by means of striving for the maximum sustainable yield (MSY) and setting TACs, and (b) optimum utilization, which obliges coastal states that cannot catch the entire TAC themselves to give other states access to the surplus. The LOS Convention acknowledges or grants rights to coastal states over marine living resources in their maritime zones and to other states on the high seas. These rights are subject to the key objectives just mentioned and many other related obligations, for instance, the obligation to take account of impacts on associated species (e.g., through bycatch) or dependent species (e.g., through predator–prey relationships) and to cooperate with relevant coastal and/or flag states on transboundary stocks/species and discrete high seas stocks (see Table 5.1). The objective of

---

[4] See, *inter alia*, UNGA Res. 46/215 (1991) and UNGA Res. 61/105 (2006), paras 80–89.

Table 5.1 Categories of fish stocks

Category	Definition
Discrete inshore stocks	Occur exclusively in the maritime zones of one single state
Joint or shared stocks	Occur within the maritime zones of two or more coastal states, but not on the high seas
Straddling stocks	Occur within the maritime zones of one or more coastal states and on the high seas
Highly migratory stocks	The fish species listed in Annex I to the LOS Convention (e.g., tuna)
Anadromous stocks	Spawn in rivers but otherwise occur mostly at sea (e.g., salmon)
Catadromous stocks	Spend greater part of life cycle in internal fresh waters but spawn at sea (e.g., eels)
Discrete high seas stocks	Occur exclusively on the high seas

optimum utilization does not apply to marine mammals and many obligations do not apply to sedentary species or to maritime zones under sovereignty (LOS Convention 1982, arts. 61–72 and 116–120). With respect to anadromous and catadromous stocks, the relevant coastal states have primary responsibility for conservation and management. For catadromous species, this specifically includes ensuring that inbound and outbound migration can take place (LOS Convention 1982, arts. 66–67).

The Fish Stocks Agreement only applies to straddling and highly migratory fish stocks. Its overarching objective is to implement the basic jurisdictional framework of the LOS Convention by means of a modernized and more elaborate and operational regulatory framework. The incorporation of an operationalised precautionary approach and a *de facto* ecosystem approach to fisheries (EAF), the clarification that RFMOs are the primary vehicles for the conservation and management of straddling and highly migratory fish stocks, and the intricate provisions on non-flag state high seas enforcement powers bear witness to that objective.

While the Fish Stocks Agreement retains MSY as a key objective, this is qualified by the need to apply the precautionary approach as operationalised in Article 6 and Annex II as well as a range of ecosystem considerations, which together constitute a *de facto* EAF. These ecosystem considerations require state parties to, among other things, minimize pollution, waste, discards, catch by lost or abandoned gear, catch of non-target species—in particular endangered species—and more generally to protect biodiversity in the marine environment (Fish Stocks Agreement 1994, art. 5).

The Fish Stocks Agreement regards RFMOs as the preferred vehicles for fisheries regulation at the regional level and imposes an obligation on state parties to the Fish Stocks Agreement to cooperate with and through them (art. 8(3)). Of crucial importance in that regard is Article 8(4), which stipulates that access to fisheries is limited to members and cooperating states. New is also the right in Article 8(3) of states with a 'real interest' to become members. Arguably, the duty to cooperate with the relevant RFMO laid down in Article 8(3) is already part of customary international law and thereby entitles the relevant members to take measures against (non-cooperating) non-members that would otherwise be in violation of international law, for instance, imposing trade-related measures. No practices of RFMOs on trade-related measures have at any rate been challenged by means of the establishment of a dispute settlement procedure under the World Trade Organization.

Article 8(5) of the Fish Stocks Agreement stipulates that RFMOs are to be established where these do not exist. This, however, only applies in the case of the presence of "a *particular* straddling fish stock or highly migratory fish stock" (emphasis added). While highly migratory fish stocks currently do not occur in the Central Arctic Ocean, this may be different for straddling fish stocks. Even though Sect. 5.3.4 below concludes that a gap in high seas coverage with RFMOs exists for most of the Central Arctic Ocean, this does not automatically mean that relevant states are obliged to ensure full coverage with RFMOs. There is nevertheless broad support in the international community to ensure that all high seas fisheries fall within the mandate of an RFMO. These developments have among other things led to the 'filling' of gaps for full high seas coverage in the Southern Indian Ocean, South Pacific, and, most recently, the Northern Pacific (Convention on the Conservation and Management of High Seas Fisheries Resources in the North Pacific Ocean 2012). As the discussion in Sect. 5.4 will reveal, several states and entities also support full high seas coverage with RFMOs in the Arctic Ocean.

5.3.4 Regional and Bilateral Fisheries Bodies and Instruments

As regards the regional and bilateral level, the following are the main regional and bilateral fisheries bodies and instruments whose spatial scope overlaps at least to some extent with the marine Arctic:

1. The bilateral (Canada and the US) International Pacific Halibut Commission (IPHC), established by the IPHC Convention (1953);
2. The bilateral (Canada and the US) Yukon River Panel of the bilateral Pacific Salmon Commission (PSC), established by the Pacific Salmon Treaty (1985)[5];
3. The North Pacific Anadromous Fish Commission (NPAFC), established by the NPAFC Convention (1992);
4. The Western and Central Pacific Fisheries Commission (WCPFC), established by the WCPFC Convention (2000);
5. The Conference of Parties (COP) to the CBS Convention (1994);
6. The North Atlantic Salmon Conservation Organization (NASCO), established by the NASCO Convention (1982);

---

[5] The Yukon River Panel was established by means of the Yukon River Salmon Agreement of 4 Dec 2002, which added Chap. 8 to the Pacific Salmon Treaty.

7. The International Commission for the Conservation of Atlantic Tunas (ICCAT), established by the ICCAT Convention (1966);
8. The Northwest Atlantic Fisheries Organization (NAFO), established by the NAFO Convention (1978);
9. The North East Atlantic Fisheries Commission (NEAFC), established by the NEAFC Convention (1980); and
10. The Joint Norwegian-Russian Fisheries Commission (Joint Commission), established by the bilateral Framework Agreement (1975).

This list can be categorized in several ways. An important distinction exists between NEAFC and the Joint Commission (Nos. 9–10) and all the other bodies, because only the spatial mandates of NEAFC and the Joint Commission indisputably extend to (part of) the Arctic Ocean. Moreover, while the bodies under Nos. 1–5 apply to certain more southerly waters of the marine Arctic in the Pacific, the bodies under Nos. 6–8 apply to certain more southerly waters of the marine Arctic in the Atlantic. The spatial mandates of ICCAT and NASCO do not clearly extend to (part of) the Arctic Ocean, even though some room for interpretation exists. Significant occurrence of tuna and tuna-like species in the Arctic Ocean is not expected in the short- or medium-term, but this may well be different for anadromous species.

As only NEAFC and the Joint Commission have clear mandates in the Arctic Ocean, this warrants some closer attention to them. While Article 1(a) of the NEAFC Convention restricts NEAFC's competence to the North-East Atlantic sector of the Arctic Ocean, the Joint Commission's constitutive instrument does not specify its spatial mandate. Fisheries for species whose distributional range extends into the (Central) Arctic Ocean therefore fall in principle within the Joint Commission's mandate, and this has also been asserted by the Joint Commission on several occasions. It is submitted, however, that this assertion relates first of all to areas of the Arctic Ocean adjacent to the Barents Sea that are part of the maritime zones of Norway and the Russian Federation. If the Joint Commission would actually exercise competence over the Central Arctic Ocean in a similar manner as with regard to the Loophole, this would not be acceptable to the other Arctic Ocean coastal states and other members of NEAFC. With respect to the Loophole, Norway and the Russian Federation have encouraged third states and entities to discontinue, or not to commence, fishing for particular species and thereby not to exercise their entitlements under international law to fish in the high seas and to be involved in high seas fisheries management. In return, they have granted fisheries access to their maritime zones and discontinued withholding benefits such as access to ports (Molenaar 2012a).

Based on the foregoing analysis, it can be concluded that, except for the area covered by NEAFC, the Central Arctic Ocean is a gap in high seas coverage with RFMOs. While Norway and the Russian Federation may not necessarily share this conclusion, other Arctic Ocean coastal states and non-Arctic states and entities do (see Sect. 5.4).

While there is a significant competence-overlap—both spatially and on species—between the two bodies, there seems to be no, or hardly any, actual conflict between their conservation and management measures. Their current relationship can therefore be regarded as complementary. As Norway and the Russian Federation form two-fifths of NEAFC's membership,[6] they are also well-positioned to withstand challenges from the three other members of NEAFC to downsize the role of the Joint Commission and enhance that of NEAFC.[7] Norway and the Russian Federation are also highly unlikely to support broader participation in the Joint Commission, as this would fundamentally alter its nature.

The following list contains other relevant regional, trilateral, and bilateral fisheries arrangements:

1. The meetings of Arctic Ocean coastal states. In addition to the two ministerial meetings held in Ilulissat in May 2008 and in Chelsea in March 2010, dedicated fisheries meetings have taken place at the level of senior officials[8] and at least one meeting of scientific experts (Anchorage Science Meeting Report 2011);
2. The trilateral Loophole Agreement between Iceland, Norway, and the Russian Federation (1999)[9];
3. Bilateral cooperation between Greenland and Norway, pursuant to an Agreement on Mutual Fishery Relations (1992), which is, *inter alia*, implemented through annual bilateral consultations;
4. Bilateral cooperation between Greenland and the Russian Federation pursuant to an Agreement on Mutual Fishery Relations (1992), presumably also implemented through annual bilateral consultations;
5. Bilateral cooperation between the Russian Federation and the US pursuant, *inter alia*, to the bilateral Intergovernmental Consultative Committee, established by an Agreement on Mutual Fisheries Relations (1988);
6. Bilateral cooperation between Canada and Greenland, which is not formalized, even though meetings are held on an annual basis (Molenaar 2012a); and
7. The numerous bilateral and multilateral agreements and arrangements establishing TACs, allocations of fishing opportunities, and mutual access to maritime zones between coastal states (including the European Union (EU)) in the North-East Atlantic (Churchill 2001; Molenaar 2012b).

It is clear from the sheer number of these regional, sub-regional, and bilateral bodies and instruments that they cannot possibly be discussed in a meaningful way in this chapter. More in-depth analyses of some of them are contained in other

---

[6] The other three are Denmark (in respect of the Faroe Islands and Greenland), the EU, and Iceland.

[7] These ratios would change if, for instance, Iceland becomes an EU Member State or Greenland becomes fully independent.

[8] One took place in June 2010, in Oslo, Norway (see the Chair's summary at <www.regjeringen. no/upload/UD/Vedlegg/Folkerett/chair_summary100622.pdf>. Accessed 27 Nov 2012).

[9] This Agreement is complemented by two Protocols between Iceland and Norway and Iceland and the Russian Federation respectively, which are currently in force.

literature (e.g., Barnes 2011; Molenaar 2012a). It is nevertheless important to highlight that many of these bodies adopt conservation and management measures that contain the types of substantive fisheries standards listed in Sect. 5.3.2.

Several of the above-mentioned bodies rely for scientific advice on other bodies, most notably the International Council for the Exploration of the Sea (ICES) and the North Pacific Marine Science Organization (PICES) (Takei 2013).

Finally, reference should be made to the fundamental disagreement that exists between Norway and most other parties to the Spitsbergen Treaty (1920) on the treaty's applicability to the maritime zones of Svalbard. As a consequence, Norway has not established an Exclusive Economic Zone (EEZ) but a Fisheries Protection Zone (FPZ) around Svalbard. Several states enjoy fisheries access to the FPZ and territorial waters of Svalbard as a result of the provisions of equal access laid down in the Spitsbergen Treaty (Molenaar 2012b).

5.3.5 Arctic Council and Arctic Council System

The Arctic Council is a high-level forum established by means of the Ottawa Declaration (1996). The choice for a non-legally binding instrument is a clear indication that the Council was not intended to be an international organization and implies that the Council cannot adopt legally binding decisions or instruments. The Arctic SAR Agreement (2011) was therefore not adopted by the Council, even though it was negotiated under its auspices and the Council's May 2011 Ministerial Meeting was also used as the occasion for its signature.

So far, the Arctic Council has not explicitly involved itself in fisheries management issues; not as the Arctic Council *per se* and not through the Arctic Council System (ACS; see further below). There is nevertheless no juridical obstacle for this; not for the Arctic Council *per se* and also not for the ACS. The mandate of the Arctic Council is very broad and relates to "common Arctic issues" with special reference to "issues of sustainable development and environmental protection in the Arctic" (Ottawa Declaration 1996, art. 1). A footnote nevertheless specifies that the Council "should not deal with matters related to military security".

In spite of this very broad mandate, however, the Council has so far avoided involvement in certain marine mammal issues (Bloom 1999) and at the November 2007 Meeting of the Senior Arctic Officials (SAOs), decided not to become involved in fisheries management issues either. The matter came up because the US drew the meeting's attention to Senate joint resolution No. 17 of 2007 (S.J. Res. 17 2007)[10] "directing the United States to initiate international discussions and take necessary steps with other Nations to negotiate an agreement for managing migratory and transboundary fish stocks in the Arctic Ocean". The ensuing

---

[10] Passed by the Senate on 4 Oct 2007. The House of Representatives voted in favor of S.J. Res. No. 17 in May 2008 and President George W. Bush signed it on 4 June 2008.

discussion at the SAO's meeting was summarized as follows: "There was strong support for building on and considering this issue within the context of existing mechanisms" (SAO 2007).

Even though the Council has not explicitly reversed its view since then, the issue of international fisheries management has come up in the context of the Arctic Ocean Review (AOR) project that is currently carried out within the Council's Protection of the Arctic Marine Environment (PAME) working group. Phase II of this project is intended to culminate in a final report adopted at the Council's May 2013 Kiruna Ministerial Meeting that will:

> summarize potential weaknesses and/or impediments in the global and regional instruments and measures for [the] management of the Arctic marine environment; outline options to address these weaknesses and/or impediments; and, make agreed recommendations to help ensure a healthy and productive Arctic marine environment in light of current and emerging trends (AOR 2011).

The AOR Phase II draft Report contains a Chap. 4 on 'Marine Living Resources', with Sect. 4.1 (Part A) on 'Fishery Resources'. Its last subsection entitled 'Opportunities' offers various proposals for policy recommendations at its very end, which are copied verbatim into Chap. 9 entitled 'Conclusions and Recommendations' and will be presented to SAOs for negotiation. These proposals consistently use either "Arctic Council States" or "Arctic Council States with coasts on the central Arctic Ocean"—and do not explicitly recommend a role for the Arctic Council as such, but do not rule that out either. Among the options mentioned in the subsection 'Opportunities' that are not specifically retained at the end—and therefore also not in Chap. 9—are a Ministerial Declaration or a statement (AOR 2013).

As alluded to above, the Council could also pursue certain options through the ACS. The concept of the ACS has been introduced by the present author (Molenaar 2012c) to clarify that legally binding instruments such as the Arctic SAR Agreement—and their institutional components—can be part of the Council's output even though they are not—and in fact could not be—formally adopted by it.

The ACS concept consists of two basic components. The first component is made up of the Council's constitutive instrument—the Ottawa Declaration, other Ministerial Declarations, other instruments adopted by the Arctic Council—for instance its Arctic Offshore Oil and Gas Guidelines (PAME 2009), and the Council's institutional structure. The second component consists of instruments 'merely' negotiated under the Council's auspices and their institutional components. The Arctic SAR Agreement and the Meetings of the Parties envisaged under its Article 10 belong to this category. Expansion of this category will occur at the 2013 Kiruna Ministerial Meeting, which will also be used as the occasion for the signature of the Agreement on Cooperation on Marine Oil Pollution Preparedness and Response in the Arctic.

While the section 'Opportunities' in Sect. 4.1 (Part A), Chap. 4 of the AOR Phase II draft Report highlights that the Arctic Council "has been the catalyst" for the adoption of the abovementioned two treaties, it does not explicitly identify this among the various 'modes of delivery' further down that could be used for a treaty on Arctic fisheries (AOR 2013). But in light of the objections by at least one Arctic Council member to the text relating to the catalyst-role of the Arctic Council—and many other parts of Sect. 4.1—it is not even certain if it will be included in the final report.[11]

5.4 National Regulation and Policy

Within the context of this chapter it is not possible to give a comprehensive overview of national regulation and policy by Arctic and non-Arctic states and entities (most notably the EU, but also Taiwan) on the conservation and management of target species and the regulation of the impacts of fishing on non-target species within the marine Arctic. A choice has therefore been made to focus on Arctic Ocean coastal states—in particular the US—and the EU.

In some parts of the marine Arctic, for instance, the North Atlantic, national regulation and policy is expected to be extensive, tailor-made, and related to all or most of the relevant capacities in which states can exercise jurisdiction, namely as flag, coastal, port, and market states and with regard to their natural and legal persons. For other parts of the marine Arctic, however, the presence of ice for most of the year may have rendered tailor-made national fisheries regulation and policy unnecessary. But as diminishing ice-coverage will attract fishing vessels looking for possible new fishing opportunities, all relevant states and entities must ensure that new Arctic fisheries in their own maritime zones and/or by their vessels, comply with applicable global and regional fisheries standards or, where these do not exist, are not conducted in such an unregulated manner that this would be "inconsistent with State responsibilities for the conservation of living marine resources under international law" (FAO 2001, para. 3.3.2).

US action with respect to its EEZ off Alaska in the Arctic Ocean clearly precludes this because it prohibits commercial fishing in the EEZ "until information improves so that fishing can be conducted sustainably and with due concern to other ecosystem components" (NPFMC 2009, sec. E.S.1.2). This proactive and precautionary action by the US is consistent with its Senate joint resolution No. 17 of 2007 mentioned earlier. The 2009 Arctic Region Policy of the US (NSPD-66 2009) contains a few paragraphs on fisheries but not a separate section (secs. III(H)(4) and (6)(b) and (c)).

Norway's laws and regulations relating to fisheries in Svalbard's maritime zones allow 'unregulated fisheries'—but not necessarily in the sense discussed above—to continue or develop, unless explicitly prohibited (Molenaar 2012b). Canada's 2001 'New and Emerging Fisheries Policy'[12] differs from both

---

[11] These written comments on Sect. 4.1 are on file with the author.

[12] The New and Emerging Fisheries Policy is one of Canada's Fisheries Management Policies and is available at <www.dfo-mpo.gc.ca>. Accessed 8 Jan 2013.

approaches as it neither freezes expansion nor allows unregulated fisheries to develop, but requires licenses for each of its three stages (feasibility, exploratory, and commercial) (Ridgeway 2010).[13] Other Canadian policies would apply to new Arctic fisheries as well.[14]

The precautionary approach as such is probably contained in the legal and policy frameworks of most Arctic Ocean coastal states and other key states and entities, even though not necessarily directly in relation to (new) Arctic fisheries. The approaches of Canada and the US are both precautionary and proactive, even though in different ways. The Kingdom of Denmark's 'Strategy for the Arctic' acknowledges that illegal, unreported, and unregulated fishing is a serious threat, that the lack of knowledge of fish stocks and fishing opportunities calls for the application of the precautionary approach, and that fisheries should not commence where a conservation and management system is not available, while explicitly mentioning the Central Arctic Ocean in this regard (Kingdom of Denmark 2011). Conversely, while Norway's 'The High North. Visions and Strategies' (Norway 2011) emphasizes the need for sustainable and science-based fisheries management and the application of the precautionary approach in a general sense, no attention is devoted to new fisheries.

As regards the EU, several of its policy statements in recent years devote specific attention to Arctic fisheries. Sect. 3.2 of the European Commission's Arctic Communication (European Commission 2008) is specifically devoted to 'Fisheries' and is among other things supportive of a temporary ban on new fisheries. Both the Council of the European Union's conclusions on Arctic issues (Council of the European Union 2009) and a 2011 European Parliament resolution (European Parliament 2011) stressed the need to avoid unregulated fishing as well.

The European Commission's 2008 Arctic Communication still viewed extension of the spatial mandate of NEAFC as the preferred option for addressing the gap in high seas coverage with RFMOs in the Central Arctic Ocean. However, the Council conclusions on Arctic issues do not contain a preferred option, but mention the possibility of extending the mandate of existing RFMOs "or any other proposal to that effect agreed by the relevant parties". The most recent EU policy document (European Commission and High Representative 2012) mentions that RFMOs "could in principle extend their geographical scope". While Iceland may prefer extending the spatial scope of NEAFC, none of the Arctic Ocean coastal states seem supportive of this option or the extension of the spatial scope of other existing RFMOs (Molenaar 2009; IAFS 2010).

---

[13] Note also the May 2010 Report of Canada's Standing Senate Committee on Fisheries and Oceans on 'The Management of Fisheries and Oceans in Canada's Western Arctic' (available at <www.parl.gc.ca>), which recommends Canada to adopt an approach for the Canadian part of the Beaufort Sea that is similar to the United States' Arctic FMP (Recommendation No. 12).

[14] Particularly relevant seem to be the 'Policy for Managing the Impacts of Fishing on Sensitive Benthic Areas' and the 'Policy on New Fisheries for Forage Species' (both available at <www.dfo-mpo.gc.ca>, accessed 8 Jan 2013).

It is not clear if other key high seas fishing states and entities such as China, Iceland, Japan, South Korea, or Taiwan have policies on (new) Arctic fisheries.

These practices and policy statements will function as points of departure in ongoing and future international discussions on the international regime for Arctic Ocean fisheries. The contrast between many of these policy statements—even the more explicit ones—and the US action with respect to its EEZ off Alaska in the Arctic Ocean is nevertheless obvious. Whereas US action already actually constrains fisheries by its nationals, most of the policy statements by other states or entities merely advocate or envisage similar action.

The action by the US is also noteworthy in light of the consensual nature of international law. Rather than awaiting agreement at the international level, the US proactively adopted more stringent domestic regulation unilaterally and thereby essentially created a competitive disadvantage for itself. As some of the fish stocks that occur in the EEZ off Alaska in the Arctic Ocean may be transboundary and in view of a more general preference for a level playing field, the US must have hoped, and must continue to do so, that other states and entities follow with similar actions or actions with similar effectiveness; both for the maritime zones of Arctic Ocean coastal states as well as for the Central Arctic Ocean (IAFS 2010). All states and entities must at any rate ensure that new Arctic Ocean fisheries are conducted in compliance with applicable international law. When policy statements are inadequate for this purpose—which is often the case—they must implement them into domestic regulation, acting in all relevant capacities; for instance, as coastal states, flag states, port states, market states, or with regard to their natural or juridical persons.

5.5 Gaps in the International Legal and Policy Framework and National Regulation and Options for Addressing Them

The existing international legal and policy framework and national regulation for Arctic fisheries contains the following main gaps:

1. Science-based and ecosystem-based fisheries management cannot be ensured due to lack of data;
2. Regulation by Arctic Ocean coastal states and other states and entities may not be adequate;
3. Gaps in Arctic Ocean coastal state fora and instruments; and
4. Gap in high seas coverage with RFMOs.

As a comprehensive discussion of all potential options for addressing these gaps is not possible in this chapter, only some comments and observations are offered here. As regards the lack of data identified in gap No. 1, some progress has already been made among Arctic Ocean coastal states (Anchorage Science Meeting Report 2011) and within ICES and PICES, among other factors due to a joint request to ICES by Norway and the Russian Federation (Molenaar 2012a; Takei 2013). Potential future options include a one-off assessment on Arctic fisheries—within or outside the Arctic Council—or the establishment of a new permanent scientific assessment and advisory body on Arctic fisheries, either self-standing, within the Arctic Council, ICES, or the International Arctic Science Committee (IASC), or established jointly by ICES/PICES. If an assessment will indeed be undertaken, it should as a minimum include plausible future scenarios and take account of the impacts of fisheries intensification and expansion on Arctic indigenous peoples. ICES might be the most likely forum for such initiatives to take place, judging by the fact that, at its 100th Meeting in October 2012, the ICES Council agreed to give its Science Committee a mandate to promote science activities in Arctic waters related to various issues, including expansion of distribution/migration of ranges of commercial fish species (ICES 2012).

As regards the regulatory gap identified in No. 2, reference is made to Sect. 5.4, which contains various options to address gaps. Gaps in Arctic coastal state fora and instruments identified in No. 3 could be addressed by formalizing existing informal cooperation, for instance that between Canada and Greenland (see Sect. 5.3.4). New fora and instruments could relate to the conservation and management of shared, straddling, or anadromous fish stocks or provide a framework for mutual fisheries access and exchange of fishing opportunities. Such fora and instruments could be bilateral, trilateral, or involve all Arctic Ocean coastal states.

The gap in RFMO coverage of the high seas identified in No. 4 has undoubtedly generated the most debate, some of which is already covered in Sect. 5.4. As noted there, insufficient support exists for extending the spatial mandate of NEAFC. A full-fledged RFMO does not seem the most likely option either, due to considerations of cost-effectiveness in light of the fact that significant commercially viable fisheries are not expected in the short term. The negotiation-process on the CBS Convention eventually decided not to push for a full-fledged RFMO for similar reasons (Balton 2001). An Arrangement[15]—whether legally binding or non-legally binding—is therefore a more likely option.

As noted in Sect. 5.3.5, there is no obstacle for such an Arrangement to be adopted by the Arctic Council—in case of a non-legally binding instrument—or through the ACS approach as previously described. If the instrument also related to the Central Arctic Ocean—where the freedom of fishing applies—its effectiveness would benefit from support by key non-Arctic states and entities. Such support could be ensured through a format or mechanism that allows them to participate in the instrument's negotiation as well as to express their consent to be bound. Involving only the current non-Arctic state observers would not work. This group consists of six EU Member States[16]—which have transferred

---

[15] See *supra* note 1.

[16] Namely France, Germany, the Netherlands, Poland, Spain, and the United Kingdom.

most of their competence on marine capture fisheries to the EU—and therefore does not include any key high seas fishing states and entities such as China, the EU, Japan, South Korea, and Taiwan. While all these except Taiwan have applied for observer status, it is by no means clear when and if these applications will be approved (Molenaar 2012c).[17] If all applications *are* approved, however, involving observers could be an important component of the aforementioned format or mechanism.

An alternative to proceeding through the Arctic Council or the ACS would be to adopt the Arrangement as a stand-alone instrument. If the instrument's spatial scope, and thereby the measures it contains, were limited to the Central Arctic Ocean—as advocated by some[18] -, similar measures or measures with similar effectiveness need to be adopted by Arctic Ocean coastal states for their own maritime zones. The need for compatibility is particularly evident as it is likely that new fishing opportunities will arise in coastal state maritime zones before arising in the high seas. An exception may nevertheless be granted in furtherance of the rights and interests of Arctic indigenous peoples.

While the assumption is that other states and entities besides the Arctic five would be allowed to participate in the negotiation of a stand-alone fisheries instrument on the Arctic Ocean or the Central Arctic Ocean, and eventually become parties thereto, this assumption could of course be proven wrong. A coastal states-only *inter se* approach would not necessarily be inconsistent with international law. This would only occur if the exercise of the right to engage in high seas fishing by other states and entities was interfered with in ways that were not consistent with international law. At-sea high seas enforcement would be an obvious example. Presumably, however, the Arctic five prefer to avoid these issues as well as the lack of legitimacy that is associated with coastal states-only approaches.[19] In a worst-case scenario, such lack of legitimacy might even prompt high seas fishing states or entities to engage in high seas fishing in the Central Arctic Ocean in order to assert their right as such, even if such fishing was not commercially viable. Fortunately, the clear commitment to peace, order, and cooperation that underlies the Ilulissat Declaration implies also a commitment by the Arctic Ocean coastal states to avoid such a scenario.

---

[17] Other non-Arctic state applicants are India (application submitted on 6 Nov 2012; information provided by N. Buvang to the author by email on 7 Feb 2013), Italy, and Singapore.

[18] For instance the Pew Environment Group's 'Oceans North' campaign, which also led to the submission of a letter signed by a large number of scientists to the International Polar Year Conference in Montréal, Canada (22–27 Apr 2012) (info at <oceansnorth.org/international>).

[19] Reference can in this context be made to the controversial 'Galapagos Agreement' (Framework Agreement for the Conservation of the Living Marine Resources on the High Seas of the Southeast Pacific, Santiago, 14 Aug 2000. Not in force, *Law of the Sea Bulletin*, 70–78, No. 45 (2001)), which never entered into force and has now essentially been replaced by the SPRFMO Convention (2009).

5.6 Potential for EU–US Cooperation

A discussion on the potential for EU–US cooperation on Arctic fisheries must acknowledge at the outset that whereas the US is an Arctic Ocean coastal state, the EU cannot rely on such a *de facto* capacity. Whereas Denmark is an Arctic Ocean coastal state with respect to Greenland and an Arctic coastal state with respect to the Faroe Islands, Denmark's EU Membership does not extend to Greenland or the Faroe Islands (TFEU 2008, arts. 204 and 355(5)(a)). But the EU can still act in various other *de facto* capacities; for instance as a flag state—including pursuant to the freedom of fishing on the high seas -, port state, market state, or with respect to the natural and legal persons of its Member States.

The "conservation of marine biological resources under the common fisheries policy" is one of the five areas listed in Article 3(1) of the TFEU (2008) in which the EU has exclusive competence, subject to some exceptions (Churchill and Owen 2010). The consequential external competence of the EU in the sphere of fisheries implies that the EU represents its Member States, for instance in negotiations with non-EU Member States and in RFMOs. In some cases, however, EU Member States can still become members to RFMOs alongside the EU. One of these exceptions relates to 'overseas countries and territories' and enables for instance Denmark to become a member of RFMOs alongside the EU in respect of the Faroe Islands, Greenland, or both; for instance, in relation to NEAFC (for both).

The differences between the EU and the US as regards their involvement in marine capture fisheries worldwide must be acknowledged as well. It is widely known that—compared to US fishing vessels—fishing vessels flying flags of EU Member States operate to a much larger extent on the high seas and within maritime zones of third states (non-EU Member States) in not only the Atlantic Ocean and adjacent seas, but also the Indian Ocean, Pacific Ocean, and Southern Ocean. The fishing fleet of the EU Member States is therefore much more a distant water fishing fleet compared to the US fishing fleet. Furthermore, it is also widely known that natural and legal persons with the nationality of EU Member States—in particular Spain—are more extensively involved as beneficial owners of fishing vessels flying the flag of third states (i.e., non-EU Member States), compared to natural and legal persons with US nationality. For these reasons, the international community views the EU in the domain of marine capture fisheries primarily as a flag state and the US primarily as a coastal state. In view of the widespread tendency to hold flag states responsible—even though not necessarily always justifiably—for failures in regional fisheries management, the US may well be quite critical of the track-record and ability of the EU and its Member States in ensuring a high standard of flag state performance.

These observations should not be interpreted as suggesting that there is no potential for cooperation between the EU and the US on Arctic fisheries. Despite their different entitlements to fisheries resources in the marine Arctic, the EU and the US share common interests in avoiding over-exploitation of target species and undesirable impacts of fisheries on non-target species, including marine mammals. Moreover, as noted in Sect. 5.4, both the EU and the US take the view that unregulated fishing in the Central Arctic Ocean must be avoided.

A suitable domain for cooperation between the EU and the US is research on the marine Arctic in general and Arctic ecosystems, commercial fish species, and fisheries in particular. Such cooperation could take place within the context of existing bodies like ICES, PICES, and the Arctic Council. As regards the Arctic Council, it seems that the success of US–EU cooperation would benefit from US support for EU observer status. The US could also consider advocating that Arctic Ocean coastal states expand participation in their fisheries science meetings with representatives of the EU.

In light of their shared concerns and positions on unregulated fishing in the Central Arctic Ocean, cooperation in that regard could be considered as well. The fact that the EU is not an Arctic Ocean coastal state and incapable of acting in such a *de facto* capacity, requires the US and the EU to proceed carefully in order not to antagonize Arctic Ocean coastal states that feel strongly about the lead role they should have in any intergovernmental consultations on international regulation of Arctic Ocean fisheries. Joint EU–US initiatives therefore run the risk of being counterproductive. Individual action in support of shared concerns and positions could avoid this. The EU could express public support for a position aligned with that of the US, for instance as laid down in the latter's Senate joint resolution No. 17 of 2007. Moreover, both the EU and the US could take individual proactive steps, for example by adopting regulations that impose a temporary ban on fishing by their vessels in the Central Arctic Ocean until such time as fishing would be permitted pursuant to an international instrument.

Even though the US and the EU should proceed carefully, their cooperation on unregulated fishing in the Central Arctic Ocean offers clear opportunities as well. The effectiveness of a future international instrument on Central Arctic Ocean fisheries would benefit from support by key non-Arctic Ocean coastal states and entities (see Sect. 5.5), and early EU support would be crucial for fostering support among them. The US could make an important contribution in this regard by advocating the inclusion of the notion of compatibility—between fisheries regulation on the high seas and fisheries regulation in coastal state maritime zones—in the future instrument. While the future instrument could also be made applicable to the Arctic Ocean as a whole, there is no indication that the US or even the EU would find this desirable; even though not necessarily for similar reasons.

Support by the EU and other key non-Arctic Ocean coastal states and entities for a future international instrument on Central Arctic Ocean fisheries also depends on their ability to participate in a meaningful way in the instrument's negotiation. Such participation would enhance the legitimacy and credibility of the negotiation-process and thereby also the instrument adopted by it. The US could contribute to these ends by undertaking efforts to ensure that preparatory consultations among Arctic Ocean coastal states have not advanced to such an extent that key non-Arctic Ocean coastal states and entities are essentially presented with a *fait accompli*. Improvements in the EU's and its Member States' performance as *de facto* flag states and coastal states in fisheries management and conservation more broadly, would also help to secure support for broader participation in negotiations. If it is decided to conduct such negotiations within the Arctic Council or by means of the ACS approach, the success of US-EU cooperation would also benefit from US support for EU observer status.

5.7 Conclusions

The unprecedented pace of change that the Arctic is currently experiencing makes it very difficult to argue that the current international legal and policy framework for Arctic fisheries conservation and management is adequate for responding to the huge challenges that lie ahead. This chapter identifies a number of gaps in the international framework as well as in national regulation and policy, and suggests various options for addressing these gaps.

Most of the suggested options to address identified gaps also offer opportunities for cooperation between the EU and the US. Cooperative initiatives could be undertaken to strengthen scientific research, 'domestic' fisheries regulation and the international regime for Arctic Ocean fisheries. The fact that the US is an Arctic Ocean coastal state and the EU is not, and is also incapable of acting in such a *de facto* capacity, requires them to proceed carefully but offers clear opportunities as well. Joint EU–US initiatives could antagonize Arctic Ocean coastal states who feel strongly about the lead role they should have in any intergovernmental consultations on international regulation of Arctic Ocean fisheries. Individual action in support of shared concerns and positions could avoid this.

Early EU support for a future international instrument on Central Arctic Ocean fisheries would be crucial for fostering support among other key non-Arctic states and entities and could thereby contribute to the future instrument's effectiveness, legitimacy, and credibility. The US could play a crucial role in this regard by advocating the inclusion of the notion of compatibility in the future instrument and by undertaking efforts to ensure that the EU and key other non-Arctic Ocean coastal states and entities can participate in a meaningful way in the instrument's negotiation.

References

ACIA (2005) Arctic Climate Impact Assessment. Cambridge University Press, New York

Agreement on Mutual Fisheries Relations between the United States and Union of Soviet Socialist Republics (1988) Agreement between the Government of the United States of America and the Government of the Union of Soviet Socialist Republics on Mutual Fisheries Relations, 31 May 1988, Treaties and other International Acts Series 11,422. Entered into force 28 Oct 1988

Agreement on Mutual Fishery Relations between Denmark and Greenland (1992) Agreement between the Government of the Kingdom of Denmark and the Local Government of

Greenland, on the one hand, and the Government of the Russian Federation, on the other hand, concerning Mutual Fishery Relations between Greenland and the Russian Federation, 7 Mar 1992, 1719 U.N.T.S. 89. Entered into force provisionally on 7 Mar 1992 and definitively on 16 Oct 1992

Agreement on Mutual Fishery Relations between Greenland/Denmark and Norway (1992) Agreement between Greenland/Denmark and Norway concerning Mutual Fishery Relations, Copenhagen, 9 June 1992, 1829 U.N.T.S. 223. Entered into force provisionally with retroactive effect from 24 Sept 1991 and definitively on 4 Mar 1994

AMAP (1997) Arctic Pollution Issues: A State of the Arctic Environment Report. Arctic Monitoring and Assessment Programme, Oslo

Anchorage Science Meeting Report (2011) Report of a Meeting of Scientific Experts on Fish Stocks in the Arctic Ocean. Anchorage, Alaska, 15–17 June 2011; on file with the author

AOR (2011) Phase I Report (2009–2011) of the Arctic Ocean Review (AOR) project. <www.pame.is>. Accessed 29 Jan 2013

AOR (2013) The Arctic Ocean review project. Phase II Report 2011–2013, consolidated version of 9 Feb 2013, on file with the author

Arctic SAR Agreement (2011) Agreement on Cooperation on Aeronautical and Maritime Search and Rescue in the Arctic, 12 May 2011, 50 I.L.M. 1119 (2011). Entered into force on 19 Jan 2013

Balton D (2001) The Bering Sea Doughnut Hole convention: Regional Solution, Global Implications. In: Stokke OS (ed) Governing High Seas Fisheries: the Interplay of Global and Regional Regimes. Oxford University Press, Oxford, pp 143–177

Barnes R (2011) International Regulation of Fisheries Management in Arctic Waters. German Yearbook Int Law 54:193–230

Bloom ET (1999) Establishment of the Arctic Council. Am J Int Law 93:712–722

CBS Convention (1994) Convention on the Conservation and Management of Pollock Resources in the Central Bering Sea, 16 June 1994, 34 I.L.M (1995). Entered into force 8 Dec 1995

Churchill RR (2001) Managing Straddling Fish Stocks in the North-East Atlantic: A Multiplicity of Instruments and Regime Linkages—but How Effective a Management? In: Stokke OS (ed) Governing High Seas Fisheries: The Interplay of Global and Regional Regimes. Oxford University Press, Oxford, pp 235–272

Churchill RR, Owen D (2010) The EC Common Fisheries Policy. Oxford University Press, Oxford

Convention on the Conservation and Management of High Seas Fisheries Resources in the North Pacific Ocean (2012) Tokyo, 24 Feb 2012. Not in force; on file with author

Compliance Agreement (1993) Agreement to Promote Compliance with International Conservation and Management Measures by Fishing Vessels on the High Seas, 24 Nov 1993, 33 I.L.M. 969 (1994). Entered into force 24 Apr 2003

Council of the European Union (2009) Council conclusions on Arctic issues, 2985th Foreign Affairs Council meeting, 8 Dec 2009

European Commission (2008) Communication from the Commission to the European Parliament and the Council. The European Union and the Arctic Region, 20 Nov 2008, COM(2008) 763

European Commission and High Representative (2012) Joint Communication to the European Parliament and the Council. Developing a European Union Policy towards the Arctic Region: progress since 2008 and next steps. 26 June 2012, JOIN(2012) 19 final

European Parliament (2011) A sustainable EU policy for the High North. European Parliament resolution of 20 Jan 2011 on a sustainable EU policy for the High North (2009/2214(INI)), 20 Jan 2011, P7_TA(2011)0024

FAO (1995) Code of Conduct for Responsible Fisheries. Adopted by the Twenty-eight Session of the FAO Conference, 31 Oct 1995

FAO (2001) International Plan of Action to Prevent, Deter and Eliminate Illegal, Unreported and Unregulated Fishing. Adopted by consensus by FAO's committee on Fisheries on 2 Mar 2001 and endorsed by the FAO Council on 23 June 2001

FAO (2008) International Guidelines for the Management of Deep-Sea Fisheries in the High Seas, 29 Aug 2008 (contained in Appendix F to the Report of the Technical Consultation on

International Guidelines for the Management of Deep-Sea Fisheries in the High Seas, 4–8 Feb and 25–29 Aug 2008 (FAO Fisheries and Aquaculture Report No. 881))

FAO (2010) International Guidelines on Bycatch Management and Reduction of Discards, Rome, 10 Dec 2010 [contained in Appendix E to the Report of the Technical Consultation to Develop International Guidelines on Bycatch Management and Reduction of Discards, 6–10 Dec 2010 (FAO Fisheries and Aquaculture Report No. 957)]

Fish Stocks Agreement (1995) Agreement for the Implementation of the Provisions of the United Nations Convention on the Law of the Sea of 10 Dec 1982 relating to the Conservation and Management of Straddling Fish Stocks and Highly Migratory Fish Stocks, 4 Aug 1995, 2167 U.N.T.S. 3. Entered into force 11 Dec 2001

Hoel AH (2011) Living marine resources in the Arctic. Trends & Opportunities. Presentation dated 20 Sept 2011 at the 2nd Arctic Ocean review (AOR) project workshop, Reykjavik, on file with author

IAFS (2010) In: Proceedings of the International Arctic Fisheries Symposium. Managing Resources for a Changing Arctic, pp 23–32 (2010). <www.nprb.org/iafs2009>. Accessed 29 Jan 2013

ICCAT Convention (1966) International Convention for the Conservation of Atlantic Tunas, 14 May 1966, 673 U.N.T.S. 63. Entered into force 21 Mar 1969. As amended by Protocols adopted in 1984 and 1992, which both entered into force

ICES (2012) Report of the 100th ICES Council Meeting, 24–25 Oct 2012. Copenhagen: ICES Headquarters. <www.ices.dk>. Accessed 29 Jan 2013

IPHC Convention (1953) Convention for the Preservation of the Halibut Fishery of the North Pacific Ocean and the Bering Sea, 2 Mar 1953, 22 U.N.T.S. 78 (1955). Entered into force 28 Oct 1953. Exchange of Notes Constituting an Agreement to Amend the [IPHC Convention], Washington, 29 Mar 1979, 1168 U.N.T.S. 380 (1980). Entered into force 29 Mar 1979

Kingdom of Denmark (2011) Denmark, Greenland and the Faroe Islands: Kingdom of Denmark Strategy for the Arctic 2011–2020. <www.nanoq.gl>. Accessed 29 Jan 2013

LOS Convention (1982) United Nations Convention on the Law of the Sea, 10 Dec 1982, 1833 U.N.T.S. 396. Entered into force 16 Nov 1994

Molenaar EJ (2007) Port State Jurisdiction: Toward Comprehensive, Mandatory and Global Coverage. Ocean Dev Int Law 38:225–257

Molenaar EJ (2009) Arctic Fisheries Conservation and Management: Initial Steps of Reform of the International Legal Framework. Yearbook Polar Law 1:427–463

Molenaar EJ (2012a) Arctic Fisheries and International Law. Gaps and Options to Address Them. Carbon Climate Law Rev 1:63–77

Molenaar EJ (2012b) Fisheries Regulation in the Maritime zones of Svalbard. Int J Marine Coastal Law 27:3–58

Molenaar EJ (2012c) Current and Prospective Roles of the Arctic Council System within the Context of the Law of the Sea. Int J Marine Coastal Law 27:553–595

NAFO Convention (1978) Convention on Future Multilateral Cooperation in the Northwest Atlantic Fisheries, 24 Oct 1978, 1135 U.N.T.S. 369. Entered into force 1 Jan 1979. 2007 Amendment, Lisbon, 28 Sept 2007. Not in force (at 27 Nov 2012), NAFO/GC Doc. 07/4

NASCO Convention (1982) Convention for the Conservation of Salmon in the North Atlantic Ocean, 2 Mar 1982, 1338 U.N.T.S. 33 (1983). Entered into force 1 Oct 1983

NEAFC Convention (1980) Convention on Future Multilateral Cooperation in the North-East Atlantic Fisheries, 18 Nov 1980, 1285 U.N.T.S. 129. Entered into force 17 Mar 1982. 2004 Amendments (Art. 18bis), London; 12 Nov 2004. Not in force (at 27 Nov 2012), but provisionally applied by means of the 'London Declaration' of 18 Nov 2005. 2006 Amendments, London (Preamble, Arts 1, 2 and 4), 11 Aug 2006. Not in force, but provisionally applied by means of the 'London Declaration' of 18 Nov 2005

Norway (2011) The High North. visions and Strategies, 18 Nov 2011

Norwegian-Russian Bilateral Framework Agreement (1975) Agreement between the Government of the Kingdom of Norway and the Government of the Union of Soviet Socialist Republics on Cooperation in the Fishing Industry, 11 Apr 1975, 983 U.N.T.S. 1975. Entered into force 11 Apr 1975

NPAFC Convention (1992) Convention for the Conservation of Anadromous Stocks in the North Pacific Ocean, 11 Feb 1992. 22 Law Sea Bull 21 (1993). Entered into force 16 Feb 1993

NPFMC (2009) Arctic Fishery Management Plan. United States North Pacific Fishery Management Council. Adopted 5 Feb 2009, effective 3 Dec 2009. 50 C.F.R. 679

NSPD-66 (2009) National Security Presidential Directive and Homeland Security Presidential Directive. Arctic Region Policy, 9 Jan 2009, NSPD-66/HSPD-25

Ottawa Declaration (1996) Declaration on the Establishment of the Arctic Council, 19 Sept 1996, 35 I.L.M. 1382 (1996)

Pacific Salmon Treaty (1985) Treaty between the Government of Canada and the Government of the United States of America Concerning Pacific Salmon, 28 Jan 1985. 1469 U.N.T.S. 358 (1987). Entered into force 18 Mar 1985. As amended

PAME (2009) Arctic Offshore Oil and Gas Guidelines. Last updated 29 Apr 2009

Port State Measures Agreement (2009) Agreement on Port State Measures to Prevent, Deter and Eliminate Illegal, Unreported and Unregulated Fishing, 22 Nov 2009. Not in force (at 27 Nov 2012)

Ridgeway L (2010) Issues in Arctic Fisheries Governance: A Canadian Perspective. In: Nordquist MH, Heidar TH, Moore JN (eds) Changes in the Arctic Environment and the Law of the Sea. Martinus Nijhoff Publishers, Canada, pp 409–446

SAO (2007) Report of the November 2007 SAOs Meeting. <www.arctic-council.org>. Accessed 29 Jan 2013

S.J. Res. 17 (2007) A joint resolution directing the United States to initiate international discussions and take necessary steps with other Nations to negotiate an agreement for managing migratory and transboundary fish stocks in the Arctic Ocean, 110th Congress, 2007

Spitsbergen Treaty (1920) Treaty concerning the Archipelago of Spitsbergen, 9 Feb 1920, 2 L.N.T.S. 7. Entered into force 14 Aug 1925

SPRFMO (2008) Draft Convention on the Conservation and Management of High Seas Fishery Resources in the South Pacific Ocean. Doc. SP/06/WP1, Revision 4 (Oct 2008). Text at <www.southpacificrfmo.org>

SPRFMO Convention (2009) Convention on the Conservation and Management of High Seas Fishery Resources in the South Pacific Ocean, 14 Nov 2009. Entered into force 24 Aug 2012

Takei Y (2013) Marine Scientific Research in the Arctic. In Molenaar EJ, Oude Elferink AG, Rothwell DR (eds) The law of the Sea and Polar Regions: Interactions between Global and Regional Regimes. Martinus Nijhoff Publishers

TFEU (2008) Treaty on the Functioning of the European Union. Consolidated Version. O.J. C 115/47

Trilateral Loophole Agreement between Iceland, Norway, and the Russian Federation (1999) Agreement between the Government of Iceland, the Government of Norway and the Government of the Russian Federation Concerning Certain Aspects of Cooperation in the Area of Fisheries, 15 May 1999. 41 Law Sea Bull 53 (1999). Entered into force 15 July 1999

WCPFC Convention (2000) Convention on the Conservation and Management of Highly Migratory Fish Stocks in the Western and Central Pacific Ocean, 5 Sept 2000, 2275 U.N.T.S. 2004 (2007) Entered into force 19 June 2004

Zeller D, Booth S, Pakhomov E, Swartz W, Pauly D (2011) Arctic fisheries catches in Russia, USA, and Canada: baselines for neglected ecosystems. Polar Biol 34:955–973

Chapter 6
Status and Reform of International Arctic Shipping Law

Erik J. Molenaar

Abstract The last years have witnessed a continuous decrease in Arctic sea ice coverage and thickness and a significant expansion in the volume of shipping traffic. This chapter assesses the adequacy of the current international legal and policy framework for Arctic shipping in view of the likelihood that these trends continue in coming years. An overview of the international legal and policy framework is provided, the main gaps therein are identified, and options for addressing these are suggested. Options could be undertaken within the International Maritime Organization (IMO) but also outside the IMO, for instance within the Arctic Council or among ad hoc groupings of states. Separate attention is devoted to the potential for cooperation between the European Union and the United States in this regard.

6.1 Introduction

As future trends in Arctic marine shipping depend to a significant extent on sea ice coverage and thickness, it is important to note that both have been steadily declining. The summer 2012 Arctic sea ice extent was the lowest on record, the

Writing this chapter was facilitated by funding from the Netherlands Polar Programme and the EU's COST Action IS1105 'NETwork of experts on the legal aspects of MARitime SAFEty and security (MARSAFENET)'. The author is grateful for comments by Carien Droppers, Henrik Ringbom, Greg O'Brien, and others on an earlier version of the chapter. This chapter builds on Molenaar EJ, Corell R (2009). Background Paper. Arctic Shipping. Arctic TRANSFORM.

E. J. Molenaar (✉)
Netherlands Institute for the Law of the Sea (NILOS), Utrecht University, Utrecht,
The Netherland
Faculty of Law, University of Tromsø, Tromsø, Norway
e-mail: E.J.Molenaar@uu.nl

E. Tedsen et al. (eds.), *Arctic Marine Governance*,
DOI: 10.1007/978-3-642-38595-7_6, © Springer-Verlag Berlin Heidelberg 2014

December 2012 sea ice was the second lowest on record, and—perhaps even more important—multi-year sea ice has declined even further after the enormous loss in the summer of 2007. For several summers now, both the Northwest Passage and the Northern Sea Route have been open.[1]

Intra- and trans-Arctic marine shipping can be interesting alternatives for the much longer routes using the Panama and Suez Canals, or for Arctic routes that are partly terrestrial and partly marine. But even though summers without sea ice in much or all of the Arctic Ocean may occur in the not too distant future, sea ice will still be widespread in winter. While much or most of this will be first-year sea ice, there may be other factors that could adversely affect shipping conditions and care must therefore be taken not to overestimate the potential growth of Arctic marine shipping (Kraska 2007; Brigham 2010; ICS 2012). However, even a limited expansion of intra- and trans-Arctic marine shipping requires an assessment of the adequacy of national and international regulation. The 1989 *Exxon Valdez* disaster has not been forgotten and the 2010 crisis with the *Deepwater Horizon* in the Gulf of Mexico painfully exposed human failure on numerous counts, including a tendency to underestimate risks. In the summer of 2010, this tendency was also illustrated by two groundings in Canadian waters: a tanker carrying fuel, close to Pangnirtung, and the cruise-vessel *Clipper Adventurer*, just east of Kugluktuk.[2]

This chapter assesses the adequacy of the current international legal framework for Arctic shipping in light of the current and expected impacts of global climate change on the marine Arctic. The focus is predominantly on the impacts of Arctic marine shipping on the Arctic marine environment and its biodiversity. The maritime safety dimension is covered where regulation serves a significant subsidiary purpose of pollution prevention. Delimiting the scope of this chapter in this way, however, is not meant to suggest that ensuring maritime safety in the marine Arctic is not as important or urgent, particularly in view of the continued and increasing interest in Arctic sea-borne tourism.

After providing some context and background information on current and future Arctic marine shipping in Sect. 6.2, an overview of the international legal and policy framework for Arctic marine shipping is provided in Sect. 6.3. Gaps in the international legal and policy framework and options for addressing them are covered in Sect. 6.4 and the potential for EU-US cooperation is examined in Sect. 6.5. Some conclusions are offered in Sect. 6.6.

For the purposes of this chapter, Arctic marine shipping is regarded as the shipping that occurs or could occur in the marine Arctic. This chapter uses the same definitions for the terms 'marine Arctic', 'Arctic Ocean', 'Arctic states', and 'Arctic Ocean coastal states' as throughout this book. It is worth noting that

---

[1] Information obtained from <nsidc.org/arcticseaicenews>, <www.arctic.noaa.gov/reportcard>, and <www.climatewatch.noaa.gov> on 11 Jan 2013.

[2] See the press releases and other information at <barentsobserver.com>, <www.arcticmonitor. net>, and <www.institutenorth.org>.

the definition of marine Arctic is a broader area than the 'Arctic waters' under the International Maritime Organization (IMO) Polar Shipping Guidelines (2009), as described in Chap. 1.

Arctic marine shipping can be intra-Arctic or trans-Arctic. Trans-Arctic marine shipping can take place by means of various routes and combinations of routes. Two of these routes are the Northwest Passage and the Northern Sea Route. The official Northern Sea Route encompasses all routes across the Russian Arctic coastal seas from Kara Gate (at the southern tip of Novaya Zemlya) to the Bering Strait (Tymchenko 2001). The Northwest Passage is the name given to the marine routes between the Atlantic and Pacific Oceans along the northern coast of North America that span the straits and sounds of the Canadian Arctic Archipelago. Pharand identified seven main routes, with minor variations (Pharand 2007). An alternative to all these routes is the Central Arctic Ocean Route, which runs straight across the middle of the Central Arctic Ocean.

6.2 Current and Future Arctic Marine Shipping

While the current volume of Arctic marine shipping is still very modest, recent years have seen steady—and sometimes significant—increases in many types of shipping, including transits through the Northwest Passage and the Northern Sea Route, destinational traffic associated with offshore resource activity, and Arctic sea-borne tourism. As regards transits, various records and first-evers have recently occurred; for instance: back-to-back crossings of both the Northwest Passage and the Northern Sea Route in a single summer by yachts; the transit of the Northwest Passage by *The World* in the summer of 2012; and—in relation to the Northern Sea Route—tankers carrying oil, gas condensate, and liquid natural gas (LNG); the first ferry (the *Georg Ots*); the first non-Russian bulk-carrier (the *MV Nordic Barents*) carrying iron-ore concentrate without stopping at a Russian port; and the first cargo vessel (the *Monchegorsk*) to transit without ice-breaker assistance. However, regular container ship operations through the Northern Sea Route have not proven commercially viable.[3]

Indications are that traffic in the Northwest Passage will grow much less compared to the Northern Sea Route (AMSA 2009; AOR 2013). This is to some extent caused by projections about the presence of sea ice and natural restraints (e.g., shallowness) in parts of the Northwest Passage. Another important factor is that Russia is much more interested in developing Arctic marine shipping than Canada, and has made large efforts in support of development (Emmerson 2011). In addition to substantial investments in infrastructure, vessels, and equipment, and reforms in domestic legislation and institutional arrangements (Solski 2013),

---

[3] See the press releases and other information at <barentsobserver.com>, <www.arcticmonitor. net>, and <www.institutenorth.org>.

reference can be made to the strategic agreement adopted in 2010 between the Sovcomflot Group and the China National Petroleum Corporation, which envisions increased usage of the Northern Sea Route.[4] The 'shale gas revolution' has also meant that Russian gas originally intended to be shipped to the east coast of the United States (US), is now shipped through the Northern Sea Route to China and other Asian states. This is a good example of the role of key variables, uncertainties, or 'wildcards' in the scenarios developed as part of the Arctic Marine Shipping Assessment (AMSA). Examples of wildcards are an accelerated Arctic meltdown, major Arctic shipping disasters, and technology breakthroughs (AMSA Scenarios 2008). The risk assessments of classification societies and the marine insurance industry are also likely to be a crucial factor for the economic viability of all Arctic marine shipping.

Due to the accelerated melting of Arctic sea ice, the Central Arctic Ocean Route may soon be an option as well (Humpert and Raspotnik 2012). The most suitable course of this latter route may vary from year to year and lead to various combinations of the Central Arctic Ocean Route on the one hand and the Northwest Passage and Northern Sea Route on the other hand. Some of the routes of which the Northern Sea Route consists already pass through the high seas area of the Central Arctic Ocean. It is finally important to note that all trans-Arctic marine shipping must pass through the Bering Strait.

Marine shipping has the following actual and potential impacts on the marine environment and marine biodiversity:

1. Shipping practices and incidents leading to accidental discharges of polluting substances (cargo or fuel) or physical impact on components of the marine ecosystem (e.g., on the benthos and larger marine mammals);
2. Operational discharges (cargo residues, fuel residues (sludge), (incineration of) garbage and sewage), and emissions;
3. Introduction of alien organisms through ballast-water exchanges or attachment to vessel hulls (e.g., in crevices); and
4. Other navigation impacts (noise pollution and other forms of impacts on, or interference with, marine species potentially causing, for instance, disruption of behaviour, abandonment, or trampling of the young by fleeing animals or displacement from normal habitat).

All these actual and potential impacts are also relevant for Arctic marine shipping. The likelihood for some of these impacts—for instance shipping incidents—to occur is higher in some parts of the marine Arctic due to the presence of ice(bergs), lack of accurate charts, and in the case of insufficient experience in navigating in ice-covered areas. In addition, cold temperatures may affect machinery, and icing can create additional loads on the hull, propulsion systems, and appendages (VanderZwaag et al. 2008). The remoteness of much of the marine Arctic, the limited available maritime safety information data, and the challenges

---

[4] See the press release of 22 Nov 2010 at <www.sovcomflot.ru>.

of navigating therein moreover mean that once shipping incidents do occur, a response will take a relatively long time and may even then be inadequate to address impacts on the marine environment and marine biodiversity.

6.3 International Legal and Policy Framework for Arctic Marine Shipping

6.3.1 Interests, Rights, Obligations, and Jurisdiction

The international legal and policy framework for vessel-source pollution seeks to safeguard the different interests of the international community as a whole with those of states that have rights, obligations, or jurisdiction in their capacities as flag, coastal, or port states or with respect to their natural and legal persons. While the term 'flag state' is commonly defined as the state in which a vessel is registered and/or whose flag it flies (LOS Convention 1982, art. 91(1)), there are no generally accepted definitions for the terms 'coastal state' or 'port state'. For the purposes of this chapter, however, the term 'coastal state' refers to the rights, obligations, and jurisdiction of a state within its own maritime zones over foreign vessels.

The term 'port state' refers to the rights, obligations, and jurisdiction of a state over foreign vessels that are voluntarily in one of its ports. In order to avoid an overlap with jurisdiction by coastal states, this chapter regards port state jurisdiction as relating to illegal discharges by foreign vessels beyond the coastal state's maritime zones as well as over violations of conditions for entry into port (Molenaar 2007).

The balance in the above-mentioned framework is first of all between the socioeconomic interests of flag states in unimpeded navigation and a minimum of globally uniform international regulation, and the environmental interests of the coastal state. The port state commonly seeks to balance its local environmental interests and the broader environmental interests that 'its' coastal state has over its maritime zones, against the socioeconomic interests of the port and its *hinterland*. States generally have interests, rights, obligations, and jurisdiction in more than one capacity. This commonly leads to a more balanced compromise position, but occasionally also to contradictory positions of the same state within different fora. There is no reason or indication to assume that Arctic states are different in this regard.

The interests of the international community—e.g., sustainable utilization, protection, and preservation of the marine environment, and conservation of marine biodiversity—normally overlap with those of flag, coastal, and port states, but are usually broader and more general. The interests of some states, however, clearly undermine those of other states and the international community, for instance, by not ensuring that their ships comply with international minimum standards or by allowing foreign vessels in their ports to be non-compliant with international minimum standards. These states, vessels, and ports thereby have a competitive advantage over states, vessels, and ports that *do* comply with international minimum standards. Such 'free riders' clearly benefit from the consensual nature of international law—meaning that a state can only be bound to a rule of international law when it has in one way or another consented to that rule.

6.3.2 Substantive Shipping Standards

The categories of substantive shipping standards below are based on the substantive focus of this chapter (see Sect. 6.1), the LOS Convention's (1982) jurisdictional framework for vessel-source pollution, and practice within IMO so far. The categories are:

1. Discharge and emission standards, including standards relating to ballast water exchange;
2. Construction, design, equipment, and manning (CDEM) standards, including fuel content specifications and anti-fouling and ballast water treatment standards;
3. Navigation standards, in the form of ships' routeing measures, ship reporting systems (SRSs), and vessel traffic services (VTS);
4. Contingency planning and preparedness standards; and
5. Liability, compensation, and insurance requirements.

This categorization is merely meant to facilitate the discussion below, however. It does not capture the entire spectrum of types of standards or requirements developed within IMO or applied by individual states acting in their various capacities. An Arctic Ocean coastal state may for instance require use of ice-breaker assistance and the payment of fees for such services.

6.3.3 Global and Regional Bodies

International regulation of vessel-source pollution by merchant ships is primarily done by global bodies. While IMO is the most prominent, the International Labour Organization (ILO) and the International Atomic Energy Agency (IAEA) adopt relevant regulation as well. The pre-eminence of global bodies is a direct consequence of the global nature of international shipping and the interest of the international community in globally uniform international minimum regulation (VanderZwaag et al. 2008; Chircop 2009; Molenaar 2010). The LOS Convention safeguards this by allowing unilateral coastal state prescription in only a few situations. These exceptions are explained in Sect. 6.3.4, which also devotes attention to so-called 'residual port state jurisdiction'.

As unilateral coastal state prescription and residual port state jurisdiction can also be exercised collectively (in concert), regional regulation of marine shipping by regional bodies or *ad-hoc* groupings of states is not inconsistent with the

LOS Convention, and thereby also not with the primary role accorded to the IMO by the LOS Convention (Stokke 2012). Regional regulation can also be pursued on an *inter se* basis to ships flying the flag of parties—for instance, Annex IV on 'Prevention of Marine Pollution' to the Antarctic Treaty's (1959), Protocol on Environmental Protection (1991).

Several IMO instruments also allow or encourage regional implementation. This has led the Arctic Council to undertake efforts to implement IMO's SAR (Search and Rescue) Convention (1979) by means of the Arctic SAR Agreement (2011), and IMO's OPRC 90 (1990) by means of the future Arctic MOPPR (Marine Oil Pollution Preparedness and Response) Agreement (2013), which is scheduled to be signed at the Arctic Council's Ministerial Meeting in Kiruna in May 2013.

Regional action within the broad spectrum of monitoring, surveillance, inspection, and enforcement is consistent with the LOS Convention as well, even though the Convention does not explicitly encourage it. IMO has frequently encouraged such regional action, for instance by means of the 1991 IMO Assembly Resolution A.682(17) 'Regional Co-operation in the Control of Ships and Discharges', which triggered the creation of a global network of regional arrangements on port state control (PSC) modelled on the then already almost decade-old Paris MOU Convention (1982).

The IMO bodies of most relevance to this chapter are the Marine Environment Protection Committee (MEPC), the Maritime Safety Committee (MSC), and the latter's Sub-Committee on Navigation (NAV) and its Sub-Committee on Design and Equipment (DE). Amendments to MARPOL 73/78 (1973/1978) are adopted by the MEPC and amendments to SOLAS 74 (1974) by the MSC. The MEPC has a coordinating role in relation to particularly sensitive sea areas (PSSAs) and the MSC has the authority to adopt mandatory SRSs and VTS pursuant to the SOLAS 74 and COLREG 72 (1972). Proposals for many of the associated protective measures (APMs) that are made applicable within PSSAs are first discussed in the NAV. The ongoing process to develop a mandatory Code for Shipping in Polar Waters (Polar Code) takes predominantly place within DE.

Other bodies relevant to Arctic marine shipping are the International Association of Classification Societies (IACS)—in particular on account of its Unified Requirements concerning Polar Class -, the Association of Arctic Expedition Cruise Operators (AECO), the Arctic Council—in particular through the efforts of its Protection of the Arctic Marine Environment (PAME) and Emergency, Prevention, Preparedness, and Response (EPPR) working groups -, and the OSPAR Commission.

6.3.4 Global and Regional Instruments

LOS Convention

Most of the LOS Convention's provisions on vessel-source pollution are laid down in its Part XII, entitled 'Protection and Preservation of the Marine Environment'. This part begins with Chap. 1, entitled 'General Provisions' and applies to all

sources of pollution. Its first provision, Article 192 lays down the general obligation for all states—in whatever capacity therefore—"to protect and preserve the
marine environment". This is elaborated in Article 194 with regard to measures
to prevent, reduce, and control pollution of the marine environment; aimed specifically at vessel-source pollution in paragraph (3)(b). Other relevant general obligations relate to rare or fragile ecosystems and the habitat of endangered species
(art. 194(5)), introduction of alien species (art. 196), cooperation on a global or
regional basis (art. 197), contingency plans against pollution (art. 199), monitoring
of the risks or effects of pollution (art. 204), and assessment of potential effects
of activities (art. 206). Sections 5 and 6 of Part XII contain separate provisions on
prescription and enforcement for all each of the sources of pollution.

The jurisdictional framework relating to vessel-source pollution laid down in
the LOS Convention is predominantly aimed at flag and coastal states. Apart from
one explicit provision (art. 218), port state jurisdiction is only implicitly dealt with
(see 'Port state jurisdiction' below).

Prescriptive jurisdiction by flag and coastal states is linked by means of rules
of reference to the notion of 'generally accepted international rules and standards'
(GAIRAS). These refer to the technical rules and standards laid down in instruments adopted by regulatory bodies, in particular IMO. It is likely that the rules
and standards laid down in legally binding IMO instruments that have entered into
force can at any rate be regarded as GAIRAS (Molenaar 1998).

The basic duty for flag states to exercise effective jurisdiction and control over
ships flying their flag laid down in Article 94 of the LOS Convention is further specified in Article 211(2), which stipulates that flag state prescriptive jurisdiction over vessel-source pollution is mandatory and must at least have the same level as GAIRAS.
It is therefore up to flag states to require their vessels to comply with more stringent
standards than GAIRAS, but this will of course impact their competitiveness.

This mandatory minimum level of flag state jurisdiction established by the LOS
Convention is balanced by according all states the following navigational rights:

1. The right of innocent passage—suspendable or non-suspendable -, in territorial
   seas, archipelagic waters outside routes normally used for international navigation or—if designated—archipelagic sea lanes, internal waters pursuant to
   Article 8(2) of the LOS Convention, and certain straits used for international
   navigation;
2. The right of transit passage in straits used for international navigation;
3. The right of archipelagic sea lanes passage within routes normally used for
   international navigation or—if designated—archipelagic sea lanes; and
4. The freedom of navigation within exclusive economic zones (EEZs) and on the
   high seas.

Coastal state prescriptive jurisdiction over vessel-source pollution is optional
under the LOS Convention but, if exercised, cannot be more stringent than the
level of GAIRAS (LOS Convention 1982, arts. 21(2), 39(2) and 211(5)). This is
the general rule even though it is subject to some exceptions that are discussed
further below.

Straits Used for International Navigation

The general rule just mentioned is also applicable to marine areas where the right of transit passage applies (LOS Convention 1982, arts. 41 and 42(1)(a) and (b)). This regime was developed for narrow international straits that would no longer have a high seas corridor once strait states would extend the breadth of their territorial seas to 12 nautical miles (nm). The applicability of the regime of transit passage is nevertheless dependent on various conditions.

One of these is laid down in Article 37 and stipulates that the regime of transit passage only applies to "straits which are used for international navigation". Diverging views exist on the words "are used", where the normal meaning points to 'actual' and not 'potential' usage. The latter is adhered to by the US, which takes the view that "the term 'used for international navigation' includes all straits capable of being used for international navigation" (Sen. Exec. Rep. 110-9 2007). Conversely, Canada and the Russian Federation take the view that the words refer to actual usage, and most commentators embrace this interpretation as well (e.g., Rothwell 1996; Pharand 2007). Close reading of the ICJ's Judgment in the *Corfu Channel* case[5]—from which the phrase originates—nevertheless reveals that it also touches on potential usage (cf. McDorman 2010).

Consistent with its above view on potential usage, the US regards the Northwest Passage and parts of the Northern Sea Route as straits used for international navigation subject to the regime of transit passage (NSPD-66 2009). None of the European Union's (EU) Arctic policy statements in recent years contain a position on the issue, even though the EU Council of the European Union's Conclusions on Arctic issues (Council of the European Union 2009) mention transit passage. However, one would assume that at least some states with large fleets engaged in international shipping or with a special interest in Arctic shipping—for instance China, Japan, Norway, South Korea, and several EU Member States—share the view of the US.

Consistent with its above view on actual usage, Canada does not regard the Northwest Passage as a strait used for international navigation. Canada combines this position with two other positions. First, the waters within its Arctic archipelago enclosed by its 1985 straight baselines[6] are internal waters based on historic title (CYIL 1987, 1988). As a corollary, the right of innocent passage pursuant to Article 8(2) of the LOS Convention does not apply (Lalonde 2004). Both the US and the then European Community (EC) Member States lodged diplomatic protests against the 1985 straight baselines, regarding them as inconsistent with international law and explicitly rejecting that historic title could provide an adequate justification (Roach and Smith 1996). The second position that Canada combines

---

[5] *Corfu Channel* case (United Kingdom of Great Britain and Ireland v. Albania), Judgment on the Merits of 9 Apr 1949, *ICJ Reports* 1949, p. 1.

[6] Territorial Sea Geographical Coordinates (Area 7) Order, S.O.R./85-872; effective on 1 Jan 1986.

with its view on the Northwest Passage is that the transit passage regime is trumped by Article 234 of the LOS Convention (see Section 'Unilateral Coastal State Prescription').

Despite their bilateral Agreement on Arctic Cooperation (1988), the dispute between Canada and the US on the legal status of the Northwest Passage and the applicable regime of navigation remains unresolved. The broad saving-clause included in its Sect. 6.4 indicates that the agreement should above all be regarded as an 'agreement-to-disagree'. The 2010 debates within IMO on Canada's mandatory Northern Canada Vessel Traffic Services (NORDREG) Regulations (SOR/2010-127)—which focus predominantly on Article 234 of the LOS Convention, however—are further proof that their disputes remain unresolved (see 'Unilateral coastal state prescription' below).

The position of the Russian Federation vis-à-vis the Northern Sea Route seems largely similar to that of Canada and consists of combined positions on actual usage, internal waters included within straight baselines pursuant to historic title, and transit passage being trumped by Article 234 (Brubaker 1999; Brubaker 2001).

General Exceptions

The above-mentioned restriction on coastal state jurisdiction applies only in relation to pollution of the marine environment, as defined in Article 1(1)(4) of the LOS Convention. Once coastal state jurisdiction is not exercised for that purpose but, for instance, for the conservation of marine living resources instead, the general rule on coastal state jurisdiction over vessel-source pollution does not apply either. As regards anchoring, this view is supported by the practice of the US and—more recently—the Netherlands on regulating anchoring beyond the territorial sea without seeking IMO approval and apparently without any objection by other states. As regards ballast water discharges, the above view is supported by the fact that, instead of an Annex to MARPOL 73/78, IMO decided to deal with ballast water management in a stand-alone treaty, namely the BWM Convention (2004). Moreover, the BWM Convention allows states individually or in concert to regulate more stringently above the minimum ballast water exchange level laid down in the Convention (BWM Convention 2004, arts. 2(3) and 13(3) and Section C of the Annex).

More stringent standards can also be adopted for special areas pursuant to Article 211(6) of the LOS Convention. But as this requires at any rate IMO approval, it gives coastal states no unilateral prescriptive authority. The PSSA Guidelines (IMO 2005) developed by IMO also implement Article 211(6) and are clearly inspired by, and consistent with, that provision (IMO 2005, para. 7.5.2.3(iii)). PSSA status is not a precondition for obtaining the majority of possible APMs as, for instance, ships' routeing measures, SRSs, or VTS can also be made applicable to the maritime zones of a coastal state upon its request by means of IMO approval.

Unilateral Coastal State Prescription

There are three exceptions to the abovementioned general rule that coastal state prescription cannot be more stringent than GAIRAS. First, as general international law does not grant foreign vessels any navigational rights in internal waters—apart from a minor exception laid down in Article 8(2) of the LOS Convention—coastal state jurisdiction is in principle unrestricted. The observations on port state jurisdiction below applies therefore *mutatis mutandis* to internal waters.

Second, a coastal state is entitled to prescribe more stringent (unilateral) standards for the territorial sea, provided they "shall not apply to the design, construction, manning or equipment of foreign ships unless they are giving effect to generally accepted international rules or standards" (LOS Convention 1982, art. 21(2)). Unilateral discharge, navigation, and ballast water management standards are, among others, allowed. The rationale of this provision is to safeguard the objective of globally uniform international minimum regulation, which would be undermined if states unilaterally prescribe standards that have significant extra-territorial effects.

A third exception is laid down in Article 234 of the LOS Convention. It is entitled 'Ice-covered areas' and provides:

> Coastal States have the right to adopt and enforce non-discriminatory laws and regulations for the prevention, reduction and control of marine pollution from vessels in ice-covered areas within the limits of the exclusive economic zone, where particularly severe climatic conditions and the presence of ice covering such areas for most of the year create obstructions or exceptional hazards to navigation, and pollution of the marine environment could cause major harm to or irreversible disturbance of the ecological balance. Such laws and regulations shall have due regard to navigation and the protection and preservation of the marine environment based on the best available scientific evidence.

Article 234 was included in the LOS Convention as a result of in particular the efforts of Canada, which sought to ensure that its Arctic Waters Pollution Prevention Act (AWPPA 1970) and underlying regulations and orders would no longer be regarded as inconsistent with international law. The negotiations on Article 234 were predominantly conducted by Canada, the Soviet Union, and the US and were closely connected to what eventually became Article 211(6) on special areas (McRae 1987; Huebert 2001; Bartenstein 2011).

While Article 234 contains a number of ambiguities—not unlike many other provisions in the LOS Convention, and in fact many treaties—the basic purpose is to provide a coastal state with broader prescriptive and enforcement jurisdiction in ice-covered areas than in maritime zones elsewhere. In particular, in contrast with Article 211(6) on special areas, Article 234 does not envisage a role for the 'competent international organization' (IMO) in case the coastal state takes the view that more stringent standards than GAIRAS are needed.

As the wording of Article 234 indicates, however, jurisdiction is subject to several restrictions and can only be exercised for a specified purpose. One such restriction follows from the words "for most of the year". Decreasing ice coverage will mean that, gradually, fewer states will be able to rely on Article 234 in fewer areas. As regards the phrase "within the limits of the exclusive economic zone", it is submitted that the better interpretation is that this is merely meant to indicate the outer limits of the EEZ but not to exclude the territorial sea (Molenaar 1998).

The purpose for which jurisdiction can be exercised pursuant to Article 234 is "the prevention, reduction and control of marine pollution from vessels". Even though 'navigation' is mentioned twice in Article 234, it does not explicitly grant jurisdiction for the purpose of ensuring maritime safety. It is nevertheless submitted that Article 234 allows regulations that have environmental protection as a primary purpose and maritime safety as a secondary purpose as well as regulations for which both purposes are more or less equally important.

The LOS Convention does not explicitly address the scenario of waters that are both ice-covered and subject to the regime of transit passage, but many commentators argue that the inclusion of the stand-alone Article 234 in the separate Sect. 8 of Part XII supports the dominance of Article 234 over transit passage (Hakapää 1981; McRae 1987; Pharand 2007). While the International Chamber of Shipping (ICS) supports the opposite view (ICS 2012), the US does not seem to have ever explicitly and publicly stated that transit passage trumps Article 234, even though this might well be its position (Roach and Smith 1996; *contra* McDorman 2010). There may be several reasons for this, including the fact that the US is not a party to the LOS Convention, awareness that its position is not very strong, and a preference for a cooperative rather than a confrontational stance.

The following states would currently be entitled to exercise jurisdiction pursuant to Article 234: Canada, Denmark (in relation to Greenland), Norway [in relation to Svalbard but subject to the Spitsbergen Treaty (1920)], the Russian Federation, and the US. So far only Canada and the Russian Federation have actually exercised such jurisdiction (Franckx 1993; Brubaker 2005; VanderZwaag et al. 2008; Solski 2013). The Kingdom of Denmark's 'Strategy for the Arctic' (2011) refers to Denmark's willingness to invoke Article 234 when adequate standards cannot be adopted within IMO.

The consistency of the national laws and regulations of Canada and the Russian Federation with international law has been questioned from time to time. For instance: the applicability of certain CDEM standards to foreign warships and other governmental vessels (re Canada); discriminatory navigation requirements, ice-breaker fees, and insurance requirements; lack of transparency; and high levels of bureaucracy (primarily re Russian Federation, even if not stated) (Molenaar et al. 2010; ICS 2012; Solski 2013).

The consistency of Canada's NORDREG Regulations with Article 234 of the LOS Convention was debated within IMO's NAV (56th Session)[7] and MSC (88th Session)[8] in 2010 (McDorman 2012). Canada introduced the voluntary NORDREG system in 1977 but decided to make it mandatory as a consequence of Canada's

---

[7] IMO doc. NAV 56/20, of 31 Aug 2010, at paras 19.21–19.24.

[8] IMO docs MSC 88/11/2, of 22 Sept 2010; MSC 88/11/3, of 5 Oct 2010; MSC 88/26, of 15 Dec 2010, at paras 11.28–11.39 and Annexes 27 and 28.

Northern Strategy (Minister of Indian Affairs and Northern Development 2009). The NORDREG Regulations became mandatory on 1 July 2010 within the extended (200 nm) scope of the AWPPA, and therefore have a much wider scope than the Northwest Passage. The cornerstone of the NORDREG Regulations is the requirement for prescribed vessels—whether domestic or foreign—to submit, prior to entering the NORDREG Zone, certain information and to obtain clearance.[9] Contravention of these requirements could lead to the vessel's detention and the imposition of a fine and/or imprisonment (c.f. Canada Shipping Act 2001, sec. 138), but none of these seem to have been imposed so far. The NORDREG Regulations are enacted pursuant to the Canada Shipping Act (2001), whose objectives include marine environmental protection (sec. 6).

At MSC 88, the debate centred mainly around the question whether or not Canada was required to seek IMO approval before imposing the NORDREG Regulations on foreign vessels. The US argued that IMO approval was necessary because in its view SOLAS 74 and associated guidelines do not provide an adequate basis for imposing the NORDREG Regulations unilaterally. The US made no references to Article 234 or even the international law of the sea, even though it made the latter references at NAV 56 and its diplomatic notes to Canada (McDorman 2013). The NORDREG Regulation's requirement to obtain clearance is probably the most troublesome for the US, among other things because it essentially amounts to the need for prior authorization and could have precedent-setting effects for other waters that the US regards as straits used for international navigation.

The US was in particular supported by interventions from Germany and Singapore at MSC 88. While the former closely followed the US position at MSC 88, the latter explicitly viewed Canada's actions as inconsistent with the LOS Convention. Prior to MSC 88, France, Germany, and the United Kingdom—and presumably other states as well—had sent *Notes Verbales* to Canada. Before the United Kingdom issued its *Note Verbale*, it approached the European Commission to verify if the Commission would be willing to issue a *Note Verbale*. The Commission declined, in part because it felt that it was not evident that Canada's actions warranted a diplomatic protest and in part also due to concerns that a diplomatic protest could compromise the EU's more important interests in cooperation with Arctic states within and outside the Arctic Council.[10]

Canada—supported among others by Norway and the Russian Federation—took the view that IMO approval was unnecessary as Article 234 provided an adequate basis. While the debates at NAV 56 and MSC 88 were inconclusive and did not resurface within IMO, they illustrate that many more states than just the US

---

[9] Cf. sec. 4 of the NORDREG Regulations (SOR/2010-127); Canada Shipping Act 2001, sec. 126(1)(a); and IMO doc. SN.1/Circ.291, of 5 Oct 2010, 'Information on the Mandatory Canadian Ship Reporting System in Canada's Northern Waters (NORDREG)'.

[10] Based on communications between the author and officials from Germany, France, the United Kingdom, and the Commission in late 2010 and early 2011.

are concerned about navigational rights and coastal state jurisdiction over shipping in ice-covered areas and potential precedent-setting effects for straits used for international navigation.

Port State Jurisdiction

As ports lie wholly within a state's territory and fall on that account under its territorial sovereignty, customary international law acknowledges that a port state has wide discretion in exercising jurisdiction over its ports. This was explicitly stated by the International Court of Justice (ICJ) in the *Nicaragua* case where it observed that it is "by virtue of its sovereignty, that the coastal state may regulate access to its ports".[11] While there may often be a presumption that access to port will be granted, customary international law gives foreign vessels no general right of access to ports (Lowe 1977). Articles 25(2), 211(3), and 255 of the LOS Convention implicitly confirm the absence of a right of access for foreign vessels to ports as well as the port state's wide discretion in exercising jurisdiction under customary international law. This so-called 'residual' jurisdiction is also recognized in several IMO instruments and has on some occasions been exercised by the US and the EU. Nevertheless, some exceptions apply—for instance in case of *force majeure* and distress—and uncertainties exist—for instance on the implications of international trade law. International law only very rarely authorizes port states to impose enforcement measures that are more stringent than denial of access or use of port (services) for extra-territorial behaviour (Molenaar 2007). Article 218 of the LOS Convention is one of these instances. This innovative provision gives port states enforcement jurisdiction over illegal discharges beyond their own maritime zones, namely the high seas and the maritime zones of other states.

IMO Instruments

In view of the focus of this chapter (see Sect. 6.1) and the main categories of substantive shipping standards listed in Sect. 6.3.2, the following are the most important legally binding IMO instruments:

1. MARPOL 73/78;
2. SOLAS 74;
3. STCW 78 (1978);
4. The Anti-Fouling Convention (2001);
5. The BWM Convention (2004);
6. COLREG 72;

---

[11] Case concerning Military and Paramilitary Activities In and Against Nicaragua (Nicaragua v. United States of America), Judgment of 27 June 1986, *ICJ Reports* 1986, p. 14.

7. OPRC 90 and its HNS Protocol (2000);
8. The various IMO instruments relating to liability, compensation, and insurance, e.g., the Civil Liability Convention (1969), the Fund Convention (1971) (each modified by several protocols), the HNS Convention (1996), and the Bunker Oil Convention (2001).

In addition, the following are the most important non-legally binding IMO instruments:

1. The General Provisions on Ships' Routeing (IMO 1985);
2. The MARPOL 73/78 Special Area Guidelines (IMO 2001);
3. The PSSA Guidelines (IMO 2005);
4. The Arctic Shipping Guidelines (2002); and
5. The Polar Shipping Guidelines (2009).

Apart from the Arctic and Polar Shipping Guidelines, all these legally binding and non-legally binding instruments have a global scope of application and therefore apply in principle to the entire marine Arctic. As is illustrated below, however, many IMO instruments allow for the adoption of more stringent measures in specified geographical areas. It should also be noted that where Arctic states are not parties to certain legally binding IMO instruments, they will not implement them in any capacity, including as a coastal state.

The remainder of this subsection will elaborate further on (a) discharge and emission standards, (b) CDEM standards, (c) navigation standards, (d) PSSA Guidelines, and (e) other standards.

Discharge and Emission Standards

MARPOL 73/78 and the BWM Convention are the only IMO instruments that contain discharge and emission standards. The Annexes to MARPOL 73/78 contain discharge standards for oil (Annex I), noxious liquid substances (Annex II), sewage (Annex IV), and garbage (Annex V), and emission standards for ozone depleting substances, nitrogen oxides (NOx), sulphur oxides (SOx), and volatile organic compounds (VOCs) (Annex VI). Annexes I, II, IV, and V allow the designation of so-called 'special areas' where more stringent discharge standards apply. Annex VI allows the designation of so-called 'Emission Control Areas' for SOx, particulate matter, and NOx. A substantial number of special areas and Emission Control Areas are currently in effect, but none of these apply to the marine Arctic.

The BWM Convention stipulates that vessels using the ballast water exchange method should not discharge ballast water within 200 nm from the nearest land or in waters less than 200 metres deep and must meet an efficiency of at least 95 % volumetric exchange (Regulations B-4 and D-1). Also, as was noted above, the BWM Convention allows states individually or in concert to regulate more stringently above this minimum level.

CDEM Standards

CDEM standards are contained in many of the main legally binding IMO instruments, in particular SOLAS 74 and STCW 78. The well-known double-hull standard—which was triggered by the *Exxon Valdez* disaster in 1989—is nevertheless laid down in Annex I to MARPOL 73/78. Arguably, the fuel content requirements in Annex VI to MARPOL 73/78 (within and beyond Emission Control Areas) and the ballast water treatment requirements in the BWM Convention must be regarded as, or treated analogous with, CDEM standards. A similar argument could be made for prescriptions on the use of certain paints or coatings pursuant to the Anti-Fouling Convention. A recent amendment to the STCW Code connected to STCW 78 concerns the inclusion of a new (voluntary) Section B–V/g on 'Guidance regarding training of Masters and officers for ships operating in Polar waters'.

As the Polar Shipping Guidelines are envisaged to be replaced by the Polar Code, it will only be briefly mentioned here. The Polar Shipping Guidelines are more elaborate and extensive than the Arctic Shipping Guidelines, for instance in relation to life-saving appliances. The Polar Shipping Guidelines contain the definition of 'ship' used in SOLAS 74 and apply to all voyages in Antarctic waters but as regards Arctic waters only to international voyages. They contain mostly CDEM standards and strong links with the IACS Unified Requirements concerning Polar Class.

The Polar Code is still under development and at least two more years beyond its original target of 2012 are needed until its adoption. The Code's maritime safety component is in a more advanced stage than the marine pollution component.[12] Slow progress on the Code is at least in part caused by complexities on the linkage between the Code and other IMO instruments, and a failure to select one of the available options early on. Apart from SOLAS 74 and MARPOL 73/78, there also seems to be support for including linkages with other IMO 'pollution' instruments such as the Anti-fouling Convention and the BWM Convention. At MSC 91 in November 2012, it was confirmed that all relevant IMO instruments "should be amended to mandate the associated provisions of the future Polar Code, as opposed to making it mandatory under the SOLAS Convention only or developing a stand-alone new Convention".[13] It is submitted that, if pursued, this approach would make the Polar Code the first genuine regional legally binding IMO instrument. Despite its being bi-polar, it would clearly be much broader and more comprehensive than, for instance, special areas and Emission Control Areas under MARPOL 73/78, or the packages of APMs under PSSAs. The preferred approach may also imply that the special area designations for Antarctica under the Annexes to MARPOL 73/78 will eventually be transferred to the Polar Code.

---

[12] The most recent publicly available version seems to be contained in IMO doc. DE 56/WP.4, of 16 Feb 2012.

[13] IMO doc. MSC 91/WP 1, of 29 Nov 2012, at para. 8.2.

Navigation Standards

As noted in Sect. 6.3.2, the category of navigation standards includes ships' routeing measures, SRSs, and VTS. These navigation standards can be adopted by the MSC based on their authority under SOLAS 74 and COLREG 72. As regards ships' routeing measures, reference should be made to the General Provisions on Ships' Routeing. Examples of routeing measures are: traffic separations schemes, deepwater routes, precautionary areas, areas to be avoided, and no anchoring areas. Apart from the regulation of anchoring for the purpose of the conservation of marine living resources, the LOS Convention does not authorize coastal states to adopt mandatory navigation standards seaward of its territorial sea without IMO approval.

While several IMO navigation standards currently apply within the marine Arctic, there is no comprehensive mandatory or voluntary IMO ships' routeing system for the Arctic Ocean or a large part thereof.

PSSA Guidelines

Designation of an area as a PSSA pursuant to the PSSA Guidelines does not bring about regulation of shipping within that area as such. This requires adoption of one or more APMs. Attention can in this context be drawn to the possibility to have special discharge standards within PSSAs (other than by means of designation as special area under MARPOL 73/78) and "other measures aimed at protecting specific sea areas against environmental damage from ships, provided that they have an identified legal basis" (IMO 2005, para. 6.1.3). Innovative standards are therefore not ruled out.

Other Standards

Reference should also be made to IMO Assembly Resolution A.999(25), 'Guidelines on voyage planning for passenger ships operating in remote areas' (IMO 2008), that was adopted a week after the tragic sinking of the *MS Explorer*—a purpose-built, ice-strengthened tourist vessel originally named *MS Lindblad Explorer*—on 23 November 2007 in Antarctic waters. IMO Assembly Resolution A.999(25) complements the more general IMO Assembly Resolution A.893(21), 'Guidelines for voyage planning' (IMO 1999). Resolution A.999(25) refers, *inter alia*, to the need to take account of shortcomings in available hydrographic data, the presence of places of refuge, and the need of experience in navigating in ice-covered areas. As regards places of refuge, IMO Assembly Resolution A.949(23), of 5 December 2003, 'Guidelines on Places of Refuge for Ships in Need of Assistance', adopted in the aftermath of the disaster with the *Prestige* in 2002, is also relevant. Finally, mention should be made of Regulation V/5 of SOLAS 74 on 'Meteorological services and warnings', Regulation V/6 on 'Ice Patrol Service', and Chapter V's Appendix on 'Rules for the management, operation and financing of the North Atlantic Ice Patrol'.

Regional PSC Arrangements

Regional PSC Arrangements for merchant shipping were established to enhance compliance with internationally agreed standards by means of commitments to carry out harmonized and coordinated inspections and to take predominantly corrective enforcement action (i.e., detention for the purpose of rectification). The instruments in which these internationally agreed standards are contained are commonly referred to as the 'relevant instruments' and include all the main IMO instruments. A participating Maritime Authority must only apply standards that are not just in force generally but also for that Maritime Authority.[14] Some applicability gaps therefore remain unavoidable.

The Arrangements are non-legally binding and, rather than states as such, maritime authorities are parties to them (Molenaar 2007). Saving-clauses have nevertheless been incorporated to ensure that nothing in them affects residual port state jurisdiction, which would include the right to take more onerous enforcement measures.[15]

The expansion in participation in the Paris MOU and the creation and expansion of eight new regional PSC Arrangements since then means that almost complete global coverage has now been achieved. However, no Arrangements have been adopted specifically for the Arctic Ocean/region or the Southern Ocean/Antarctic region. Some advantages and disadvantages of an Arctic Ocean/region MOU will be discussed under Sect. 6.4.2, among other things in view of the likelihood that practically all ships engaged in either intra- or trans-Arctic marine shipping will make use of ports subject to either the Paris MOU or Tokyo MOU Tokyo (1993). None of the other Arrangements seem therefore relevant for Arctic marine shipping. However, when considering amendments to the Paris MOU it is,—in light of the EU's Directive on Port State Control (2009) and the need of convergence between the Directive and the Paris MOU[16]—essential to obtain prior agreement within the EU.

The maritime authorities of the following 27 states currently participate in the Paris MOU:

Belgium	Estonia	Ireland	Norway	Spain
Bulgaria	Finland	Italy	Poland	Sweden
Canada	France	Latvia	Portugal	United Kingdom
Croatia	Germany	Lithuania	Romania	
Cyprus	Greece	Malta	Russian Federation	
Denmark	Iceland	Netherlands	Slovenia	

---

[14] Cf. Paris MOU, sec. 2.3, and Tokyo MOU (1993), sec. 2.4.

[15] E.g., of the Paris MOU, secs. 1.7 and 9.1.

[16] See the 13th preambular paragraph of the Directive.

The participation by the Danish Maritime Authority extends to Greenland as well. Moreover, even though the US Coast Guard merely has observer status, it has been cooperating with the Paris MOU since at least 1986—when it first attended meetings within the Paris MOU—and its PSC system is more or less compatible with that of the Paris MOU.[17]

The Paris MOU does not contain a provision that explicitly defines its spatial coverage. However, Sect. 9.2 stipulates that adherence is open for "A Maritime Authority of a European coastal state and a coastal state of the North Atlantic basin from North America to Europe". This has facilitated the participation or cooperation of the Maritime Authorities of all Arctic states, even though the description is not intended to encompass the entire marine Arctic.

As the Maritime Authorities of both Canada and the Russian Federation also participate in the Tokyo MOU (see below)—and, in addition, the Maritime Authority of the Russian Federation also participates in the Black Sea MOU (2000)-, clarity is needed as to which of their ports are subject to which Arrangement. In 2009, Canada decided to also subject its Pacific ports to the Paris MOU. The Pacific ports of the Russian Federation are currently still subject to the Tokyo MOU, even though some Paris MOU requirements, for instance on training for PSC officers, are applicable throughout the Russian Federation.[18]

The Maritime Authorities of the following states currently participate in the Tokyo MOU:

Australia	Fiji	Republic of Korea	Philippines	Thailand
Canada	Hong Kong, China	Malaysia	Russian Federation	Vanuatu
Chile	Indonesia	New Zealand	Singapore	Vietnam
China	Japan	Papua New Guinea	Solomon Islands	

Sections 1.2 and 8.2 of the Tokyo MOU and Sect. 1.1 of its Annex 1, entitled 'Membership of the Memorandum', stipulate that the Tokyo MOU applies to the "Asia–Pacific region", a term that is not further defined. The US Coast Guard has observer status with the Tokyo MOU and cooperates in a similar way as with the Paris MOU.

Output of the Arctic Council

The Arctic Council is a high-level forum established by means of the Ottawa Declaration (1996). The choice for a non-legally binding instrument is a clear indication that the Council was not intended to be an international organization

---

[17] Information provided to the author by C. Droppers, Paris MOU Secretariat, on 25 Jan 2013.

[18] Information provided to the author by C. Droppers, Paris MOU Secretariat, on 25 Jan 2013.

and implies that the Council cannot adopt legally binding decisions or instruments. The Arctic SAR Agreement was therefore not adopted by the Council, even though it was negotiated under its auspices and the Council's May 2011 Ministerial Meeting was also used as the occasion for its signature.

The mandate of the Arctic Council is very broad and relates to "common Arctic issues" with special reference to "issues of sustainable development and environmental protection in the Arctic" (Ottawa Declaration 1996, art. 1). A footnote nevertheless specifies that the Council "should not deal with matters related to military security". Marine shipping falls squarely under this broad mandate and this is also subscribed by the fact that the Arctic Council has produced output that relates specifically to marine shipping as well less specific or more indirectly relevant output.

The latter includes the 2004 Arctic Marine Strategic Plan (AMSP; Arctic Council 2004), which was developed under PAME and is currently under revision, with adoption scheduled for the 2014 Deputy Ministerial Meeting (PAME 2012b). Both the Arctic SAR Agreement and the future Arctic MOPPR Agreement—scheduled to be signed at the 2013 Kiruna Ministerial Meeting—could be regarded to belong in this category as well. While both implement global IMO instruments—namely the SAR Convention and OPRC 90 -, the Arctic SAR Agreement also implements the ICAO Convention (1994), and neither deals exclusively with shipping incidents, but also with incidents relating to air traffic and offshore installations. Finally, much of the output of EPPR belongs in this category as well, as evidenced by EPPR's role in the negotiation process of the future Arctic MOPPR Agreement by developing Operational Guidelines that will be appended to the Agreement (EPPR 2012).

The most important Arctic Council output that focuses specifically on Arctic marine shipping is the AMSA, completed by PAME in 2009. The AMSA contains a considerable number of recommendations categorized under the headings 'Enhancing Arctic Marine Safety', 'Protecting Arctic People and the Environment', and 'Building the Arctic Marine Infrastructure'. At least three of these recommendations have already been implemented, namely Recommendation I(B), which includes support for the updating and mandatory application of the Arctic Shipping Guidelines; Recommendation I(E), which supports the negotiation of an Arctic SAR instrument; and Recommendation III(C), which supports, inter alia, the development of circumpolar agreements on environmental response capacity.

As Recommendation I(B) eventually shaped to a considerable extent—in addition to actions undertaken within the Antarctic Treaty System (ATS)—the decision to develop the IMO Polar Code, it is a good example of the Arctic Council's so-called 'decision-shaping' function (Molenaar 2012). This function continues to be relevant through PAME's continuous monitoring of progress with the implementation of the AMSA Recommendations.

New Arctic Council initiatives in the domain of Arctic marine shipping will probably arise from the Arctic Ocean Review (AOR) project that is currently carried out within PAME. Phase II of this project is intended to culminate in a final report adopted at the Council's May 2013 Kiruna Ministerial Meeting that will:

> summarize potential weaknesses and/or impediments in the global and regional instruments and measures for [the] management of the Arctic marine environment; outline options to address these weaknesses and/or impediments; and, make agreed recommendations to help ensure a healthy and productive Arctic marine environment in light of current and emerging trends (AOR 2011).

The AOR Phase II draft Report (AOR 2013) contains a Chap. 3 on 'Arctic Marine Operations and Shipping', with specific opportunities (A-L) in Sect. 3.4.3, some of which build on the AMSA recommendations. Section 6.4 below incorporates some of these.

As the Polar Code will ultimately be adopted by the IMO, it will be regarded as that body's output and not as the Council's. The connection between the Polar Shipping Code and the Council is clearly very different from the connection between the Council and the Arctic SAR Agreement and the future Arctic MOPPR Agreement. This author has introduced the concept of the Arctic Council System (ACS) to clarify that legally binding instruments such as the Arctic SAR Agreement and the future Arctic MOPPR Agreement—and their institutional components—can be part of the Council's output even though they are not—and in fact could not be—formally adopted by it (Molenaar 2012).

The ACS concept consists of two basic components. The first component is made up of the Council's constitutive instrument—the Ottawa Declaration, other Ministerial Declarations, other instruments adopted by the Arctic Council— for instance its Arctic Offshore Oil and Gas Guidelines (PAME 2009), and the Council's institutional structure. The second component consists of instruments 'merely' negotiated under the Council's auspices and their institutional components. The 2011 Arctic SAR Agreement and the Meetings of the Parties envisaged under its Article 10 belong to this category and the Arctic MOPPR Agreement and, if included, its institutional component, will soon be as well.

The AOR Phase II draft Report proposes recourse to the ACS approach in order to amend or complement the future Arctic MOPPR Agreement to ensure coverage of other pollutants—in particular noxious liquid substances—, as well as on the collection and sharing of Arctic marine traffic data (AOR 2013).

Acts of the OSPAR Commission

The spatial competence of the OSPAR Commission extends to the 'OSPAR Maritime Area', which includes areas within and beyond national jurisdiction (OSPAR Convention 1992, art. 1(a)). The OSPAR Maritime Area roughly overlaps with the Atlantic sector of the marine Arctic, but about half extends further south. Nothing in the OSPAR Convention or the Acts of the OSPAR Commission challenges IMO's primacy in the regulation of international merchant shipping, but also does not entirely preclude action in relation to merchant shipping. Article 4(2) of

Annex V to the OSPAR Convention stipulates that Members of the OSPAR Commission can raise the need for regulatory action within IMO and requires them to cooperate among other things on regional implementation of IMO instruments. An example of action by the OSPAR Commission in the domain of shipping is the voluntary interim application of certain standards of the BWM Convention adopted in 2007. In 2012, this action was replaced by joint action between the OSPAR Commission and the regional seas bodies for the Baltic and Mediterranean Seas.[19]

6.4 Gaps in the International Legal and Policy Framework and Options for Addressing them

6.4.1 Gaps

Before identifying gaps in the international legal and policy framework for Arctic marine shipping, it is only fair to note that much progress has been made in addressing such gaps in the last five years or so. IMO managed to adopt the Polar Shipping Guidelines in 2009 and is working hard on the adoption of the Polar Code. The Arctic Council finalized AMSA in 2009 and has made good progress in implementing several of its recommendations, most notably culminating in the signature of the Arctic SAR Agreement in 2011 and the future Arctic MOPPR Agreement, scheduled to be signed in May 2013.

As the Polar Code is still to be adopted and enter into force, the following appear to be the main gaps in the international legal and policy framework for Arctic marine shipping:

1. Insufficient participation in relevant international instruments, for instance the LOS Convention—to which the US is not a party—and instruments such as the BWM Convention and the Arctic SAR Agreement, which are still to enter into force;
2. No dedicated, legally binding IMO standards for the marine Arctic, for instance in relation to discharge, emission, or ballast water exchange standards and CDEM (including fuel content (e.g., heavy fuel oil (HFO), anti-fouling, and ballast water treatment) standards;
3. No comprehensive mandatory or voluntary IMO ships' routeing system for all or part of the Arctic Ocean; and
4. No dedicated pan-Arctic mechanisms on monitoring, surveillance, inspection, and enforcement.

---

[19] Joint Notice to Shipping from the Contracting Parties of the Barcelona Convention, OSPAR and HELCOM on: 'General Guidance on the Voluntary Interim Application of the D1 Ballast Water Exchange Standard by Vessels Operating between the Mediterranean Sea and the North-East Atlantic and/or the Baltic Sea' (Annex 17 to 2012 OSPAR Summary Record).

6.4.2 Options

This subsection contains various options for addressing the gaps identified in the previous subsection. They are discussed under the following subsections: options for action within IMO; options for action outside IMO; and options for PSC initiatives. Some of the options for action outside IMO also highlight a potential role for the Arctic Council or the ACS approach.

Options for Action Within IMO

As the Polar Code is still under negotiation, the most obvious option for action within IMO is inclusion of the commonly used IMO standards mentioned under Gap No. 2 in Sect. 6.4.1 above. The desirability of restrictions on the use and carriage of HFO in the marine Arctic is under consideration within PAME as well. The negotiations on the Polar Code may also provide opportunities to include new types of standards, for instance compulsory pilotage and ice-breaker or tug assistance.

So far, there are no indications that navigation standards (ships' routeing measures, SRSs, and VTS) will be included in the Polar Code. Navigation standards could be adopted on a case-by-case basis as stand-alone standards, for instance a SRS or VTS for the Bering Strait, or speed restrictions for certain areas in order to avoid ship strikes of marine mammals or to reduce emissions.[20] An alternative would be to develop a comprehensive ships' routeing system for part or all of the Arctic Ocean, which may be desirable in view of the continuous expansion of Arctic marine shipping. As the main shipping routes described in Sect. 6.2 resemble somewhat archipelagic sea lanes established pursuant to Article 53 of the LOS Convention, the procedure laid down in Article 53—implemented by Annex 2 to the IMO's General Provisions on Ships' Routeing—may be suitable as a model for developing 'Arctic Sea Lanes'. It is not a problem if one or more of these sea lanes would be partially on the high seas, as consensus-based IMO approval reflects support by the entire international community.

Designating one or more PSSAs with APMs in the marine Arctic is also an option that could be pursued in parallel with the Polar Code. Many of the above-mentioned standards could be adopted as APMs. Area-based measures for the marine Arctic are also under consideration by PAME as part of the implementation of AMSA Recommendations II(A), (C), and (D) and the AOR project. It is disappointing—but to some extent understandable—that PAME decided to limit its efforts on PSSAs and MARPOL 73/78 special areas to the high seas of the Arctic Ocean (PAME 2012a; AOR 2013).

---

[20] See IMO doc. MEPC 60/4/24, of 15 Jan 2010, at para. 8, in the context of black carbon emissions. The MSC has adopted at least one speed restriction—albeit recommendatory and primarily for the purpose of human safety—, namely in the TSS "Between Korsoer and Sprogoe" (see IMO doc. MSC 78/26/Add.2, of 4 June 2004, at Annex 21, p. 8, n. 3 to the TSS).

Options for Action Outside IMO

As was pointed out in Sect. 6.3.3, the LOS Convention does not preclude action outside IMO.

1. All states—in their capacities as flag states, port states, coastal states, or with regard to their tour operators—can encourage self-regulation by the (cruise) shipping industry;
2. All states, whether individually or collectively, can in their capacities as flag states impose standards on their vessels for shipping in the marine Arctic that are more stringent than GAIRAS, for instance special discharge, emission, and ballast water exchange standards or HFO standards. Such proactive steps can also be taken in anticipation of the entry into force of the Polar Code and other relevant IMO instruments and standards;
3. Arctic Ocean coastal states and other (Arctic) states—whether within or outside the scope of the Arctic Council or by pursuing the ACS approach—can, in their capacities as port states, develop a collective multifaceted strategy on port state jurisdiction for Arctic marine shipping. This strategy could consist of the following elements:
   a. PSC initiatives (see 'Options for PSC initiatives' further below in this subsection);
   b. Coordinated and optimized use of port state jurisdiction, for instance by implementing Article 218 of the LOS Convention in concert, exercising 'departure state jurisdiction', or using criminal or administrative law to impose charges such as furnishing false information or obstruction of inspection in connection with behaviour prior to entry into port (Molenaar 2007); and
   c. Exercise of port state residual jurisdiction in concert in case the Polar Code takes too long to enter into force or its stringency level is deemed insufficient.
4. Arctic Ocean coastal states or Arctic states—whether within or outside the scope of the Arctic Council or by pursuing the ACS approach—could, in their capacities as coastal states, collectively
   a. Amend or complement the future Arctic MOPPR Agreement to ensure coverage of other pollutants, in particular noxious liquid substances (AOR 2013);
   b. Develop a new regional instrument on the collection and sharing of Arctic marine traffic data (AOR 2013);
   c. Ensure regional implementation of IMO's Guidelines on Places of Refuge;
   d. Develop a regional mechanism for coordinated aerial and satellite surveillance of intentional and accidental marine pollution;
   e. Harmonize relevant domestic laws, regulations, and policies, including in relation to enforcement; and
   f. Take other action consistent with international law, including by relying on Article 234 of the LOS Convention, in case the Polar Code takes too long to enter into force or its stringency level is deemed insufficient.
5. Canada, the Russian Federation, and key flag states could convene multilateral consultations on Arctic marine shipping in order to exchange views and address concerns on navigation in the Northwest Passage and the Northern Sea Route.

Options for PSC Initiatives

PSC initiatives can either be undertaken within existing regional PSC Arrangements or by establishing a new Arrangement, namely an Arctic Ocean/region MOU.

As regards possible initiatives on Arctic marine shipping within existing Arrangements, one approach would be to bring as much Arctic marine shipping as possible under the scope of the Paris MOU. This would be based on the assumption that the stringency level and performance of the Paris MOU is the highest of all the regional PSC Arrangements. Accordingly, the Russian Federation could follow Canada's example (see Sect. 6.3.4 'Regional PSC Arrangements') by subjecting all its Pacific ports to the Paris MOU. The Paris MOU would thereby cover all intra-Arctic shipping and a sizeable part of trans-Arctic shipping, in particular if relatively much use would be made of transhipment ports in the high North Atlantic and the high North Pacific.

Further initiatives could be developed within the Paris MOU as well. As highlighted earlier, these would not relate to the prescription of new standards, but rather be concerned with harmonized and coordinated inspection, and corrective enforcement action with respect to existing standards. Initiatives should be specifically tailored to ships that have engaged in Arctic marine shipping since their last port visit and those that will do so before their next port visit. As regards the Paris MOU, adjustments could for instance be made to one or more Port State Control Committee Instructions (e.g., 'Guidance on Type of Inspections') to include special guidance/instructions for inspections of ships that have engaged or will engage in Arctic marine shipping, as well as specific requirements for the qualification and training of PSC officers in that regard. Such guidance could also be developed by, and made applicable to, a sub-set of the maritime authorities that participate in, or cooperate with, the Paris MOU.

But unless trans-Arctic shipping would make extensive use of transhipment ports in the high North Pacific, departure or destination ports in the Asia–Pacific region would still constitute a significant gap. Similar dedicated guidance/instructions on Arctic marine shipping could therefore be developed within the Tokyo MOU.

An alternative to developing initiatives under the Paris and Tokyo MOUs is the development of an Arctic Ocean/region MOU. As participation in regional PSC Arrangements is always reserved for maritime authorities of the region's coastal states, it follows that the maritime authorities from the following states would be participants: Denmark (Greenland), Canada, Norway, the Russian Federation, the US and—especially in case ships involved in Arctic marine shipping are expected to make extensive use of Icelandic ports—Iceland.

As noted in the discussion on Regional PSC Arrangements in Sect. 6.3.4, the maritime authorities from these states either already formally participate in, or cooperate with, both the Paris and Tokyo MOUs (Canada, the Russian Federation,[21] and the US) or just the Paris MOU (Denmark (Greenland), Iceland, and Norway).

---

[21] See also *supra* note 64 and accompanying text. Memorandum of Understanding on Port State Control in the Black Sea Region, Istanbul, 7 Apr 2000. In effect 19 Dec 2000, as regularly amended. Most recent text at <www.bsmou.org>.

While the cost-effectiveness of all regional PSC Arrangements as a whole would not necessarily be negatively affected by further overlaps in participation, the six maritime authorities will have to weigh the costs of participating in, or cooperating with, yet another MOU against the benefits that its establishment would bring. This would seem to depend, among other things, on their views as to the need and urgency of dedicated PSC initiatives for Arctic marine shipping; the extent to which Arctic marine shipping is expected to be composed of intra-Arctic shipping and ships using transhipment ports in the high North Atlantic and the high North Pacific; and the prospects of adopting satisfactory dedicated PSC initiatives for Arctic marine shipping within the Paris or Tokyo MOUs (Stokke 2012).

6.5 Potential for EU–US Cooperation

A discussion on the potential for EU-US cooperation on Arctic marine shipping must acknowledge at the outset that whereas the US is an Arctic Ocean coastal state, the EU cannot rely on such a *de facto* capacity. Denmark is an Arctic Ocean coastal state with respect to Greenland and an Arctic coastal state with respect to the Faroe Islands, but Denmark's EU Membership does not extend to Greenland or the Faroe Islands (TFEU 2008, arts. 204 and 355(5)(a)). However, the EU can still act in various other *de facto* capacities; for instance as a flag state—pursuant to the various navigational rights applicable within the marine Arctic -, port state, or with respect to natural and legal persons of its Member States.

'Transport' and 'environment' are among the areas listed in Article 4(2) of the TFEU where the EU and its Member States share competence. This shared competence in shipping is among things reflected in the fact that the EU is not a member of IMO. The European Commission nevertheless has observer status with IMO.

Both the US and several EU Member States—in particular Cyprus, Denmark, Germany, Greece, and Malta—are important actors in the domain of international merchant shipping on account of the number of vessels and cumulative deadweight tonnage registered under their flags or in terms of (beneficial) ownership over such vessels (UNCTAD 2011). More in general, seaborne trade is vital to the economies of the US and EU Member States and they also have a range of other interests that are closely associated with shipping (Raspotnik and Rudloff 2012). The high priority accorded to safeguarding navigational rights from undue interference also results from the naval capability of the US and several EU Member States and concerns that restrictions on merchant shipping may spill over to warships and other government ships. It is worth noting that the US Arctic Region Policy NSPD-66 (2009) discusses navigation rights and interests in Section III(B), entitled 'National Security and Homeland Security Interests in the Arctic'.

As regards merchant shipping in the marine Arctic, the US seems so far mainly involved in intra-Arctic traffic. Among the EU Member States, Denmark, Finland, and Germany are involved in both intra- and trans-Arctic shipping. Other EU Member States may also have an interest in Arctic marine shipping on account or their ports, for instance the Netherlands on account of Rotterdam.

In addition to their shared shipping interests, the US and the EU share interests and obligations in relation to the Arctic marine environment and its marine biodiversity, including marine mammals. All these rights and interests of the EU and the US related or associated with Arctic shipping are reflected in various recent policy statements including the US Arctic Region Policy, the European Commission's Arctic Communication (European Commission 2008) and the EU Council conclusions on Arctic issues Council of the European Union (2009). For the purpose of this section, they are grouped together below under three headings, followed by various opportunities for bilateral cooperation derived in some cases from the options identified in Sect. 6.4.2:

1. Protection and preservation of the Arctic marine environment and its marine biodiversity:
    a. Joint and coordinated engagement and support for the negotiations on the IMO Polar Code;
    b. Joint efforts to ensure that the US and EU Member States have a reputation as responsible and high-performance shipping states, including by ensuring compliance with international obligations in all their capacities; and
    c. Joint pro-active steps in their (*de facto*) capacities as flag states (see No. 2 under 'Options for action outside IMO' in Sect. 6.4.2).

2. Safeguarding navigational rights from undue interference:
    a. Cooperation on monitoring the laws, regulations, and practices of Canada, the Russian Federation, and other Arctic states to verify consistency with the international law of the sea; and
    b. Joint or coordinated diplomatic protests in case laws, regulations, or practices are not consistent with the international law of the sea.

3. Promoting multilateral regulation of Arctic marine shipping:
    a. Joint and coordinated engagement within relevant international bodies, including the IMO, the Paris and Tokyo MOUs, and the Arctic Council;
    b. Joint actions to initiate multilateral consultations on the Northwest Passage and the Northern Sea Route (see No. 5 under 'Options for action outside IMO' in Sect. 6.4.2);
    c. Joint action to discourage unnecessary reliance on Article 234 of the LOS Convention, among other things by working towards a high stringency level of the IMO Polar Code and its speedy entry into force; and
    d. Joint action to encourage regional harmonization of relevant laws, regulations, and practices, including in relation to enforcement.

6.6 Conclusions

In coming years, Arctic sea ice coverage and thickness are highly likely to gradually decrease and Arctic marine shipping to increase. In view of these trends, the international legal and policy framework for Arctic marine shipping cannot be assumed to be adequate. This is in fact broadly acknowledged as much progress has been made in addressing gaps in this framework during the last five years or so. The IMO managed to adopt the Polar Shipping Guidelines in 2009 and is working hard on the adoption of the Polar Code, even though this will not be achieved before 2014. The Arctic Council finalized AMSA in 2009 and has made good progress in implementing several of its recommendations, most notably culminating in the signature of the Arctic SAR Agreement in 2011 and the future Arctic MOPPR Agreement, scheduled to be signed in May 2013.

Many of the options suggested in Sect. 6.4.2 to address the gaps identified in Sect. 6.4.1 also offer opportunities for cooperation between the EU and the US. These opportunities can be grouped together under headings that reflect the rights and interests of the EU and the US related or associated with Arctic shipping in line with their recent policy statements, namely: protection and preservation of the Arctic marine environment and its biodiversity; safeguarding navigational rights from undue interference; and promoting multilateral regulation of Arctic marine shipping.

References

Agreement on Arctic Cooperation (1988) Agreement between the Government of Canada and the Government of the United States of America on Arctic Cooperation. Canada Treaty Series 1988, No. 29. Entered into force 11 Jan 1988

AMSA (2009) Arctic Marine Shipping Assessment 2009 Report. Arctic Council, Apr 2009

AMSA Scenarios (2008) The Future of Arctic Marine Navigation in Mid-Century. Scenario Narratives, Prepared for the Arctic Marine Shipping Assessment. May 2008

Antarctic Treaty (1959) The Antarctic Treaty, 1 Dec 1959, 402 U.N.T.S. 71. Entered into force 23 June 1961

Anti-Fouling Convention (2001) International Convention on the Control of Harmful Antifouling Systems on Ships, 5 Oct 2001, IMO Doc. AFS/CONF/26, of 18 Oct 2001. Entered into force 17 Sept 2008

AOR (2011) Phase I report (2009–2011) of the Arctic Ocean Review (AOR) project. <www.pame.is>. Accessed 1 Feb 2013

AOR (2013) The Arctic Ocean Review Project. Phase II report 2011–2013, consolidated version of 9 Feb 2013. On file with author

Arctic Council (2004) Arctic Marine strategic Plan. 24 Nov 2004. Available at <http://www.pame.is/images/stories/AMSP_files/AMSP-Nov-2004.pdf>. Accessed 5 Feb 2013

Arctic MOPPR Agreement (2013) Agreement on Cooperation on Marine Oil Pollution Preparedness and Response in the Arctic, scheduled to be signed at the Arctic Council's Ministerial Meeting in Kiruna, May 2013

Arctic SAR Agreement (2011) Agreement on Cooperation on Aeronautical and Maritime Search and Rescue in the Arctic, 12 May 2011, 50 I.L.M. 1119 (2011). Entered into force on 19 Jan 2013

Arctic Shipping Guidelines (2002) Guidelines for Ships Operating in Arctic Ice-Covered Waters, IMO MSC/Circ. 1056, MEPC/Circ. 399, 23 Dec 2002

AWPPA (1970) Arctic Waters Pollution Prevention Act 1970, R.S.C. 1985, Canada

Bartenstein K (2011) The "Arctic Exception" in the Law of the Sea Convention: a Contribution to Safer Navigation in the Northwest Passage? Ocean Dev Int Law 42:22–52

Black Sea MOU (2000) Memorandum of Understanding on Port State Control in the Black Sea Region, Istanbul, 7 Apr 2000. Entered into force 19 Dec 2000, as regularly amended

Brigham LW (2010) Think again: The Arctic. Foreign Policy, Sept/Oct. 2010
Brubaker RD (1999) The Legal Status of the Russian Baselines in the Arctic. Ocean Dev Int Law 30:191–233
Brubaker RD (2001) Straits in the Russian Arctic. Ocean Dev Int Law 32:263–287
Brubaker RD (2005) The Russian Arctic Straits. Martinus Nijhoff, Leiden
Bunker Oil Convention (2001) International Convention on Civil Liability for Bunker Oil Pollution Damage, 23 Mar 2001, LEG/CONF.12/19. Entered into force 21 Nov 2008
BWM Convention (2004) International Convention for the Control and Management of Ships' Ballast Water and Sediments, 13 Feb 2004. Not in force, IMO Doc. BWM/CONF/36, of 16 Feb 2004
Canada Shipping Act (2001) S.C. 2001, c. 26
Chircop A (2009) The Growth of International Shipping in the Arctic: Is a Regulatory Review Timely? Int J Marine Coastal Law 24:355–380
Civil Liability Convention (1969) International Convention on Civil Liability for Oil Pollution Damage, Brussels, 29 Nov 1969, 9 I.L.M. 45 (1970). Entered into force 19 June 1975. Replaced and entered into force 30 May 1996
COLREG 72 (1972) Convention on the International Regulations for Preventing Collisions at Sea, 20 Oct 1972, 1050 U.N.T.S. 16. Entered into force 15 July 1977
Council of the European Union (2009) Council Conclusions on Arctic issues, In: 2985th Foreign Affairs Council Meeting, 8 Dec 2009
CYIL (1987) Canadian Yearbook of International Law, Vol. 25. University of British Columbia Press, Victoria
CYIL (1988) 26 Canadian Yearbook of International Law, Vol. 26. University of British Columbia Press, Victoria
Directive on Port State Control (2009) Directive 2009/16/EC of the European Parliament and of the Council of 23 Apr 2009 on port State control, 2009 O.J. (L 131/57) (23 Apr 2009)
Emmerson C (2011) Russia's Arctic Opening. Foreign Policy 30 Mar 2011. <http://www.foreignpolicy.com/articles/2011/03/30/russias_arctic_opening>
EPPR (2012) EPPR Progress report to Senior Arctic Officials; ACSAO-SE03 Haparanda, Doc 6.1
European Commission (2008) Communication from the Commission to the European Parliament and the Council. The European Union and the Arctic Region, 20 Nov 2008, COM(2008) 763
Franckx E (1993) Maritime Claims in the Arctic: Canadian and Russian Perspectives. Martinus Nijhoff Publishers, Dordrecht
Fund Convention (1971) International Convention on the Establishment of an International Fund for Compensation for Oil Pollution Damage, 18 Dec 1971, 11 I.L.M. 284 (1972). Entered into force 16 Oct 1978
Hakapää K (1981) Marine Pollution in International Law: Material Obligations and Jurisdiction. Suomalainen Tiedeakatemia, Helsinki
HNS Convention (1996) International Convention on Liability and Compensation for Damage in Connection with the Carriage of Hazardous and Noxious Substances by Sea, 3 May 1996, 35 I.L.M. 1406. Not in force
HNS Protocol (2000) Protocol on Preparedness, Response and Co-operation to Pollution Incidents by Hazardous and Noxious Substances, 15 Mar 2000, IMO Doc. HNS-OPRC/CONF/11/Rev.1, of 15 Mar 2000. Entered into force 14 June 2007
Huebert R (2001) Article 234 and Marine Pollution Jurisdiction in the Arctic. In: Oude Elferink AG, Rothwell DR (eds) The Law of the Sea and Polar Maritime Delimitation and Jurisdiction. Martinus Nijhoff Publishers, New York, pp 249–267
Humpert M, Raspotnik A (2012) The Future of Arctic Shipping Along the Transpolar Sea Route. Arctic Yearbook 2012:281–307
ICAO Convention (1994) Convention on International Civil Aviation, Chicago, 7 Dec 1944, 15 U.N.T.S. 295 (1948). Entered into force 4 Apr 1947
ICS (2012) Position Paper on Arctic Shipping. International Chamber of Shipping. <www.ices-shipping.org>. Accessed 1 Feb 2013

IMO (1985) General Provisions on Ships' Routening. IMO Resolution A.572(14). Adopted on 20 Nov 1985, as amended

IMO (1999) Guidelines for voyage planning. IMO Doc. A 21/Res.893, 4 Feb 2000. Adapted on 25 Nov 1999

IMO (2001) Guidelines for the Designation of Special Areas under MARPOL 73/78', as set out in Annex 1 to IMO Assembly Resolution A.927(22) 2001, to be revoked in 2013 by means of an IMO Assembly Resolution (see IMO Doc. 63/23/Add.1, 14 Mar 2012, Annex 27)

IMO (2005) IMO Assembly Resolution A. 982(24) of 1 Dec 2005, 'Revised Guidelines for the Identification and Designation of Particularly Sensitive Sea Areas' (IMO doc. A 24/Res.982, 6 Feb 2006)

IMO (2008) Guidelines on voyage planning for passenger ships operating in remote areas. IMO Doc. A 25/Res.999, 3 Jan 2008. Adapted on 29 Nov 2007

Kingdom of Denmark (2011) Denmark, Greenland and the Faroe Islands: Kingdom of Denmark Strategy for the Arctic 2011–2020. <www.nanoq.gl>. Accessed 29 Jan 2013

Kraska J (2007) The Law of the Sea Convention and the Northwest Passage. Int J Marine Coastal Law 22:257–282

Lalonde S (2004) Increased Traffic Through Canadian Arctic Waters: Canada's State of Readiness. Revue Juridique Themis 38:49–124

LOS Convention (1982) United Nations Convention on the Law of the Sea, 10 Dec 1982, 1833 U.N.T.S. 396. Entered into force 16 Nov 1994

Lowe AV (1977) The Right of Entry into Maritime Ports in International Law. San Diego Law Rev 14:597–622

MARPOL 73/78 (1973/1978) International Convention for the Prevention of Pollution from Ships, 2 Nov 1973, 2 I.L.M. 1319 (1973), as modified by the 1978 Protocol Relating to the International Convention for the Prevention of Pollution from Ships (17 Feb 1978, 17 I.L.M. 546 (1978)) and the 1997 Protocol to Amend the International Convention for the Prevention of Pollution from Ships (26 Sept 1997) and as regularly amended. Entry into force varies for each Annex. At the time of writing Annexes I-VI were all in force

McDorman TL (2010) The Northwest Passage: International Law, Politics and Cooperation. In Nordquist MH, Heidar TH, Moore JN (eds) Changes in the Arctic Environment and the Law of the Sea, Martinus Nijhoff Publishers, Dordrecht, pp 227–250

McDorman TL (2012) National Measures for the Safety of Navigation in Arctic Waters: NORDREG, Article 234 and Canada. In Nordquist MH, Moore JN, Soons AHA, Kim H (eds) The Law of the Sea Convention. US Accession and Globalization. Martinus Nijhoff Publishers, Leiden

McDorman TL (2013) Canada, the United States and International Law of the Sea in the Arctic Ocean. In VanderZwaag D, Stephens T (eds) Polar Oceans Governance in an Era of Environmental Change. Edward Elgar, Cheltenham

McRae D (1987) The Negotiation of Article 234. In: Griffiths F (ed) Politics of the Northwest Passage. McGill-Queen's University Press, Kingston and Montreal, pp 98–114

Minister of Indian Affairs and Northern Development (2009) Canada's Northern Strategy: Our North, Our Heritage, Our Future, Ottawa

Molenaar EJ (1998) Coastal State Jurisdiction Over Vessel-Source Pollution. Kluwer Law International, The Hague

Molenaar EJ (2007) Port State Jurisdiction: Toward Comprehensive, Mandatory and Global Coverage. Ocean Dev Int Law 38:225–257

Molenaar EJ (2010) Arctic Marine Shipping. Overview of the International Legal Framework, Gaps and Options. J Trans Law Policy 18:289–325

Molenaar EJ (2012) Current and Prospective Roles of the Arctic Council System within the Context of the Law of the Sea. Int J Marine Coastal Law 27:553–595

Molenaar EJ, Hodgson S, VanderZwaag D, Heidar Hallsson H, Henriksen T, Holm-Peterson L, Vladimirovich Korel'skiy M, Kraska J, Már Magnússon B, Rolston S, Serdy A (2010) Legal Aspects of Arctic Shipping. Report commissioned by the European Commission, Directorate-General Maritime Affairs & Fisheries, Feb 2010. On file with author

NSPD-66 (2009) National Security Presidential Directive and Homeland Security Presidential Directive. Arctic Region Policy, 9 Jan 2009, NSPD-66/HSPD-25

OPRC 90 (1990) International Convention on Oil Pollution Preparedness, Response, and Cooperation, 30 Nov 1990. 30 I.L.M. 733 (1991). Entered into force 13 May 1995

Ottawa Declaration (1996) Declaration on the Establishment of the Arctic Council, 19 Sept 1996, 35 I.L.M. 1382

PAME (2009) Arctic Offshore Oil and Gas Guidelines. Last updated 29 Apr 2009

PAME (2012a) PAME Working Group Meeting Report No: I-2012. 26–27 Mar 2012, Sweden

PAME (2012b) Record of Decisions and Follow-up Actions. PAME II-2012. 18–20 Sept 2012

Paris MOU (1982) Memorandum of Understanding on Port State Control, Paris, 26 Jan 1982. Entered into force 1 July 1982, as regularly amended. This chapter uses the version that includes the 34th amendment and entered into force on 1 July 2012

Pharand D (2007) The Arctic Waters and the Northwest Passage: a Final Revisit. Ocean Dev Int Law 78:3–69

Polar Shipping Guidelines (2009) Guidelines for ships operating in polar waters, IMO Assembly Resolution A.1024(26), 2 Dec 2009

Protocol on Environmental Protection (1991) Protocol on Environmental Protection to the Antarctic Treaty. Annexes I–IV, 4 Oct 1991. Entered into force 14 Jan 1998; Annex V (adopted as Recommendation XVI-10), 17 Oct 1991. Entered into force 24 May 2002; Annex VI (adopted as Measure 1(2005)), 14 June 2005. Not in force

Raspotnik A, Rudloff B (2012) The EU as a shipping actor in the Arctic. characteristics, interests and perspectives. Working Paper FG 2, 2012/Nr. 4 Dec 2012, SWP Berlin

Roach JA, Smith RW (1996) United States Responses to Excessive Maritime Claims, 2nd edn. Martinus Nijhoff Publishers, The Hague

Rothwell DR (1996) The Polar Regions and the Development of International Law. Cambridge University Press, Cambridge

SAR Convention (1979) International Convention on Maritime Search and Rescue, 27 Apr 1979. 1405 U.N.T.S. 118. Entered into force 22 June 1985

Sen. Exec. Rep. 110-9 (2007) United States Senate Executive Report 110-9: Convention on the Law of the Sea. In: 110th Congress, 19 Dec 2007

SOLAS 74 (1974) International Convention for the Safety of Life at Sea, 1 Nov 1974, 1184 U.N.T.S. 278. Entered into force 25 May 1980

Solski J (2013) New Legal Developments in the Russian Regulation of Navigation in the Northern Sea Route. Arctic Rev Law Politics, pp 90–120

Spitsbergen Treaty (1920) Treaty concerning the Archipelago of Spitsbergen, 9 Feb 1920, 2 L.N.T.S. 7. Entered into force 14 Aug 1925

STCW 78 (1978) International Convention on Standards of Training, Certification and Watchkeeping for Seafarers, 1 Dec 1978, 1361 U.N.T.S. 2 Entered into force 28 Apr 1984. As amended and modified by the 1995 Protocol

Stokke OS (2012) Regime interplay in Arctic shipping governance: explaining regional niche selection. Int Environ Agree Politics Law Econ 13(1):65–85

TFEU (2008) Treaty on the Functioning of the European Union. Consolidated Version. O.J. C 115/47

Tokyo MOU (1993) Asia-Pacific Memorandum of Understanding on Port State Control in the Asia-Pacific Region, Tokyo, 1 Dec 1993. Entered into force effect 1 Apr 1994, as regularly amended. This chapter uses the version that includes the 12th amendment and in effect on 1 June 2012

Tymchenko L (2001) The Northern Sea Route: Russian Management and Jurisdiction Over Navigation in Arctic Seas. In: Oude Elferink AG, Rothwell DR (eds) The law of the sea and polar maritime delimitation and jurisdiction. Martinus Nijhoff Publishers, The Hague, pp 269–291

UNCTAD (2011) Review of Maritime Transport. In: United Nations Conference on Trade and Development. doc. UNCTAD/RMT/2011

VanderZwaag D et al (2008) Governance of Arctic Marine Shipping. Report prepared for the AMSA. Marine & Environmental Law Institute, Dalhousie University, Halifax

Chapter 7
Understanding Risks Associated with Offshore Hydrocarbon Development

Kamrul Hossain, Timo Koivurova, and Gerald Zojer

Abstract Arctic sea ice is rapidly reducing due to climatic changes occurring in the region, allowing for easier access to vast amounts of undiscovered oil and gas resources. In recent years, growing interest in exploitation of Arctic hydrocarbon resources has led to an increase in exploration activity. Nevertheless, because of the Arctic's harsh conditions, activities remain costly and are linked to serious environmental risks for vulnerable and unique Arctic ecosystems. Clean-up of potential oil spills would be highly complicated, if not impossible, and routine operational activities connected to hydrocarbon development, such as drilling or increased shipping traffic, have adverse consequences on marine flora and fauna. This chapter examines past, current, and potential future hydrocarbon activities in the Arctic, associated environmental impacts from accidents as well as normal operations, and possible cooperation between the European Union (EU) and United States (US) in mitigating the adverse environmental consequences of oil and gas development. Possibilities for transatlantic cooperation regarding hydrocarbon development in the Arctic are considered, including the use of legal and institutional frameworks to which both the EU and US have commitments.

Based on Koivurova T, Hossain K (2008).

K. Hossain · T. Koivurova · G. Zojer (✉)
Arctic Centre, University of Lapland, Rovaniemi, Finland
e-mail: info@gerald-zojer.com

K. Hossain
e-mail: kamrul.hossain@ulapland.fi

T. Koivurova
e-mail: timo.koivurova@ulapland.fi

E. Tedsen et al. (eds.), *Arctic Marine Governance*,
DOI: 10.1007/978-3-642-38595-7_7, © Springer-Verlag Berlin Heidelberg 2014

7.1 Introduction

Offshore hydrocarbon development is anticipated to be a major future economic activity in the Arctic. One of the primary driving factors behind this is climate change, from which access to the Arctic marine area is gradually becoming easier. The Arctic Climate Impact Assessment (ACIA) estimates that there are significant oil and gas reserves in the Arctic marine area, most of which are located in Russian territory, with additional fields in Canada, Alaska (United States (US)), Greenland (Denmark), and Norway (ACIA 2004). The US Geological Survey (USGS) suggests that the area north of the Arctic Circle holds approximately 30 % of the world's undiscovered gas and 13 % of undiscovered oil, with most resources located under less than 500 metres of water (Gautier et al. 2009). Approximately 84 % of this undiscovered oil and gas, representing about 90 billion barrels of technically recoverable oil, is believed to be offshore (Bird et al. 2008). While the Eurasian side of the Arctic has more natural gas reserves, the North American Arctic is more oil-prone: The North American Arctic is estimated to have 65 % of undiscovered Arctic oil, but only 26 % of undiscovered Arctic natural gas (Hong 2012).

Despite the economic opportunities that a surge in offshore hydrocarbon activity may bring, the resulting environmental impacts will contribute to instability in Arctic ecosystems. The end result of the activity, leading to increased use of fossil fuels, will further accelerate climate change, making the Arctic's climate, weather, and ice conditions seemingly less predictable. Although the long-term trends are clear,[1] there will be large variations in ice from year to year, with some seasons colder and having more ice than has been 'normal' in recent years.[2] These variations may have the effect of limiting possibilities for moving oil and gas operations further out to sea, even if the ice edge retreats. Thus, industry cannot necessarily count on areas remaining ice-free and in the case of fixed installations, operators need to set up infrastructure in consideration of maximum extent of ice (Rottem and Moe 2007).

There are four main stages of hydrocarbon development: geological and geophysical survey, exploration, development and production, and decommissioning. Each of these stages has associated environmental impacts. This chapter focuses on the potential risks associated with offshore hydrocarbon development in the Arctic, as well as on potential cooperation between the European Union (EU) and US in the context of hydrocarbon activities.

---

[1] A recent study by the Arctic Monitoring and Assessment Program (AMAP) suggests that the Arctic Ocean will be seasonally ice-free during the summer within the next thirty to forty years (AMAP 2011).

[2] The National Snow and Ice Data Center announced a new record low in Arctic sea ice extent in Sept 2012, following the previous minimum in summer 2007 (NSIDC 2012).

7.2 Offshore Hydrocarbon Potential in the Arctic

7.2.1 Development of Oil and Gas Extraction in the Arctic

Oil and gas exploration and extraction in the Arctic began in the 1920s, but the second half of the twentieth century has witnessed a rapid growth in activities. In recent years, both energy industries and Arctic states have found oil and gas exploration in the Arctic to be increasingly attractive. In many regions within the circumpolar North, exploitation of oil and gas is already a major economic driver (AMAP 2007). Due to estimated increases in global oil demand[3] and even greater natural gas demand,[4] it is projected that hydrocarbon extraction will continue to expand in the future (IEA 2011a).

A number of reasons are arguably behind this trend: First, the price of oil on the world market is anticipated to remain on a relatively high level, supporting investment in hydrocarbon development, even when extracted in costly regions. Second, resource exploitation may become increasingly feasible in the Arctic with continued advances in ship design and drilling equipment. Third, in comparison to many other hydrocarbon-rich regions, the Arctic can be viewed as a relatively safe region, as there are no ongoing conflicts that would potentially disrupt production.[5] Nonetheless, Arctic offshore drilling is more expensive than in other regions of the world, due to harsh Arctic conditions, which require the use of advanced technologies and enhanced safety measures. The resulting high cost of doing business in the Arctic suggests that only the world's largest oil and gas companies may have the financial, technical, and managerial strength to meet the costs and long lead-times for projects that are dictated by challenging Arctic conditions (Hong 2012). From an investment standpoint, the risks in the Arctic are considered to be far greater than in other regions, which is one reason why hydrocarbon extraction in the Arctic remains in the early phases of development, with limited activity to date. The short drilling season due to sea ice onset, and in some cases tighter environmental regulations than in other regions, could slow the development of new fields.

---

[3] The International Energy Agency (IEA) projects in a current policies scenario, as well as in a new policies scenario (with more oil use efficiency and switching to other fuels), an absolute global primary oil use increase (e.g., reaching 107 MMb/d in 2035 in the current policies scenario, compared to 84 MMb/d in 2009), even if the share of oil in total primary energy demand is expected to decrease (IEA 2010b).

[4] For natural gas, the IEA expects an absolute increase in demand as well as an increase in the share of total primary energy demand in all scenarios (e.g., in the current policies scenario, the share grows 1.6 % per year, attaining 4.9 tcm in 2035, compared to 3.2 tcm in 2008) (IEA 2010b).

[5] This fear stems from the October 1973 world oil crisis, when Arab members of petroleum producing countries announced a ban on oil shipment to countries supporting Israel in the 1973 Arab–Israeli war (EIA n.d.).

7.2.2 Arctic Resources and Activities

Within the US, more than half of the current Alaskan oil and gas production rate of approximately 0.55 Bb/d[6] originates onshore in the Prudhoe Bay area (AOGCC 2012). The most important *offshore* extracted oil fields[7] in the North of Alaska are the fields of Endicott, Point McIntyre, Oooguruk, and Northstar; from the latter, oil is transported to shore by the first Arctic subsea pipeline, operating since 2001 (Piepul 2001). These fields are, however, near-shore and in shallow waters, using causeways or infrastructure on artificial islands (ADNR 2009), and thus in relatively easy conditions for the Arctic. The results of past exploration, as well as recent surveys, suggest that the chances of discovering large volumes of oil and gas in Alaska's Beaufort and Chukchi Seas are promising. The USGS estimates that 72,766 MMboe of undiscovered oil-equivalent are located in the Alaskan Arctic (Bird et al. 2008). For the US, concerns after the Deepwater Horizon disaster in the Gulf of Mexico in April 2010 led to a temporary moratorium on deepwater offshore drilling in the Outer Continental Shelf and to tighter regulations. In the US Department of Interior's Five Year Outer Continental Shelf Oil and Gas Leasing Program 2012–2017, which was finished in the aftermaths of the Deepwater Horizon accident, new leasing sales for the Beaufort and Chukchi Seas were shifted towards the end of this five-year period. Shell Oil, holding pre-existing leases, was given permission to initiate drilling in 2012, but was forced to postpone completion of wells for another year, after a spill containment dome was damaged during testing in September 2012 (Krauss 2012). In February 2013, Shell Oil announced it would suspend planned offshore drilling in 2013. Under optimistic estimates, assuming high oil and gas prices, development of new oil and gas resources in the offshore US Arctic might compensate for a current decline in Alaskan production (EIA 2012).

The Russian Federation is one of world's leading oil and natural gas producers, producing 10.28 MMb/d oil and 670 Bcm gas in 2011 (Watkins 2012), and holds the world's largest natural gas reserves with approximately 474.6 Tcm. The majority of Russia's estimated 88.2 billion barrels proven oil reserves (BP 2012) are located in Western Siberia. Current production from oil and natural gas fields in Western Siberia is expected to decline, thus in order to maintain high levels of production, "a new generation of higher-cost fields need to be developed, both in the traditional production areas of Western Siberia and in the new frontiers of Eastern Siberia and the Arctic" (IEA 2011a). With an estimated 132,572 MMboe in the West Siberian Basin and 61,755 MMboe in the Eastern Barents Basin (Bird et al. 2008), Russia has strong prospects for maintaining its position as an oil and gas supply leader. Nonetheless, high exploration costs have delayed utilization of many of these resources. For example, even after years of negotiations, the Shtokman project, one of the world's biggest undeveloped gas fields, located in the Barents Sea, was put on hold due to its high development costs (Chazan 2012). On the other hand, some projects have moved forward in the Russian Arctic: The Prirazlomnaya platform in the Nenets Autonomous

---

[6] Average daily production rate for Sept 2012.

[7] Other offshore fields are extracted from land, using directional drilling technology (ADNR 2009).

Okrug was put in place in August 2011 and, despite several delays, "will be the first ever offshore field in the Russian Arctic put in production" (Pettersen 2012). An agreement between Russia and Norway on the demarcation of their maritime border in the Barents Sea in February 2011 increased stability in the region, opening the door for new exploration in both Norwegian and Russian waters (IEA 2011b).

In 2010, Norway produced 2.16 MMb/d crude oil, and had an estimated natural gas production of 105.9 Bcm in 2009. A share of 3.4 Bcm originated from the Snøhvit LNG plant (IEA 2011b). The Snøhvit gas field is the only field in the Norwegian Arctic where production has taken place. The Snøhvit field is situated in the Barents Sea along with the Goliat field, where, according to the operator Eni, production is scheduled to commence in 2014 (Eni Norge n.d.). Since initial gas field discoveries in the 1970s, hydrocarbon extraction has been the main driver of Norway's economy, with most resources exploited from offshore platforms off Norway's western coast on the continental shelf. Since 2002, however, oil production has declined, and without significant new discoveries, peak Norwegian oil production may have already been reached (IEA 2011b). Such new discoveries are anticipated in Arctic territories, where vast untapped oil and gas reserves are expected. USGS estimates an undiscovered 7,322 MMboe in the Norwegian Margin and 6,704 MMboe in the Barents Platform (Bird et al. 2008).

Combining both conventional and non-conventional oil reserves, Canada is the world's second largest oil-resource holder behind Saudi Arabia with an estimated 267 billion barrels and rising production (IEA 2010a). The Canadian National Energy Board estimates an oil production rate of 3.45 MMb/d for 2012 (NEB 2012a). Canada is also the world's third largest natural gas producer, with an estimated production of 145.7 Bcm in 2012 (NEB 2012b). However, the vast majority of this gas, approximately 98 %, originates *onshore* in the Western Canada Sedimentary Basin and only 2 % is produced—mainly offshore—in Atlantic Canada (IEA 2010a). In the Canadian Arctic, oil and gas activities have been carried out since the 1960s. The first significant discoveries that led to production, such as the Drake Point gas field, were located on Canada's Arctic islands and the Beaufort-Mackenzie area. In total, 65 fields were discovered until the late 1980s, when activities began to decline (McCracken et al. 2007) due to problems with transporting hydrocarbons to the markets, tight environmental regulations, and unsettled land claims (AMAP 2010). Canada's first Arctic offshore wells were drilled in the 1970s in the Beaufort Sea, but no field development took place until the (now ceased) Cohasset Panuke started production in 1992. In recent years, activities have begun increasing again, showing a rise in interest in undiscovered hydrocarbon deposits in the Canadian Arctic. Currently, production takes place off of Nova Scotia (Sable project, 235,067 b/d[8]; production from Deep Panuke was expected to start in 2012), Newfoundland, and Labrador[9] (Hibernia, Terra Nova, White Rose, and its satellite

---

[8] Average of monthly production rates in 2011. CNSOPB. <http://www.cnsopb.ns.ca/pdfs/production_report.pdf.> Accessed 5 July 2012.

[9] Following the determination of the Arctic Human Development Report (AHDR), parts of the North Atlantic Ocean, including the Labrador Sea, belong to the Arctic marine area (Young and Einarsson 2004).

North Amethyst; together: 271,791 b/d; CNLOPB 2011), only contributing a moderate share to Canada's overall production. However, USGS estimates potential undiscovered resources of, for example, 17,063 MMboe in West Greenland-East Canada, or 5,108 MMboe in the Amerasian Basin (Bird et al. 2008).

Even though exploration for hydrocarbons in Greenland was first initiated in the early 1970s, to date, no economically feasible amounts of oil or gas have been discovered. Nevertheless, the area offshore of Northwest Greenland is expected to hold 17 BBoe of undiscovered oil and gas resources, and another 31.4 Bboe are predicted to exist in Northeast Greenland (BMP 2011). Greenland's new self-government is active in inviting foreign partners to invest in offshore oil and gas exploration. British Cairn Energy, one of the most prominent of these, finished eight exploration wells in a number of basins from 2010 to 2011 and discovered reservoir-quality sands in the Atammink block (BMP 2012).

7.3 Risks Associated with Hydrocarbon Development in the Arctic

Hydrocarbon development in the Arctic marine area could result in potentially devastating damage to the environment, particularly were a serious accident to occur at any stage of extraction activities, although normal operational activities affect the environment as well. Pollution, such as from oil discharges during drilling operations, releases of drilling mud and chemicals used during the development and production phases, and operational air emissions (e.g., carbon dioxide, nitrous oxide, sulfur dioxide, VOCs, methane, black carbon) all have negative impacts on marine flora and fauna abundance, health, and diversity (Casper 2009).

Oil spills are more dramatic events and cause both short- and long-term adverse effects on the marine environment. Oil spills and releases can occur as the result of blowouts during exploration or production activities, from slow releases of oil from sub-sea pipelines and on-land storage tanks or pipelines travelling to water, or from accidents involving oil transportation vessels or vessels carrying fuel oil (Casper 2009). While intermittent oil spills can occur quickly and remain in specific areas (such as from tanker accidents), persistent oil spills occur mainly in the phase of exploration or production, for example, by means of a blowout or through leaking pipelines. In the case of a persistent oil spill, oil releases continuously and may spread over a larger area if it cannot be embanked in time (Belanger et al. 2010). Spills from drilling platforms normally last for longer periods of time, resulting in immediate and drastic consequences to the environment and wildlife within the marine area, depending on factors such as the type of crude oil spilled, environmental conditions, time of year, currents, and more (Belanger et al. 2010). The impacts of spilled oil, petroleum by-products, and dispersants used for clean-up are of great concern for marine organisms (Muhling et al. 2012).

While true that the Arctic has not experienced any major spills from oil drilling activities to date, it is likewise true that extensive offshore development in the Arctic has yet to commence and would increase the risk of spills. The 1989 *Exxon*

Valdez spill, the largest sub-Arctic oil spill, occurred in significantly more accessible and favourable conditions, but nonetheless left a severe footprint in the region after a huge amount of oil was released in a short time and gradually spread along the coastline (Pew Environment Group n.d.). Following the (non-Arctic) 2010 *Deepwater Horizon* accident, leaking of oil into water column continued persistently for over ninety days and an estimated 4.9 million barrels of oil were discharged into the ocean environment (Muhling et al. 2012), causing shocking consequences to a large marine area. The official estimate suggests that despite clean-up operation efforts—in conditions that were far more accessible than in the Arctic and with greater infrastructure at hand—26 % of residual oil remained in the seawater, whereas 24 % was naturally or chemically dispersed (Maltrud et al. 2010). Should any such accident occur in the Arctic Ocean, the potential impacts would presumably be much greater considering the fact that an effective clean-up operation in Arctic conditions would face greater challenges, not to mention high financial costs.

The main concerns related to offshore development in the Arctic are the potential environmental impacts on its fragile ecosystems. The unique environmental conditions in the Arctic include extended periods of darkness, reduced visibility, ice-covered ocean areas, severe cold, high winds, and extreme storms (Casper 2009). The Arctic marine environment also has a unique seasonal shoreline and oceanographic changes. The Arctic shore consists of ice shelves, glacier margins, ice foot features, and tundra coast; the unique seasonal oceanographic and shoreline changes are due to open water, freeze-up, frozen conditions, and break-up (EPPR 1998). If a large-scale oil spill were to occur in the Arctic, the marine environment would undoubtedly suffer serious adverse impacts and the impacts would be severe for the region's species and ecosystems, causing long-term contamination that would affect populations and ecosystems for decades (Kaczynski and Brosnan 2008; AMAP 2007; Carpenter 2009). For example, twelve years after the (sub-Arctic) *Exxon Valdez* accident, a survey found sixty-one tons of undecayed oil in the subsurface sediments of Prince William Sound's intertidal shorelines and an almost equal amount of only minimally decayed subsurface oil, representing only a 20–26 % per year decay rate. Direct exposure to oil, oil by-products, and dispersants almost certainly results in increased rates of mortality for many organisms. The effects of incorporation of oil into marine food webs are not yet fully understood (Muhling et al. 2012), but would arguably cause contamination. Moreover, increased infrastructure development in the Arctic to facilitate transportation of potential oil and gas, both on land and at sea, will further accelerate adverse consequences on the Arctic environment, which is already vulnerable due to climate change.[10]

The potential environmental impacts of oil and gas production are also relevant for probable costs and insurance issues. Given the difficulty of handling an accident, risk criteria will be set higher than in other offshore areas, such as the North Sea, and thus managing and insuring risk in the offshore Arctic is likely to be costly.

---

[10] Climate change may allow for increased transport and greater access to Arctic resources (particularly fossil fuels) which would not only create potential environmental consequences, but the burning of extracted fuels to meet global energy demand would further accelerate climate change. See Koivurova T, Hossain K (2008).

Allowing investors without sufficient funds for potential clean-up operations to drill in the Arctic is essentially a risk transfer towards the public sector (Emmerson and Lahn 2012).

7.3.1 Oil Spill Pollution

It is estimated that at present, 80–90 % of petroleum hydrocarbons entering the Arctic come from natural seeps. Increases in oil and gas exploitation, however, will correspondingly heighten the risk of oil contamination from spills and leaks. Oil spills during the exploration, production, and transportation phases of hydrocarbon development are the most serious and direct sources of oil pollution (AMAP 2002). Near-shore facilities and tanker routes near land pose the risk of coastal damage, with spills possibly dispersing over wide stretches (AMAP 2007).

Arctic conditions significantly enhance routine oil spill risks. Ice keels—large pieces of ice that gouge the seafloor—may pummel undersea supply pipes (Wolf 2007). In remote Arctic conditions, a pipeline leak could go undetected for months, leaving extensive pollution behind. Further, oil trapped under ice could be difficult to reach and could potentially travel long distances and towards land, fouling bays, estuaries, and inlets and harming coastal and near-coast species, or travelling further out to sea, placing offshore species at risk and contaminating the sea floor. The persistence of oil, even in small amounts, harms wildlife by reducing species' survival rates, slowing reproduction, and stunting growth (Carpenter 2009).

There are a number of ways in which oil spills could contaminate the Arctic marine area. First, there is a risk of losing well control during the exploration phase, resulting in discharges of oil or releases of gas and creating the potential for fires. Second, other accidental spills and releases may occur from operating activities, both at sea and on land during storage and shipping.[11] Third, oil may be released during offloading in harbours and terminals. Finally, undersea oil pipelines could leak, which poses a special spill risk in the Arctic, as this might not immediately get detected and clean-up operations would be difficult if surface-level sea ice is present (Carpenter 2009).

As mentioned, oil spills are not the only harmful releases that might occur during drilling operations. The offshore oil and gas industry generates huge amounts of dirty water; this water may contain chemicals, and unless it is re-injected, may cause contamination of sea water if discharged into the ocean.[12]

---

[11] Smaller, diffused spills might occur from increased transportation by ships in the Arctic (AMAP 2007).

[12] See OSPAR Recommendation 2001/1 for the Management of Produced Water from Offshore Installations. Under this recommendation, each contracting party was to ensure that the total quantity of oil in produced water discharged into the sea in the year 2006 was reduced by a minimum of 15 % compared to the equivalent discharge in the year 2000. The means used by most of the contracting parties to achieve the goal of a 15 % reduction is the re-injection of produced water (OSPAR Commission 2007).

7.3.2 Clean-Up Challenges

One major reason why the Arctic is considered to be exceptionally vulnerable to oil spills is because of the potentially slow recovery from a spill in cold and highly seasonal ecosystems and the difficulty of clean-up in the remote, cold region, especially where sea ice is present. The ability to cope with spills and to conduct clean-up efforts would vary greatly depending on the location, time of year, weather conditions, volume and characteristics of the oil spilled, and more (Belanger et al. 2010). According to the ACIA report, ice movement, even in the open sea, would hinder some clean-up operations (ACIA 2004). A recent study commissioned by Canada's National Energy Board suggests that clean-up efforts for an offshore oil spill in the Arctic could be impossible at least one day in five because of bad weather or sea ice (Weber 2011).

One of the main methods of clean-up for ocean spills is through the burning of oil slicks. In the Arctic Ocean, these would be subject to high winds that can reach over 10 metres per second, potentially making it impossible to burn the slicks. Even in June, the most favourable Arctic month, weather and ocean conditions would likely prevent conventional clean-up methods from being effectively used about 20 % of the time. These conditions deteriorate over time through the summer until October, when traditional clean-up would likely be impossible 65 % of the time (SLR 2011). Spills in broken sea ice and under sea ice are the most difficult conditions to respond to and cannot be cleaned up effectively (NRC 2003; AMAP 2007). Effective spill response strategies are still being developed for Arctic ice-filled waters (Carpenter 2009). There have been no successful oil spill response tests in ice-covered waters[13] that demonstrate the actual difficulties in clean-up operations.

7.3.3 Oil Spill Impacts on Marine Living Resources

Unlike in ecosystems with less extreme climatic conditions, most organisms in the fragile Arctic depend on limited sources of food supply. The rates of biological factors—such as productive season lengths, generational turnover time, and age of maturity—which determine how quickly an ecosystem would recover from a spill are much slower in the Arctic. Therefore, a single serious oil spill could destroy entire populations and greatly endanger unique species, particularly were the event to overlap with the presence of migratory species, which often congregate

---

[13] See, e.g., Pedersen (2012), which states that the machinery Shell planned to use for oil spill response has not yet successfully been used in Arctic waters. Wolf (2007) quotes Michael Macrander, a biologist with Shell Oil, as saying that, to date, there are no methods to clean up oil in ice-laden conditions.

in relatively small areas. In addition, flora in Arctic terrestrial environments tends to be much more susceptible than in less extreme environments (Kaczynski and Brosnan 2008).

The living resources in the Arctic that would be most severely affected by oil spills are fish stocks in the embryonic stage, and feathered and fur-bearing animals, which are harmed if they inhale or ingest oil (Carpenter 2009). Seasonal aggregations of animals may be particularly vulnerable (e.g., marine mammals in open water areas in sea ice, seabirds in breeding colonies or feeding sites, or fish at spawning time). Near-coast species are affected by even small oil spills and animals far offshore are at risk if a spill moves out to sea rather than along the coast (Wolf 2007).

Oil spills pollute seawater and adult fish readily take up oil components; however, it is unlikely that high concentrations of these components would accumulate in the fish, which are able to metabolize and excrete them. Massive fish kills caused by oil spills have not been documented in the open sea, although this is largely because toxics concentrations in the wider and deeper open seas seldom reached significant levels. Another reason for this could be avoidance behaviour, whereby adult fish are mobile and can escape contaminated areas, as is witnessed among salmon and cod (Mosbech 2002). Nonetheless, oil spills do directly contribute to fish kills and lead to the gradual reduction of fish stocks. Spawn and fish larvae are particularly sensitive to the effects of petroleum products, as eggs and larvae cannot move to avoid spilled oil, unlike adult fish, and even greater mortality may occur (Mosbech 2002). Hydrocarbons poison the larvae of many aquatic organisms and can kill them during the initial days following an oil spill (Lesikhina et al. 2007). Low levels of dissolved oil hydrocarbons may also slow larval growth rates and affect swimming and feeding behaviours (Muhling et al. 2012). An oil spill in spawning areas could severely reduce that year's recruitment. The risk is greater in the Arctic conditions where effective oil spill clean-up operation is difficult. Dispersants, if used for spill clean-up, could also expose fish eggs and larvae to harmful concentrations of oil components, as dispersants' low evaporation rate increases aquatic exposure.

In ice-free waters, marine mammals may be able to avoid oil, but frozen sea ice may limit open water areas on which marine mammals rely. White whales, narwhals, bowhead whales, ringed seals, walrus, and bearded seals are particularly at risk from oil exposure, as these species' primary habitat is ice-covered waters. Polar bears, seals, and walrus are the most commonly occurring species in the Arctic waters during the icy period. Marine mammals with fur, such as sea otters, polar bears and seals, are more vulnerable to oil spills than other sea mammals, as fur contaminated with oil mats and loses its ability to retain heat and repel water. Oiling may disrupt fur's insulating effect, which species like polar bears depend upon. Oil is, as for other marine species, toxic to the bears and studies suggest that ingestion results in lethal poisoning (Boertmann and Aastrup 2002). Oil can additionally cause irritation to animals' skin and eyes and impede their normal ability to swim (AMAP 1997). Whales and most seals, which rely on blubber rather than fur for insulation, are generally less vulnerable to oiling, but ingested oil can still result in cause gastrointestinal bleeding, renal failure, liver poisoning, and blood pressure disruption. Fumes resulting from the evaporation of oil lead to problems in the respiratory organs of mammals near to, or in the immediate vicinity of, large-scale oil spills. Moreover, oil spills and contamination result in loss of the mammals' food supply (OGP 2002).

Seabirds are among the immediate indicators of wildlife and environmental damage during marine spill events and spilled oil contaminates birds' food supplies, eggs, and habitat (Montevecchi et al. 2012). Birds that feed at sea throughout or for part of the year are considered sensitive to oil spills (Mosbech 2002). Exposure to oil destroys plumage, mats feathers, and causes eye irritation. Oily feathers hinder birds from flying and deprive them of their ability to retain warmth, eventually causing death from hypothermia. Arctic seabirds, which live in cold water, are especially vulnerable to the destruction of the insulating capacity of plumage. Long-term exposure to toxic oil may also hamper bird's reproductive capacity (AMAP 2007).

7.3.4 Other Operational Impacts

Oil and gas activities have a number of other operational impacts on the marine environment. For example, the construction of gravel islands and causeways can impede fish migrations and near-shore water flow. Drill cutting piles accumulating near rigs can disturb bottom-dwelling animals.

The use of ice-breakers can affect ice habitats and also create considerable noise, as can air traffic noise that may occur during transport of logistics supplies to the offshore installations and can frighten animals, causing displacement and disrupting feeding schedules. Large increases in ocean vessel traffic to support hydrocarbon development will raise the number of bird and animal strikes and disturb wildlife (Wolf 2007). Fish and marine mammals both are affected by noise, the effects of which can extend tens of kilometres from the source, particularly by sounds generated from seismic exploration (NRC 2003). For instance, in the Alaskan Beaufort Sea, bowhead whales have been observed to change swimming direction in response to noise sources up to 30 kilometres away. Whale hunters in northern Alaska report that they must travel farther offshore to find whales, a change attributed to the displacement of whales from near-shore areas by industrial noise (AMAP 2007). Species such as whales, walruses, and seals are sensitive to man-made sounds and research shows they move away from industrial noises (AMAP 2007), even though such avoidance behaviour is often temporary. Further, since marine mammals rely on hearing to locate prey, seismic activities could drive animals away from important feeding sites (Wolf 2007).

Hydrocarbon-related transportation and other activities create pressures for improving infrastructure, which may cause fragmentation of both maritime and terrestrial habitats. Many animals have dense seasonal aggregations on breeding grounds, along migratory pathways, or along the ice edges and in open water polynyas in the sea ice, making them temporarily vulnerable to even localised incidents. Rigs, drill ships, and offshore pipelines also tend to impair migration routes. Even without pollution or accidents, oil and gas activities can reduce the wilderness character of a region (AMAP 2007).

Oil and gas development in the Arctic is expected to exacerbate global climate change by increasing the availability of oil and gas to be consumed, contributing to increased greenhouse gas emissions (see, e.g., Casper 2009). Offshore installations and gravel islands, infrastructural development, transportation facilities, and industrial activities related to offshore development will lead to significant new emissions sources, perpetuating the impacts of climate change and Arctic sea ice melt (Hossain 2010). A rough estimate suggests that a barrel of crude oil produces 300 kg of carbon dioxide after refining and combustion processes. If the Arctic's resources comprise 90 billion barrels of technically recoverable oil (based on the USGS 2008 estimate), the region's reserves could eventually produce 27 billion tons of carbon dioxide emissions, an amount comparable to current world total annual emissions, further hindering efforts to mitigate climate change (Greenpeace 2010).

7.4 Potential for EU-US Cooperation

Energy is one of the EU's significant Arctic interests: Although EU member states do not have direct access to Arctic offshore areas, a major share of the EU's energy demands—currently 50 % of total consumption—is imported, 53 % of which comes from the Arctic (Russia and Norway). It has been estimated that over the next twenty years, EU energy imports will rise approximately 65–70 % (Airoldi 2010). Energy development in the Arctic is an EU concern in terms of climate change and sustainability policies as well as energy security. Still, without territorial jurisdiction over the Arctic Ocean, the EU does not have direct authority in regards to hydrocarbon developments in the offshore Arctic. Denmark, which is an EU Member State, does not provide EU territorial jurisdiction over Arctic waters as Greenland withdrew itself from the EU by referendum in 1982. While not coastal, the EU does have Arctic territory through Finland and Sweden, both members of the Arctic Council. Iceland and Norway, coastal Arctic states, are not members of the EU, but are committed to many EU regulations through the EEA Agreement and Iceland has applied for EU membership. Thus, even while the EU lacks direct jurisdiction over offshore activities in the Arctic, it nonetheless has competences and influence that do intersect with its Arctic interests (Jerome et al. 2009).

The US, on the other hand, holds a considerably different position by virtue of its Arctic coastal territory in Alaska. The US has already engaged in Arctic offshore hydrocarbon production and holds notable offshore oil and gas resources within its territorial jurisdiction in accordance with the law of the sea. Further, as a member of the Arctic Council, the US has an effective means of participation and strong voice in any joint decision making processes, including issues directly concerning hydrocarbon development. The US, along with the other Arctic countries and the European Parliament, is also a member country of the Conference of Arctic Parliamentarians. Accompanying its jurisdiction over offshore resources in the Arctic Ocean, the US has legal and political commitments, both domestic and international, to ensure environmental standards in resource development. In addition, the US is bound to observe a number of contingency plans concluded with Russia and Canada. In 1986, the US and Canada, for example, concluded the Canada-United States Joint Marine Pollution Contingency Plan, which was later revised in 2003 (JPC Canada-US 2003). A similar contingency plan—the US-Russia Joint Contingency Plan against Pollution in the Bering and Chukchi Seas—was created between the US and Russia (JPC US-Russia 1989). Moreover, the US also takes part in other regional and multilateral bodies, such as International Regulators Forum for global offshore safety (IRF), that work for the promotion of safe and sustainable offshore hydrocarbon development.

Due to the vast potential hydrocarbon reserves underneath the Arctic Ocean seabed and increasing global demand for energy resources, including in both EU and US markets, extensive offshore exploitation is likely to occur sooner rather than later. The critical factor for EU-US cooperation in this arena will be maintaining high environmental standards so that demand is not met at the expense of environmental sustainability.

Both the EU and the US view the Arctic Council as a high-level forum that can enhance cooperation between states on Arctic issues. The US is a member of the Arctic Council, unlike the EU, which nevertheless supports Arctic Council agendas through various channels such as through financing Arctic research in order to promote science-based knowledge on environmental challenges and sustainable Arctic development. The EU has applied to be an observer to the Arctic Council.

The Arctic Council, through its working groups, has produced a number of scientific documents with a view to promoting Arctic environmental protection and which can support better understanding of offshore oil and gas risks and impacts. In 2009, the PAME (Protection of the Arctic Marine Environment) working group adopted the Arctic Offshore Oil and Gas Guidelines (PAME 2009), a comprehensive document identifying fundamental issues to be taken into account in all stages of offshore oil and gas activities in the Arctic region. These issues include, *inter alia*, environmental impact assessment (EIA), environmental monitoring, safety and environmental management, operating practices, decommissioning, and site clearance. Although the guidelines are not legally binding, they provide important guidance for best practices in offshore activities and have been endorsed by all eight Arctic states, demonstrating wide support. The Arctic Council's EPPR (Emergency Prevention, Preparedness, and Response) working group also supports marine oil pollution preparedness and response.

Currently, a new legally binding agreement on oil spill response is being concluded under the auspices of the Arctic Council in order to implement effective response mechanisms. The new Agreement on Cooperation on Marine Oil Pollution Preparedness and Response (MOPPR) in the Arctic has been drafted and negotiated by a task force, as mandated at the Arctic Council Ministerial Meeting in Nuuk in May 2011, led by Norway, the Russian Federation, and the US, and is intended for completion and signature in May 2013. Following the OPRC Convention (1990), the agreement looks to achieve prompt and effective cooperation amongst Arctic states to undertake actions to minimize any damage or threat of damage resulted from oil spills. Steps taken to combat oil spill in the Arctic marine area are to include strengthened cooperation, coordination, and mutual assistance among the parties and the establishment of national contingency plans and contract points. Voluntary operational guidelines are being developed as well.

In light of the Arctic Council's successful history of cooperation and focus on offshore oil and gas activities, it follows that EU-US cooperation on hydrocarbon development in the offshore Arctic should primarily be conducted through Arctic Council initiatives. Arctic Council initiatives have provided science-based policy documents, founded on extensive research within the auspices of Arctic Council to which the Arctic states have political commitment. Additionally, it may be argued that the Arctic Offshore Oil and Gas Guidelines should be made a legally binding convention, creating legal obligations for actors in Arctic hydrocarbon development.

In addition to the Arctic Council, the EU has engaged with the region's development through other bilateral and multilateral arrangements such as through Barents Euro-Arctic Cooperation (BEAC) and Northern Dimension Policy (NDP). Cooperation between the EU and the US on mitigation of risks associated with offshore hydrocarbon development should additionally be viewed in consideration of these institutional arrangements. Bridging BEAC's initiatives to enhance regional sustainable development, which the EU is a part of, with the Arctic Council's agenda could further enhance EU-US cooperation. Transatlantic partnership between the EU and US could also be developed through the NDP, a common policy of the EU, Iceland, Norway, and Russia that promotes partnership between the EU and other northern non-Member States with regard to sustainable development. Both Canada and the US participate as observers in Northern Dimension initiatives, providing an opportunity whereby the forum could serve as a venue for transatlantic cooperation on sustainable offshore activity.

EU-US cooperation can also address the implementation of existing international legal mechanisms, such as the LOS Convention (1982). Despite not being a party to the LOS Convention, the US has expressed its commitment to following the law of the sea in the Arctic in the Ilulissat Declaration (2008). Implementation of other pertinent regulations, such as IMO (International Maritime Organization) regulations and the OPRC Convention, may contribute to enhancing EU-US cooperation in offshore development as far as these regulations deal with oil discharges from sources such as ships and offshore installations, environmental impacts of offshore development, and preparedness and response in the case of potential oil spills. The EU is not a signatory to the OPRC Convention (while most of its Member States are), but is an observer in the IMO (while all of its 27 Member States are IMO members) as well as a party to the Espoo Convention on Environmental Impact Assessment (1991). The US is a party to the OPRC Convention and is a signatory to but has not ratified the Espoo Convention.

7.5 Conclusion

On balance, there are many promising aspects for EU-US cooperation to help achieve sustainable offshore hydrocarbon development in the Arctic. Both the EU and the US have highlighted the importance of regional knowledge-building in their respective Arctic policies. Sharing information, new knowledge, and industry best practices can facilitate effective management of offshore energy development. The Arctic Council's initiatives to conclude legally binding instruments, such as the Agreement on Cooperation on Aeronautical and Maritime Search and Rescue in the Arctic (Arctic SAR Agreement 2011) and forthcoming oil spill prevention and response agreement, are major developments to which both the EU and US have shown strong support and commitment. In addition, both the EU and US, because of their commitments to sustainable hydrocarbon development in the Arctic, support the types of measures suggested by the guidance provided by the Arctic Council Offshore Oil and Gas Guidelines. The EU and US therefore could further cooperate to support implementation of and adherence to the Guidelines, perhaps even including a push to make the guidelines legally binding. Additionally, cooperation could lend support to effective utilization of existing mechanisms, such as implementation of relevant provisions of the LOS Convention, the OPRC Convention within the auspices of IMO, the Espoo Convention, and more. While cooperation on offshore hydrocarbon development may best be promoted under the auspices of the Arctic Council, the EU and US could also utilize available institutional arrangements, such as cooperation through BEAC and NDP initiatives.

References

ACIA (2004) Impacts of a Warming Arctic: Arctic Climate Impact Assessment. Cambridge University Press, Cambridge

ADNR (2009) Beaufort Sea area wide oil and gas lease sale: Final finding of the Director. Alaska Department of Natural Resources. 9 Nov 2009

Arctic SAR Agreement (2011) Agreement on Cooperation on Aeronautical and Maritime Search and Rescue in the Arctic, 12 May 2011, 50 I.L.M. 1119 (2011). Entered into force on 19 Jan 2013

Airoldi A (2010) The European Union and the Arctic. Main developments July 2008–2010. Nordic Council of Ministers, Copenhagen

AMAP (1997) Arctic Pollution Issues: a State of the Arctic Environment Report. Arctic Monitoring and Assessment Programme, Oslo

AMAP (2002) Arctic Pollution Issues: a State of the Arctic Environment Report. Arctic Monitoring and Assessment Programme, Oslo

AMAP (2007) Arctic Oil and Gas 2007. Arctic Monitoring and Assessment Programme, Oslo

AMAP (2010) Assessment 2007: Oil and Gas Activities in the Arctic: Effects and Potential Effects, vol 1. Arctic Monitoring and Assessment Programme, Oslo

AMAP (2011) Snow, Water, Ice and Permafrost in the Arctic (SWIPA). Executive Summary. Oslo: Arctic Monitoring and Assessment Programme. <http://amap.no/swipa/SWIPA2011ExecutiveSummaryV2.pdf>. Accessed 6 Aug 2011

AOGCC (2012) Alaska Average Daily Oil and NGL Production Rates. Alaska Oil and Gas Conservation Commission, Alaska Department of Administration. <http://doa.alaska. gov/ogc/ActivityCharts/Production/2012_09-ProdChart.pdf>. Accessed 8 Nov 2012

Arctic MOPPR Agreement (2013) Agreement on Cooperation on Marine Oil Pollution Preparedness and Response in the Arctic, scheduled to be signed at the Arctic Council's Ministerial Meeting in Kiruna, May 2013

Belanger M, Tan L, Askin N, Wittnich C (2010) Chronological effects of the Deepwater Horizon Gulf of Mexico oil spill on regional seabird casualties. J Marine Animals Ecology 3(2):10–14

Bird KJ, Charpentier RR, Gautier DL, Houseknecht DW, Klett TR, Pitman JK, Moore TE, Schenk CJ, Tennyson ME, Wandrey CJ (2008) Circum-Arctic Resource Appraisal: Estimates of Undiscovered Oil and Gas North of the Arctic Circle. U.S. Geological Survey, Fact Sheet 2008–3049

BMP (2011) Report to Inatsisartut on mineral resource activities in Greenland. Bureau of Minerals and Petroleum. <http://www.bmp.gl/images/stories/about_bmp/publications/Report_ to_inatsisartut_on_mineral_reousrce_activities_in_2011.pdf>. Accessed 13 June 2012

BMP (2012) Report to Inatsisartut, the Parliament of Greenland, concerning mineral resources activities in Greenland. Bureau of Minerals and Petroleum

Boertmann D, Aastrup P (2002) Impacts on Mammals. In A. Mosbech (Ed.), Potential environmental impacts of oil spills in Greenland. An assessment of information status and research needs.(pp 113–117). NERI (National Environmental Research Institute), Technical Report 415

BP (2012) Statistical Review of World Energy, June 2012. <http://www.bp.com/statisticalrev iew>. Accessed 27 July 2012

JPC Canada-US (2003) Canada-United States joint marine pollution contingency plan, Revised 2003. <http://www.nrt.org/production/NRT/NRTWeb.nsf/AllAttachmentsByTi tle/A-403CANUSJCPEnglish/$File/CANUS%20JCP%20English.pdf?OpenElement>

JPC US-Russia (1989) Agreement between the government of the United States of America and the government of the union of soviet socialist republics concerning cooperation in combating pollution in the Bering and Chukchi Seas in emergency situations, with joint contingency plan against pollution in the Bering and Chukchi. Agreement, Moscow, 11 May 1989. Plan, London, 17 Oct 1989. TIAS 11446

Carpenter B (2009) Warm is the new cold: global warming, oil, UNCLOS Article76, and how an Arctic treaty might stop a new Cold War. Environ Law Rev 39(1):215–252

Casper KN (2009) Oil and Gas Development in the Arctic: Softening of Ice Demands Hardening of International Law. Nat Res J 49(3/4):825–882

Chazan G (2012) 31 Aug Gazprom puts Shtokman project on ice. Financial Times. <http://www. ft.com/intl/cms/s/0/604b9b38-f359-11e1-9ca6-00144feabdc0.html>. Accessed 3 Oct 2012

CNLOPB (2011) Annual Report 2010/2011. Canada-Newfoundland and Labrador Offshore Petroleum Board

OPRC Convention (1990) International Convention on Oil Pollution Preparedness, Response, and Cooperation, 30 Nov 1990. 30 I.L.M. 733 (1991). Entered into force 13 May 1995

EIA (2012) Annual Energy Outlook 2012. US Energy Information Administration. DOE/EIA-0383(2012)

EIA (n.d.) 25th Anniversary of the 1973 oil Embargo. US Energy Information A   dministration. <http://www.eia.doe.gov/emeu/25opec/anniversary.html>. Accessed 5 Feb 2013

Emmerson C, Lahn G (2012) Arctic Opening: Opportunity and Risk in the High North. Chatham House, Lloyd's. <http://www.lloyds.com/news-and-insight/risk-insight/reports/arctic-report-2012>. Accessed 6 Feb 2013

Eni Norge (n.d.) Field development facts. <http://www.eninorge.com/en/Field-development/Goliat/ Facts>. Accessed 26 Aug 2012

EPPR (1998) Field Guide for Oil Spill Response in Arctic Water. <http://eppr.arctic-council.org/ content/fldguide/index.html>. Accessed 09 Aug 2011

Espoo Convention (1991) Convention on Environmental Impact Assessment in a Transboundary Context, 25 Feb 1989, 1989 U.N.T.S. 309. Entered into force 10 Sept 1997

Gautier DL, Bird KJ, Charpentier RR, Grantz A, Houseknecht DW, Klett TR, Moore TE, Pitman JK, Schenk CJ, Schuenemeyer JH, Sørensen K, Tennyson ME, Valin ZC, Wandreym CJ

(2009) Assessment of Undiscovered Oil and Gas in the Arctic. Science 324(59321):1175–1179. doi:10.1126/science.1169467

Greenpeace (2010) The risks and potential impacts of oil exploration in the Arctic. Media Briefing, Apr 23. <http://www.greenpeace.org.uk/files/pdfs/climate/arctic_briefing_gp.pdf>. Accessed 09 Aug 2011

Hong N (2012) The energy factor in the Arctic dispute: A pathway to conflict or cooperation? J World Energy Law Bus 5(1):13–26. doi:10.1039/jwelb/jwr023

Hossain K (2010) International Governance in the Arctic: The Law of the Sea Convention with a Special Focus on Offshore Oil and Gas. In G. Afredsson, T. Koivurova (Ed.), Yearbook of Polar Law, 2, 139–169

IEA (2010a) Oil and Gas Security. Emergency Response of IEA countries: Canada. <http://www.iea.org/papers/security/canada_2010.pdf>. Accessed 13 June 2012

IEA (2010b) World Energy Outlook 2010. International Energy Agency, Paris OECD/IEA

IEA (2011a) World Energy Outlook 2011. Executive Summary. Paris: OECD/IEA. <http://www.iea.org/Textbase/npsum/weo2011sum.pdf>. Accessed 13 June 2012

IEA (2011b) Oil and Gas Security. Emergency Response of IEA Countries: Norway. <http://www.iea.org/papers/security/Norway_2011.pdf>. Accessed 13 June 2012

Ilulissat Declaration (2008) Arctic Ocean Conference. Ilulissat, Greenland. 27 May 2008, 48 I.L.M. 382 (2009)

Jerome D, Hossain K, Koivurova T (2009) Canadian Arctic Offshore Oil and Natural Gas and European Union Energy Diversification: Towards a New Perspective? In T. Koivurova, A. Chircop, E. Franckx, E.J. Molenaar and D.L. VanderZwaag (Eds.), Understanding and strengthening European union-Canada relations in law of the sea and ocean governance. Juridica Lapponica, 35 (227–251), University of Lapland Printing Centre, Rovaniemi

Kaczynski V, Brosnan M (2008) Management of Arctic Resources: Economic, Environmental, Legal and Policy Considerations. In: Proceedings of 4th international conference on globalization, energy and environment, Warsaw School of Economics, May 29–30, 2008

Koivurova T, Hossain K (2008) Background Paper: Offshore Hydrocarbon—Current Policy Context in the Marine Arctic. Arctic transform

Krauss C (2012) Shell Delays Arctic Oil Drilling Until 2013. New York Times, 17 Sept. <http://www.nytimes.com/2012/09/18/business/global/shell-delays-arctic-oil-drilling-until-next-year.html>. Accessed 31 Oct 2012

Lesikhina N, Rudaya I, Kireeva A, Krivonos O, Kobets E (2007) Offshore Oil and Gas Development in Northwest Russia: Consequences and Implications. Bellona. <http://www.bellona.org/reports/report/russian_arctic_shelf>. Accessed 15 Nov 2012

Maltrud M, Peacock S, Visbeck M (2010) On the possible long-term fate of oil released in the Deepwater Horizon incident, estimated using ensembles of dye release simulations. Environ Res Lett 5(3):035301. doi:10.1088/1748-9326/5/3/035301

McCracken AD, Poulton TP, Macey E, Monro Gray JM, Nowlan GS (2007) Arctic Oil and Gas. Geological Association of Canada. <http://www.gac.ca/PopularGeoscience/factsheets/Arctic OilandGas_e.pdf>. Accessed 4 July 2012

Montevecchi W, Fifield D, Burke C, Garthe S, Hedd A, Rail JF, Robertson G (2012) Tracking long-distance migration to assess marine pollution impact. Biol Lett 8(2):218–221. doi:10.1098/rsbl.2011.0880

Mosbech A (2002) Impacts of oil spill on fish. In: A. Mosbech (ed.), Potential environmental impacts of oil spills in Greenland. An assessment of information status and research needs. (pp 79–92) NERI (National Environmental Research Institute), Technical Report 415

Muhling BA, Roffer MA, Lamkin JT, Ingram GW, Upton MA, Gawlikowksi G, Muller-Karger F, Habtes S, Richards WJ (2012) Overlap between Atlantic bluefin tuna spawning grounds and observed Deepwater Horizon surface oil in the Northern Gulf of Mexico. Mar Pollut Bull 64:679–687. doi:10.1016/j.marpolbul.2012.01.034

NEB (2012a) Estimated Production of Canadian Crude Oil and Equivalent. National Energy Board. <http://www.neb.gc.ca/clf-nsi/rnrgynfmtn/sttstc/crdlndptrlmprdct/stmtdprdctn-eng.html>. Accessed 14 Nov 2012

NEB (2012b) Marketable Natural Gas Production in Canada. National Energy Board. <http://www.neb.gc.ca/clf-nsi/rnrgynfmtn/sttstc/mrktblntrlgsprdctn/mrktblntrlgsprdctn-eng.html>. Accessed 14 Nov 2012

NRC (2003) Cumulative Environmental Effects of Oil and Gas activities on Alaska's North slope. National Research Council: National Academies Press, Washington, DC

NSIDC (2012) Press Release: Arctic sea ice shatters previous low records; Antarctic sea ice edges to record high. National Snow and Ice Data Center. <http://nsidc.org/news/press/20121002_MinimumPR.html>. Accessed 30 Oct 2012

OGP (2002) Oil and gas exploration and production in arctic offshore regions. Guidelines for environmental protection. Report No. 2.84/329. International Association of Oil and Gas Producers

OSPAR Commission (2007) Annual Report of the OSPAR Commission 2006/07

PAME (2009) Arctic Offshore Oil and Gas Guidelines. Last updated 29 Apr 2009

Pedersen S (2012) Shell's Arctic Drilling Mission Begins Without Oil Recovery Vessel. The international, 15 July. <http://www.theinternational.org/articles/221-shells-arctic-drilling-mission-begins-wi>. Accessed 13 Aug 2012

Pettersen T (2012) More delays at Prirazlomnoye. Barents Observer, 24 May. <http://barentsobse rver.com/en/energy/more-delays-prirazlomnoye>. Accessed 16 June 2012

Pew Environment Group (n.d.) Oil Spills. Oceans North. <http://www.oceansnorth.org/oil-spills>. Accessed 6 Feb 2013

Piepul R (2001) Northstar field begins producing through first subsea Arctic pipeline. Oil Gas J. <http://www.ogj.com/articles/2001/11/northstar-field-begins-producing-through-first-subsea-arctic-pipeline.html>. Accessed 27 Oct 2012

Rottem SV, Moe A (2007) Climate Change in the North and the Oil Industry. Input to Strategic Impact Assessment. Barents region 2030. Fridtjof Nansens Institute Report 9/2007

SLR (2011) Spill Response Gap Study for the Canadian Beaufort Sea and the Canadian Davis Strait. SL Ross Environmental Research Limited, National Energy Board. <https://www.neb-one.gc.ca/ll-eng/livelink.exe?func=ll&objId=702903&objAction=Open>. Accessed 10 Nov 2012

Watkins E (2012) CGES: Russia's 2011 output sets new post-Soviet production record. Oil Gas J 110(1b), 17–18

Weber B (2011) Arctic oil spill cleanup impossible one day in five: energy board report. The globe and mail. <http://www.theglobeandmail.com/news/national/arctic-oil-spill-cleanup-impossible-one-day-in-five-energy-board-report/article2116242/>. Accessed 9 Aug 2011

Wolf E (2007) Oil and water. The Arctic Seas Face Irreversible Damage. Earth I J, 22(2)

Young OR, Einarsson N (2004) Introduction. In: AHDR (Arctic Human Development Report. Steffanson Arctic Institute, Akureyri, pp 15–26

Part III
Improving Marine Governance

Chapter 8
Impact Assessments and the New Arctic Geo-Environment

Pamela Lesser and Timo Koivurova

Abstract This chapter discusses impact assessment on both sides of the Atlantic, focusing on its current use in, and potential for, the Arctic region. There is a large body of information regarding the use of impact assessments, primarily as a tool for environmental analysis, but also in a lesser capacity for multinational collaboration. As a result of climate change, regardless of whether impact assessment as a regulatory mechanism becomes legally enforceable, this instrument has an opportunity for an increasingly useful role in pan-Arctic cooperation. The chapter concludes with a discussion on how impact assessments can be used as a mechanism to view the Arctic and address environmental problems in a new 'geo-environmental' context.

8.1 An Overview of Environmental Impact Assessment and the EU and US

According to the International Association for Impact Assessment (IAIA), 'impact assessment', simply defined, is the process of identifying the future consequences of a current or proposed action (IAIA 2013). The requirements for, and implementation of, impact assessments vary widely from country to country. Much of the Arctic is considered to be the sovereign territory of the eight Arctic states, and therefore, it is primarily national and sub-national authorities that implement environmental impact assessments (EIAs).

P. Lesser (✉)
Ecologic Institute, Berlin, Germany
e-mail: pamela.lesser@ecologic.eu

T. Koivurova
Arctic Centre, University of Lapland, Rovaniemi, Finland
e-mail: timo.koivurova@ulapland.fi

E. Tedsen et al. (eds.), *Arctic Marine Governance*,
DOI: 10.1007/978-3-642-38595-7_8, © Springer-Verlag Berlin Heidelberg 2014

The European Union's (EU) EIA Directive (1985) provides the legal framework for EIAs in the EU. These assessments may be either mandatory—when projects fall into a designated category automatically triggering preparation of an EIA—, or discretionary —when project impacts are not yet known but are considered likely to result in significant impacts and where Member States determine whether an EIA is needed. The Directive has a limited role for the marine Arctic. Of the Arctic states, Denmark, Finland, and Sweden are EU Member States and are clearly subject to the Directive. Iceland and Norway, as European Free Trade Association (EFTA) countries and parties to the European Economic Area (EEA) agreement are also bound to the EIA Directive. Denmark, via Greenland, has Arctic coastal waters, however, Greenland withdrew from the European Economic Community (EEC) in 1985 and is therefore not subject to the EIA Directive. With respect to the marine Arctic, the EIA Directive's reach is limited due to Greenland's withdrawal from the EEC, the fact that neither Finland nor Sweden have Arctic coastal waters, and that only the mainland of Norway (not the Svalbard Islands which are excluded from the EEA agreement) is subject to the Directive.

In general, preparation and implementation of EIAs are left to the individual EU Member States as well as to sub-national authorities (Koivurova 2008). The European Commission has issued Guidelines for EIAs and Environmental Impact Statements (EIS) with the "aim...to provide practical help to those involved in [stages of screening, scoping, and review of the EIA process] drawing upon experience from around Europe and worldwide" (European Commission 2001). The Guidelines note that while "designed to be useful across Europe [they] cannot reflect all the specific requirements and practice of EIA in different countries [and] cannot substitute for Member State guidance on EIA which should always be referred to first."

While such guidelines are helpful, they remain precisely that—guidelines–, having little to no regulatory power behind them and often, at best, are used in a piecemeal fashion, and at worst, may be forgotten. In terms of the Arctic, the Guidelines for Environmental Impact Assessment (AEPS 1997) adopted as part of the Arctic Environmental Protection Strategy (AEPS 1991) are legally non-binding; as a result, they have not been fully incorporated into the national legislation of the eight Arctic states (Koivurova 2008).

The United States (US) has federal regulations for impact assessments and several individual states also have their own regulations applicable to projects that complement the federal regulations. The National Environmental Policy Act (NEPA 1969) requires federal agencies to undertake an assessment of the environmental effects of proposed actions prior to making decisions. Federal government agencies have a responsibility to implement NEPA and are required to determine if their proposed actions have significant environmental effects and to consider the environmental and related social and economic effects of the proposed actions (CEQ 2007). The range of actions considered is broad and includes issuing regulations, providing permits for private actions, funding private actions, making federal land management decisions, constructing publicly-owned facilities, and more. US state environmental impact assessment laws require additional EIA requirements for state-level agencies, often adding more substantive, rather than procedural, requirements. It should be noted, however, that Alaska (the only US state with Arctic territory) does not have a state-level environmental impact assessment law in place. While the information that must be included in analyses under these laws is specific, the rigor with which that analysis has been undertaken is far from consistent, the result being that the quality of the assessments is often challenged by opponents of a particular project with the courts becoming the final arbiters of what qualifies as an adequate environmental assessment.

8.2 Environmental Assessments in the Marine Environment

Although there are marked differences between marine and terrestrial ecosystems, the most obvious being the prevalence of water versus air, there is also an important legal difference in that the marine environment is generally common property, rather than a freehold or owned by an individual, as on land; nonetheless, the conventional approach to EIA for both is still typically the same. Some have argued that using the same analytical framework for both actually diminishes the influence EIA has to control development and to protect the environment (Smith 2008). There is no conclusive evidence to validate this assertion, and there are of course many similar pressures on both marine and terrestrial environments, but it is worth noting that marine environments, and in particular the marine Arctic, may benefit from more tailored EIAs that account for the many differences.

There are few developments that have either a positive or neutral impact on the marine environment. EIAs are most often used in the marine environment for offshore wind farms, hydrocarbon production, aquaculture, dredging, laying cable lines, hydropower projects, and for permitting desalination plants (OSPAR 2010). While there has been some good work regarding legislation, planning, and application of the EIA process to the marine environment, there is still much additional work that could be done to ensure accurate evaluation of impacts and effective implementation of mitigation measures (Smith 2008).

National guidance has been—and is being—developed in many countries to assist developers and regulators in adapting the environmental impact assessment process to the marine environment. One example is the German *Standards for Environmental Impact Assessments of Offshore Wind Turbines on the Marine Environment* (BSH 2007). The United Kingdom has also been very active in this area, having prepared several guidance documents, including *Offshore Wind-Farms—Guidance Note for Environmental Impact Assessment In Respect of FEPA and CPA requirements* (CEFAS 2004) and the *Nature Conservation Guidance on Offshore Wind Farm Development* (Defra 2005).

8.3 The Role of Impact Assessments (EIAs, SEAs, and TEAs) in the Arctic

8.3.1 Environmental Impact Assessments

For over 20 years, environmental protection in the Arctic has been a catalyst for multilateral cooperation among the eight Arctic states (Canada, Denmark/Greenland, Finland, Iceland, Norway, Sweden, Russia, and the US). The Arctic Environmental Protection Strategy (AEPS), signed in 1991, was followed by, and eventually enveloped under, the Arctic Council which was established five years later. One of the Arctic Council's key objectives has been to promote sustainable development, which is described in the Terms of Reference for the Sustainable Development Program (Arctic Council 2000) as including opportunities to protect and enhance the environment and the economies, culture, and health of indigenous communities and of other inhabitants of the Arctic, as well as to improve the environmental, economic, and social conditions of Arctic communities as a whole. In order to begin implementing this goal, the Guidelines for Environmental Impact Assessment in the Arctic ('Arctic EIA Guidelines'; AEPS 1997) were developed. Finland volunteered to take on the lead role, and the Arctic EIA Guidelines take an arguably holistic approach to looking at the Arctic's unique natural, historical, and cultural environment.

The objectives of an Arctic EIA, following the Arctic EIA Guidelines, are similar to those of non-Arctic environmental assessments: i.e., the nature and likelihood of environmentally damaging events should be accounted for to provide a basis for decision making; the location, technical solutions, construction, operation, and decommissioning aspects of a project should be identified if they are likely to cause adverse environmental effects; alternative options to the project should be identified with the goal of balancing environmental protection and the conservation of natural resources with other social, health, and economic considerations; and to devise and implement remedial measures for eliminating or minimizing undesirable impacts. A unique facet of the Arctic EIA Guidelines is the objective "to provide for the incorporation of traditional knowledge and consultations with the developer, the public, regulatory, and non-regulatory authorities to guide decision making." Later in this chapter, case studies illustrating the role of indigenous peoples in Arctic impact assessments show how traditional knowledge can fuse with contemporary science.

While there is genuine opportunity to enhance environmental protection by using the Arctic EIA Guidelines as a tool for evaluation, they are legally non-binding and therefore have not been directly incorporated into national legislation (Koivurova 2008). That said, all Arctic countries also have national EIA procedures that may mirror some of the procedures in the Arctic EIA Guidelines, despite not having directly adopted them (Koivurova 2012).

The role of the Arctic EIA Guidelines is to offer a framework for preparing EIAs, but, as explained, as they are not mandatory, it is often national EIA laws that determine the information contained in an EIA. In the marine Arctic, there have been relatively few EIAs, and of those, the comprehensiveness of topics covered and analytical robustness vary considerably. Nevertheless, guidelines generally can be useful, an example of which is the Arctic Offshore Oil and Gas Guidelines (PAME 2009), prepared by the Arctic Council's Protection of Arctic Marine Environment (PAME) working group. These, too, are not mandatory, but as they have already been revised two times, most recently in 2009, this demonstrates that the Arctic Offshore Oil and Gas Guidelines continue to be a living document and presumably have some continuing utility, in contrast to the Arctic EIA Guidelines.

Box 8.1 Example of EIAs in the Arctic: Oden Arctic Technology Research Cruise

The Swedish Polar Research Secretariat (SPRS) is a government agency promoting and coordinating Swedish polar research, including planning research and development and organizing and leading research expeditions to the Arctic and Antarctic regions. The Norwegian University of Science and Technology (NTNU) hosts a centre for research-based innovation, Sustainable Arctic Marine and Coastal Technology (SAMCoT), which works with the development of technologies necessary for the sustainable exploration and exploitation of the Arctic region (Swedish Polar Research Secretariat 2012).

In 2012, the NTNU and the SPRS established a collaboration known as the 'Nordic Cooperation in Polar Research'. The first step in this collaboration was performing a research cruise to the waters northeast of Greenland on the Swedish icebreaker *Oden* in the summer of 2012. The cruise, called the 'Oden Arctic Technology Research Cruise 2012' (OATRC), was a SAMCoT-associated project with financial support from Statoil (Swedish Polar Research Secretariat 2012).

Oden's first scientific expedition was actually carried out in 1991 and an EIA prepared in 1993 to respond to concerns about the effects of underwater noise, interference with marine mammals, and exhaust emissions (Swedish Polar Research Secretariat 2012). The EIA verified that the environmental impacts for *Oden* during polar operations would be minor and any adverse effects were likely to be the result of the use of fossil fuels and resultant air emissions. Emissions were expected to be transitory and in negligible concentrations. All waste would be retrograded back to Sweden for proper disposal. *Oden* would not visit any terrestrial areas, and the cumulative impacts also were considered negligible and to have less than minor or transitory impacts (Swedish Polar Research Secretariat 2012).

The main modification of *Oden* since the 1993 EIA was the installation of a multi-beam echo sounder with an integrated sub-bottom profiler, to be operated during transits and in the survey areas on the continental shelf in the Arctic Ocean. Debate over underwater sound and its effect on marine mammals became the catalyst for another EIA. Similar to the 1993 EIA, the EIA undertaken in 2012 noted that the unavoidable environmental impacts from the OATRC would be transitory and less than significant. At most, minor or transitory impacts would result from logistic activities. Thus, from an environmental point of view, there were no reasons not to perform OATRC, assuming the expedition was conducted within the framework described in the EIA (Swedish Polar Research Secretariat 2012). Interestingly, the actual EIA noted that "[p]resently there is a lack of coherent policy framework for evaluating the environmental impacts of underwater noise pollution. There are no regulations that specifically address the operations of sonar or other ship-borne transmissions of sound in the Arctic" (Swedish Polar Research Secretariat 2012). Given the lack of either a policy or regulatory framework against which to evaluate the project's impact, i.e., underwater noise pollution, the conclusions of significance may be seen as challenging to evaluate and perhaps arbitrary.

8.3.2 *Strategic Environmental Assessments*

Strategic environmental assessments (SEAs) are often referred to as the environmental impact process applied to policies, plans, and programmes. Although the Arctic EIA Guidelines only briefly address SEAs, with their utility described as a regional and sectoral planning tool, useful in facilitating identification of general sustainability issues and setting the context for more specific project EIAs (AEPS 1997), SEAs are becoming increasingly relevant with respect to EIAs in the region.

SEA Protocol to the Espoo Convention

The SEA Protocol (2011) of the Espoo Convention (1991) entered into force in July 2010 (see Sect. "Espoo Convention" below on the Espoo Convention). The impetus for adding the Protocol stemmed from the need to include environmental and health issues in the preparation of plans, programmes, policies and legislation. Transboundary environmental assessments (TEAs) are only discretionary under the Espoo Convention; therefore, a specific protocol on SEA was developed (Azcarate et al. 2011). The new SEA Protocol is directly linked to TEAs in that the SEA Protocol contains rules requiring a transboundary SEA in certain cases of transboundary environmental effects.

The Protocol focuses on creating national SEA procedures and augments the Espoo Convention by ensuring that individual parties integrate environmental assessment into their plans and programmes at the earliest stage, thus helping to lay the groundwork for sustainable development. SEA is undertaken much earlier in the decision making process than project EIAs, and is therefore seen as a key tool for sustainable development. The SEA Protocol also provides for extensive public participation in government decision making processes. In practice, the applicability of the SEA Protocol to the Arctic marine area is limited given the fact that of the Arctic states, only Finland, Denmark, Norway, and Sweden have ratified it; Canada, Iceland, Russia, and the US are not even signatories.

EU SEA Directive

Under the EU's SEA Directive (2001), SEA is mandatory for plans and programmes (but not policies) which are prepared for agriculture, forestry, fisheries, energy, industry, transport, wastewater management, telecommunications, tourism, town and country planning, or land use and which set the framework for future development consent or have been determined to require an assessment under the EU's Habitats Directive. In general, for plans and programmes not included in the above, Member States must carry out a screening procedure to determine whether they are likely to have significant environmental effects; if there are significant effects, an SEA is needed.

The European Commission summarizes the SEA procedure as follows: An environmental report is prepared in which the likely significant effects on the environment and the reasonable alternatives of the proposed plan or programme are identified and public and environmental authorities are then consulted on the draft documents. Regarding plans and programmes likely to have significant effects on the environment of another Member State, the Member State in whose territory the plan or programme is being prepared must consult the potentially affected Member State(s). The environmental report and the results of consultations are taken into account before adoption. Once the plan or programme is adopted, the environmental authorities and public are informed and relevant information made available. In order to identify unforeseen adverse effects at an early stage, significant environmental effects of the plan or programme are to be monitored (see European Commission 2013).

The EU's SEA Protocol was largely inspired by the SEA Directive and the influence can be seen when comparing the Protocol and Directive, particularly where each requires consultation with potentially affected states/Member States. The Protocol does, however, go beyond the scope of the Directive in that it proposes SEAs be discretionally applied to legislation and policies (Azcarate et al. 2011). Commonalities to both include a provision on transboundary consultations and both explicitly apply to offshore hydrocarbon exploitation.

Box 8.2. Example of SEAs in the Arctic: Dreki maritime area

In 2007, a SEA for the development of Iceland's Dreki maritime area for oil and gas exploitation was undertaken. The proposed plan entailed granting exclusive licenses for exploration and production of oil and gas in the northern part of the Dreki area, which covers approximately 42,700 km^2,. An agreement between Iceland and Norway on the continental shelf between Iceland and Jan Mayen applies to almost 30 % of this area (Ministry of Industry 2007). The possible environmental effects associated with issuing licenses for prospecting, exploration, and production of oil and gas were evaluated along with the risk of accidents. As part of the analysis of the plan, the SEA included a review of the existing legal framework, health, safety, and environmental issues (geology, biota, and climate in the Dreki area), an evaluation of potential hydrocarbon resources in the Dreki area, and also a gap analysis. No major obstacles were discovered to initiating exploration and production of hydrocarbons in the area by granting exclusive licenses.

The SEA made note of its limited scope by stating that "[i]t is important to keep in mind that a strategic environmental assessment of a plan is never as exhaustive as an environmental impact assessment of the environmental effects of construction that may be required because of particular aspects of these activities," the implication being that an EIA would be required for the actual exploratory and production drilling activities in accordance with Icelandic law on EIA (Ministry of Industry 2007).

Iceland's government provisionally awarded Faroe Petroleum the offshore exploration licenses under the country's second licensing round, without requiring any mitigation measures (Offshore 2012). In January 2013, the final licenses were issued. It is interesting to note that the SEA states that the main environmental impact during the prospecting phase is noise from an air gun used during measurements of the ocean floor taken with the help of seismic surveys, finding that "research has shown that whales avoid the noise and alter their diving pattern for a distance from the source of up to 20 kilometres. Since this involves a limited and demarcated activity, no special measures are deemed necessary because of this" (Ministry of Industry 2007). As mentioned previously in this chapter, the analytical rigor between impact assessments can vary greatly and decisions on whether or not to impose mitigation measures remain highly subjective.

The analysis and determination of alternatives are also often highly subjective. For example, the Dreki SEA lists three possible alternatives: to offer licenses throughout the northern part of the Dreki area (42,700 km^2) in accordance with the plan; to restrict the size of the licensing area from what the plan provides to the 5,000–10,000 km^2 considered most promising for finding oil and gas; and not to offer any licenses in the area—either temporarily or for the indefinite future (Ministry of Industry 2007). None of these alternatives were specifically analyzed, but the conclusion was that even without issuing any licenses: the impacts of climate change will result in greater pollution (i.e., more oil tankers transiting the Arctic will increase exposure to oil pollution and emissions); Norway might begin exploring for oil in the neighbouring region, thus exposing the Dreki ecosystem to adverse impacts; and anthropogenic climate change has a large impact on the Arctic region that will likely cause other diverse changes in the environment to occur. While these conclusions are technically not incorrect, an analysis of alternatives could have been conducted in a more robust manner. Impact assessments in general, when sufficiently rigorous, can be valuable tools for marine governance, but can also be used merely as procedural tools to justify political and business priorities.

8.3.3 Transboundary Environmental Assessments

The Arctic EIA Guidelines address the need for looking at transboundary issues, but there is little discussion of actual methods, or of their use as a possible tool for implementing environmental strategies. Rather, they point to the Espoo Convention (see below Sect. "Espoo Convention") as the best framework in which to address transboundary issues, suggesting that these be later coupled with either bilateral or multilateral agreements. Rather than focusing on implementation, the discussion centres more on building trust (especially with indigenous peoples); although, the linkage is later made in the statement that building trust would also, over the long term, produce better EIAs and ultimately a better environmental state. The Arctic EIA Guidelines note that this would be a particularly relevant and important aspect of TEAs as it is sovereign states that need most to cooperate. For any TEA to be successful, "it is essential that individual states understand it will be a long process, that the need to discuss issues early is essential, and that the knowledge of likely transboundary impacts is built together and is followed up together" (AEPS 1997).

Espoo Convention

The main international instrument on transboundary EIA is the Espoo Convention (1991). Adopted in 1991 and entered into force in 1997, the Convention now has a total of 30 signatories and 45 parties.[1] Of the Arctic states, Canada, Denmark, Finland, Norway, and Sweden have all ratified the Convention. Iceland, the Russian Federation, and the US are signatories. While this provides a good foundation for a pan-Arctic legal framework for transboundary EIA, it is not a

---

[1] As of 10 Feb 2013. See <http://www.unece.org/env/eia/ratification.html>.

comprehensive, multilateral legal framework. Still, because transboundary issues are especially prevalent in the Arctic, the influence of the Espoo Convention as a catalyst for Arctic cooperation, in particular during the stage of the AEPS, should be acknowledged (Koivurova 2009).

Activities requiring a TEA include development activities that are expected to cause significant impacts to the environment of another state; however, TEAs for strategic actions above the project planning level are discretionary, as described in Sect. 8.3.3. TEA processes follow the standard procedures of EIAs and SEAs, but the transboundary issues dealt with in TEAs add administrative, political, and regulatory complexities (Azcarate et al. 2011). The origin state is only obliged to notify a potentially affected state if the planned activity (e.g., offshore hydrocarbon extraction) is likely to cause significant adverse transboundary environmental impacts. If the concerned states disagree on the likelihood of impacts, Appendix IV of the Convention provides for an inquiry commission procedure. It should be noted that the Espoo Convention does not apply to cases of potential harm to global commons, such as the high seas, but only when proposed activities are likely to cause harm to the environment located in another state's maritime zone (Koviurova and Molenaar 2009).

Box 8.3 Example of transboundary environmental assessments in the Arctic: Tornio Stainless Steel Works and Tornio Fairway

The Tornio Stainless Steel Works plant in Finland is situated at the northern end of the Gulf of Bothnia and at the mouth of the river Tornionjoki. The plant has produced ferrochrome since 1968 and steel since 1976. The first EIA for the facility was prepared from 1996 to1997, and the second in 2005. The procedures of the two were identical with the exception that the latter required the draft EIA report to both be submitted for public comment in Finland and submitted to the affected party—in this case, Sweden.

The project to enlarge the Steel Works plant was multi-faceted and included an increase in capacity, enlargement of the nearby harbour, sulphuric acid installation, and a recycling installation for metal dust. According to national Finnish legislation, the proposed project required preparation of an EIA. The enlargement project triggered the requirements of the Espoo Convention as it was determined that transboundary impacts from chemical installation and waste-disposal installation of dangerous wastes could easily impact Sweden, only 2 km away.

Procedures under the second EIA encountered some difficulty as the initial draft of the EIA report sent to Sweden was only partially translated; Sweden asserted that the translated version did not include all of the information that had been stated at an earlier stage during the scoping process. The additions were subsequently provided, which then allowed Finland to receive

---

comments from Sweden and transmit the document to the Lapland Regional Environmental Centre (Rantakallio 2005).

Public involvement in both projects was substantial. For the steel works component, a combined public information meeting for Sweden and Finland with interpretation was provided and the developer also arranged a separate meeting for the media. In the case of the Tornio fairway project, public information meetings were held separately in Sweden and Finland (Rantakallio 2005). Cooperation with the authorities in both projects consisted of a steering group and a follow-up group with relevant authorities from both sides of the border, including the Lapland Regional Environmental Centre, the Swedish provincial environmental administration, the towns of Tornio (Finland) and Haaparanta (Sweden), the Swedish fisheries administration, and fishermen's associations from Sweden and Finland (for the fairway project).

Lessons learned included that procedures should be well-planned in advance, responsibilities should be clear with training regional competent authorities and providing written guidance, and meetings should be held with points of contacts during the process to agree on practicalities and improve procedure (Rantakallio 2005).

8.3.4 Other Application Under International Law

In addition to the Espoo Convention and SEA Protocol, a duty to conduct an EIA is found under other international treaties such as the LOS Convention (1982) (art. 206) and the Convention on Biological Diversity (1992) (art. 14). Principle 17 of the Rio Declaration (1992) calls for an EIA to be undertaken for "proposed activities that are likely to have a significant adverse impact on the environment".

Transboundary EIA under the LOS Convention is of particular interest for the Arctic marine area. Under the Convention, when there are reasonable grounds for believing that planned activities within the jurisdiction or control of a state may cause substantial pollution of or significant harmful changes to the marine environment, states must assess the potential effects of the activities on the marine environment. States are required to conduct an assessment of the effects of activities taking place within their maritime jurisdiction on the marine environment located in other states' jurisdiction as well as on areas beyond national jurisdiction. Unfortunately, it is unlikely that an assessment of transboundary impacts on the marine environment located in another state's jurisdiction would be evaluated in a systematic manner given the fact that there is no guidance on how potentially affected states can contribute to an assessment. More importantly, the duty of assessment is qualified by the phrase "as far as practicable", giving the origin state a fair amount of discretion. The results of assessments are to be communicated to competent international organizations "which should make them available to all states", and it is in this manner that a potentially affected state can obtain information.

The International Court of Justice (ICJ) has recognised a requirement under customary international law to carry out assessments where there is a risk that a proposed industrial activity may have a significant adverse impact in a transboundary context, in particular, on shared resources. In the *Pulp Mills on the river Uruguay* case,[2] the ICJ held that conducting an EIA can be considered a requirement under general international law and is part of exercising due diligence; this judgment does not necessarily extend to a requirement for SEAs. The ICJ did not specify the requisite content of assessments, but did clarify that content be determined "having regard to the nature and magnitude of the proposed development and its likely adverse impact on the environment" (para. 205), that impact assessments be carried out prior to implementation of activities, and that continuous monitoring is needed. As a rule of customary international law, this development may require additional clarification. The ICJ also suggested that the content of this obligation might evolve over time.

In its Advisory Opinion for the *Nauru* case,[3] the International Tribunal on the Law of the Sea with its specialised section for the settlement of seabed disputes elaborated on aspects of due diligence with regard to EIAs. It concluded that there exists an obligation of conduct—not, however, an obligation of result—and understood due diligence as a variable concept. It also remains to be seen if the transfer of this principle to areas beyond national jurisdictions will be confirmed in the future.

8.3.5 Do Any of These Impact Assessments and Guidelines Matter and are They Effective?

Although specific and easy to understand, in practice, the Arctic EIA Guidelines are often not applied due to both a lack of awareness of their existence coupled with the fact they are not mandatory (Koivurova 2012). While many project-level EIAs have been prepared, there is little consistency regarding methodology and rigor of analysis, or even what substantively is to be covered.

A similar lack of methodology is also a persistent problem in SEAs. In theory, by having adopted various regulations for EIA and SEA within national legal systems, the Arctic states are obligated to carry out environmental assessments for projects, plans, and programmes, and, in some cases, for policies that could potentially harm the environment. However, as variations exist between the environmental assessment systems of the different Arctic countries, and between rigor of analysis

---

[2] Pulp Mills on the River Uruguay (Argentina v. Uruguay), Judgment of 20 Apr 2010, *ICJ Reports* 2010.

[3] Responsibilities and obligations of States sponsoring persons and entities with respect to activities in the Area (Request for Advisory Opinion submitted to the Seabed Disputes Chamber), Case No. 17, Advisory Opinion, 1 Feb 2011, see paras. 110ff., 117.

and implementation, the application of EIA and SEA has varied considerably in the circumpolar region (Azcarate et al. 2011).

Regarding TEAs, planned activities for the Arctic have not yet taken place on a large scale, and where activities have occurred, they have been far away from national borders (Koivurova 2008). Moreover, there seems to be a lack of capacity and knowledge on how to implement TEAs, demonstrated by the low number of national assessments and TEAs that have been implemented in the Arctic (Azcarate et al. 2011).

Thus, while most of the Arctic is covered by national provisions on EIA and SEA, and international treaties such as the Espoo Convention and its SEA Protocol require certain Arctic states to carry out TEAs, gaps still remain, suggesting that some activities with environmental impacts in the Arctic are not assessed.

8.4 Climate Change in the Arctic and the EIA process

With climate change, Arctic glaciers have drastically receded, but it is the Arctic Ocean itself that is most changed. From the 1970s to the 1990s, the minimum extent of polar pack ice fell by around 8 % per decade. In 2007, record sea ice lows left the Northwest Passage ice-free for the first time in memory (see Astill 2012). Scientists found that in 2007 every natural variable, including warm weather, clear skies, and warm currents, had lined up to reinforce the seasonal melt. But in 2012, under less 'primed' conditions, sea ice reached a new record low. Past climate models predicted that the Arctic Ocean could be ice-free in summer by the end of this century; an analysis published in 2009 in *Geophysical Research Letters* Wang and Overland (2009) suggested it might happen as early as 2037. Many now think it will be sooner.

8.4.1 The Arctic Climate Impact Assessment

During the US chairmanship of the Arctic Council from 1998 to 2000, the Council and the International Arctic Science Committee (IASC) commenced work on the Arctic Climate Impact Assessment (ACIA), and their findings were released in a publicly accessible format in 2004 (ACIA 2004). The ACIA concluded that ocean warming and loss of ice is expected to accelerate, exacerbating the major physical, ecological, social, and economic changes already underway in the Arctic marine environment.

Even though a 2001 report by the Intergovernmental Panel on Climate Change (IPCC) had already noted that warming is more intense in the Arctic, the Arctic Council-sponsored ACIA report established the Arctic as an early warning region for climate change observation. In addition to revealing serious impacts on the environment, ecosystems, and local communities, the ACIA led to important changes in the way the Arctic is perceived. The ACIA findings demonstrated that the Arctic is a region undergoing dramatic transformation. The public view developed from an understanding of the Arctic as being naturally guarded from human activity to an image of the region as dynamic and rich with economic potential, and therefore requiring stricter governance measures.

8.4.2 Climate Change and a Stronger Arctic Council

Until recently, Arctic cooperation has functioned for over fifteen years in a fairly consistent mode of operation with work primarily conducted by the six working groups focused on environmental protection and sustainable development (see Chap. 1). But in response to alarming climate change, and the new changes and challenges that are rapidly manifesting in the region, the Council has recently strengthened the way it functions. Given the enormity of challenges posed by climate change in the Arctic, the Arctic Council has gradually started to support the adoption of binding legal responses, rather than sticking only to traditional soft law regulation through guidelines, best practices, and manuals. These soft law measures serve their purpose in some policy arenas, especially since the region's indigenous peoples' organizations can participate in their drafting and some soft law guidance has likely made its way into practice, although this is often difficult to verify. One case that indicates the limitations of soft law regulation, particularly where Arctic Council member states have viewed guidance as an intrusion upon their sovereignty, is the EIA Guidelines, which as noted, appear to have been gradually forgotten (Koivurova 2012).

8.4.3 Analyzing Climate Change in EIAs

There are no written guidelines for the Arctic that provide a framework for assessing climate change in impact assessments. However, factors to consider when analyzing climate change effects in the Arctic include rising temperatures, increased extreme weather, loss of permafrost affecting facilities and transportation design and performance, environmental and social changes as a result of climate changes, and effects arising from increased pressure to develop resources given easier Arctic access.

8.4.4 The Canadian Example

At the prompting of the IAIA and the Canadian Environmental Assessment Agency, the *Practitioner's Guide to Incorporating Climate Change into the*

Environmental Impact Assessment Process—ClimAdapt EIA Guide (Bell et al. 2003) was developed in 2003 with the intent to provide the following:

• An understanding of the implications of climate change in relation to the preparation of an EIA;
• On a project-specific basis, to give direction on how to determine whether or not climate change should be considered;
• Sources of information for use in assessing climate change implications; and
• Guidance in incorporating climate change consideration into the EIA process.

Part of the guidance includes reversing the normal thought process, which looks at the impacts of the project on the environment. Conversely, the question was asked—*what are the effects of the environment on the project?* To answer this question, issues such as the following were addressed: human health and safety; how the operation and productivity of a project may be adversely affected; how the cost of development may rise and project design need to be modified; how maintenance requirements could increase; and the increasing importance of monitoring (Bell et al. 2003).

It also promoted the consideration of cumulative effects within the context of climate change, involving issues such as the increased transport of physical or chemical constituents; an increase or decrease in habitat area for a species or species group that is already affected by the project; and the secondary effects related to climate change modification to the environment or its effects on the project (Bell et al. 2003). Six existing EIAs in Canada were used as case studies, for example:

Gahcho Kue Diamond Mine—2007: The MacKenzie Valley Review Board has jurisdiction for EIAs in a large portion of Canada's Northwest and was faced with the proposal for a new proposed Debeers diamond mine. Given scientific consensus that the North is particularly vulnerable to impacts from a changing climate, the Board found that the EIS must examine and evaluate the development as a potential greenhouse gas contributor, as well as examine potential climate change effects on the proposed development (Bell et al. 2003).

Beaufort Sea Gas Development Project (Canadian Arctic)—1982: A detailed EIS described the impacts of a project to recover Arctic oil and gas and deliver it to southern markets through either a MacKenzie River pipeline or tanker. While current and historic climate conditions were reviewed in detail in the EIA, there was no consideration or integration of climate change into the assessment. The decision making panel did, however, recommend the possibility that climate change be considered in the design and construction of a pipeline and other fixed facilities in areas of permafrost (Bell et al. 2003).

Voisey's Bay Nickel Mine (Canadian province of Newfoundland and Labrador near the bay of Voisey)—1997: The project was located in Labrador where climate change predictions at the time were for reductions in temperatures rather than increases. The assessment revealed that Valued Environmental Components warranted ongoing monitoring and observation, and the preparation of a risk assessment that looked at caribou, birds, Arctic char, sea ice with respect to shipping and its interaction with local coastal activity, and changes to social activities of local aboriginal peoples. The last two are possibly the most likely to involve interactions of the project and the effects of climate change together (Bell et al. 2003).

Although the Guide was prepared in 2003, it appears that little has changed in EIA practice. The emphasis on addressing climate change impacts remains focused on the reduction of greenhouse gas emissions and the potential effects of change on the project. As noted by Collins (2010), the question should be whether we can agree that climate change is the largest global cumulative impact that we need to address. While effort is increasing to mainstream climate change into existing sustainable development management processes, integration in EIA has been largely limited to greenhouse gas contributions and effects on the project. There is a world of difference between consideration of greenhouse gas reductions and climate change contributions, and planning for and taking into consideration the impacts of climate change.

8.5 Progressiveness in Using Traditional Knowledge

One important aspect of impact assessments in the Arctic is the role and contribution of indigenous peoples. As explained the Arctic EIA Guidelines emphasize the inclusion of indigenous peoples into the EIA process, reflecting the status of indigenous peoples as permanent participants in the Arctic Council.

Comparing impact assessments between indigenous and non-indigenous peoples provides a revealing contrast in terms of environmental approaches and resultant analyses. For example, the Sámi perceive landscape in a cultural way, as part of the community's history, language, myths, values, and livelihoods. In short, the Sámi cultural landscape, heritage, and biodiversity are inextricably linked (see Chap. 4).

As a result of this cultural landscape having been formed through traditional livelihoods, traditional ecological knowledge (TEK) and respect for the land can result in a more holistic approach in terms of environmental assessment. In Canada, this can be seen in a number of instances, and through the following example of the Baffinland project, it is clear that TEK helps provides a foundation for a more integrated approach.

8.5.1 Baffinland Project

The Mary River Project is a proposed iron ore mine on North Baffin Island, in the Nunavut Settlement Area (NSA) of the Mackenzie Valley, in conjunction with the development of a railway that is planned to transport 18 million tons per year of ore from an open pit mine to an all-season deepwater port. The project would be the largest in Nunavut history, involving technological challenges such as the construction of a railway on permafrost and cumulative impact concerns such as land use changes, mining, land transportation, and marine shipping.

The Nunavut Impact Review Board (NIRB) is responsible for the environmental assessment of project proposals in the NSA [Collins N (2003)]. An EIA was requested by the NIRB in June 2008, designed to include preparation of a climate change assessment addressing project impacts to permafrost and soils with high ice content, the hydrological regime, marine ice flow regimes, and the long-term impacts of such changes on the project. Multiple impact assessment scenarios spanning the range of possible future climates were to be designed and applied, and the long-term effects of climate change were to be discussed up to the projected closure phase of the project (Collins 2010).

The NIRB's guidelines (NIRB 2008) took a more holistic view that required not only a detailed evaluation of climate change impacts, as discussed above, but in addition, that an *ecosystem-based approach* to EIA be adopted, that socioeconomic issues—such as the economic development within the region—be included, that past and potential future environmental, economic, and social trends be taken into account, and that the well-being of residents of Canada outside of the Nunavut Settlement Area also be considered (NIRB 2008; see also Collins 2010).

In addition to seeing how climate change impacts are addressed in both the Canadian and Nunavut cases, what is interesting about the Baffinland example is the extent to which TEK underlay many of the other environmental baseline assumptions against which impacts were later measured. For example, five local communities were consulted regarding their knowledge of terrestrial wildlife, caribou in particular. Traditional knowledge of caribou habitat and abundance cycles was then collected and summarized in a spatial database. The traditional knowledge and survey data were ultimately used as a baseline to conduct an impact assessment to determine practical options to mitigate the effects of the mine, road, and railway on caribou habitat and movement (Environmental Dynamics n.d.).

The need to survive extremely harsh conditions for generations has built a repository of local knowledge that easily melds the understanding of how local and regional ecosystems operate with the more specific interrelationships between animals, plants, and the physical environment. In fact, it has been documented that in remote parts of the world, the ecological knowledge of indigenous peoples is often 'geographically and temporally more extensive than scientific knowledge' (Ferguson and Messier 1997).

The final public hearings on the Mary River Project by the NIRB occurred in July 2012. The first phase of the mine is expected to last 21 years and generate revenues of 3–5 billion CND while creating 950 jobs. Baffinland representatives at the hearing stated that they are committed to hiring Inuit and creating economic opportunities (Murphy 2012).

In spite of these professed benefits, the NIRB, Nunavut Tunngavik Inc., and Qikiqtani Inuit Association repeatedly voiced concerns regarding Baffinland's preparedness in the event of a spill. Baffinland insisted the risk of fuel spills associated with the project's transportation system are low and that emergency clean-up plans are in place in the "unlikely" event that one will occur; nonetheless representatives admitted their current techniques for dealing with a spill are "not optimal for recovery on the ice" (Murphy 2012).

8.6 Toward 'Geo-Environmental' Cooperation in the Arctic'

The Espoo Convention initially held much promise as the best hope for transboundary, pan-Arctic environmental cooperation in the Arctic region. Since the late 1990s, however, that promise has waned as Iceland, Russia, and the US have been unwilling to accept the Convention as legally binding; however, recent developments indicate that Russia may be moving closer to acceptance of the Convention. Iceland's reasons for not ratifying the Convention are said to be the impracticality of TEA procedures due to the country's location and the fact that priorities need to be established for a country with such a small civil service (Koivurova 2008). The US, while a signatory, is not interested in becoming a party. Even so, the Espoo Convention is still generally cited as being a standard for TEA in the Arctic, as exemplified by the voluntary application by Finland of Espoo procedures towards Russia in the hopes of inducing Russia to finally ratify the Convention.

It has been more than 15 years since Espoo has come into force, and in the interim, the ramifications of climate change, and the projections of future adverse impacts from climate change, continue to grow. Could this changing reality be the catalyst for a new type of environmental cooperation—one that takes climate change fully into account and addresses not only traditional environmental issues, but also, in an integrated fashion, those relating to environmental security and socioeconomic considerations? Could this new reality be the catalyst for a new form of Arctic cooperation (i.e., the 'geo-environmental' dimension) where geopolitics—including relationships based on shipping and trade, offshore oil and gas development, and energy security—are increasingly shaped by environmental issues? Perhaps transboundary environmental assessments will build stronger relations among the Arctic nations by encouraging environmental cooperation in the recognition of a common goal: sustainable use of the Arctic.

Perhaps most important of all is the potential for new cooperation in Arctic governance. Whether or not an EIA or SEA with actual legal authority will ever be a reality, can the Arctic, in all its complexity, be looked at from a geo-environmental perspective, and if not, what are the ramifications for the Arctic and the world?

The key question is: Can new and better approaches to impact assessments in the Arctic help to inform and shape a more effective approach for the region's environment? The limits to this will likely be set by Arctic politics and sovereignty concerns, but the demand for more rigorous and enforceable EIA guidelines will almost certainly grow over the coming years. The growing significance of Arctic natural resources has a direct impact on events in the global economy and politics, and with climate change, globalization, and technological developments, the

Arctic region is no longer seen as a peripheral area, but is gaining a place at the centre of international affairs. Arctic change is inevitable, and still largely unpredictable, but of huge—'geo-environmental'—dimensions.

References

ACIA (2004) Impacts of a Warming Arctic: Arctic Climate Impact Assessment. Cambridge University Press, Cambridge

AEPS (1991) Arctic Environmental Protection Strategy, 14 Jan 1991, 30 I.L.M. 1624

AEPS (1997) Guidelines for Environmental Impact Assessment (EIA) in the Arctic. Sustainable Development and Utilization, Arctic Environment Protection Strategy. Finnish Ministry of the Environment, Finland

Arctic Council (2000) Framework Document (Chapeau) for the Sustainable Development Programme, Arctic Council, 13 Oct 2000

Astill J (2012) The vanishing north, The Economist, 16 June

Azcarate J, Balfors B, Destouni G, Bring A (2011) Shaping a Sustainability Strategy for the Arctic. Royal Institute of Technology and Stockholm University. IAIA11 Conference on Impact Assessment and Responsible Development for Infrastructure, Business and Industry

Bell A, Collins N, Young R (2003) Practitioner's Guide to Incorporating Climate Change into the Environmental Impact Assessment Process: Clim Adapt EIA Guide

BSH (2007) Standards for Environmental Impact Assessments of Offshore Wind Turbines on the Marine Environment. Bundesamt für Seeschifffahrt und Hydrographie, Feb 2007

CEFAS (2004) Offshore Wind-Farms—Guidance Note for Environmental Impact Assessment in Respect of FEPA and CPA requirements (United Kingdom), Version 2 June 2004. Prepared by the Centre for Environment, Fisheries and Aquaculture Science on behalf of the Marine Consents and Environment Unit

CEQ (2007) A Citizen's Guide to the NEPA: Having your voice heard. Council on Environmental Quality, Executive Office of the President. Dec 2007

Collins N (2003) Incorporating Climate Change into Cumulative Effects Assessment. Presentation for the 2003 IAIA Calgary Conference, CEF Consultants Ltd., Halifax

Collins N (2010) Climate Change and Canada's North. Presentation for the IAIA conference in Washington DC, CEF Consultants Ltd., Halifax

Convention on Biological Diversity (1992) Convention on Biological Diversity, 5 June 1992, 1760 U.N.T.S. 79. Entered into force 29 Dec 1991

Defra (2005) The Nature Conservation Guidance on Offshore Wind Farm Development—A guidance note on the implications of the EC Wild Birds and Habitats Directives for developers undertaking offshore wind-farm developments. Mar 2005

EIA Directive (1985) Council Directive 85/of 27 June 1985 on the assessment of the effects of certain public and private project son the environment, 27 June 1985, O.J. (L 175/40), as amended as amended by Directives 97/11/EC and 2003/35/EC

Environmental Dynamics (n.d.) Mary River Terrestrial Wildlife Baseline and Impact Assessment. <http://www.edynamics.com/assets/files/Baffinland_MaryRiver_Project.pdf>

Espoo Convention (1991) Convention on Environmental Impact Assessment in a Transboundary Context, 25 Feb 1989, 1989 U.N.T.S. 309. Entered into force 10 Sept 1997

European Commission (2001) Guidance on EIA—EIS review. Prepared by Environmental Resources Management, June 2001

European Commission (2013) Additional tools. Strategic Environmental Assessment—SEA. <http://ec.europa.eu/environment/eia/sea-legalcontext.htm>. Accessed 20 Feb 2013

Ferguson M, Messier F (1997) Collection and Analysis of Traditional Ecological Knowledge about a Population of Arctic Tundra Caribou. Arctic 50(1):17–28

IAIA (2013) International Association for Impact Assessment. <www.iaia.org>. Accessed 8 Feb 2013

Koivurova T (2008) Transboundary environmental assessment in the Arctic. Impact Assess Project Appraisal 26(4):265–275

Koivurova T (2009) Transboundary EIA in the Arctic. UNECE Seminar. <http://www.unece.org/fileadmin/DAM/env/eia/documents/WG12_may2009/seminar_arctic.pdf>. Accessed 10 Feb 2013

Koivurova T (2012) New ways to Respond to Climate Change in the Arctic. insights. American Society of International Law, Insights, 16(33)

Koivurova T, Molenaar EJ (2009) International Governance and Regulation of the Marine Arctic: Overview and Gap Analysis. WWF International Arctic Programme, Oslo

LOS Convention (1982) United Nations Convention on the Law of the Sea, 10 Dec 1982, 1833 U.N.T.S. 396. Entered into force 16 Nov 1994

Ministry of Industry (2007) Strategic Environmental Assessment (SEA) of the proposed plan to offer exclusive exploration and production licenses for oil and gas in the Dreki (Dragon) area on the Jan Mayen Ridge, Northeast of Iceland. Ministry of Industry, Iceland Mar 2007

Murphy D (2012) Baffinland reps field questions on risk, socio-economics, July 18. Nunatsiaq Online. <http://www.nunatsiaqonline.ca/stories/article/65674baffinland_reps_field_questions_on_risk_socio-economics/>. Accessed 11 Feb 2013

NEPA (1969) National Environmental Policy Act of 1969, 42 U.S.C. 4321 et seq

NIRB (2008) Guide to the NIRB Review Process. Guide 5. Nunavut Impact Review Board, Sept 2008. <http://www.nirb.ca/04-GUIDES/NIRB-F-Guide%205-The%20NIRB%20Review%20Process-OT5E.pdf>. Accessed 11 Feb 2013

Offshore (2012) Iceland issues offshore licenses, 4 Dec <http://www.offshore-mag.com/articles/2012/12/iceland-issues-offshore-licenses.html>. Accessed 20 Feb 2013

OSPAR (2010) Quality Status Report 2010. Assessment report on the environmental impact of offshore wind farms. Publication No. 385/2008

PAME (2009) Arctic Offshore Oil and Gas Guidelines. Last updated 29 Apr 2009

Rantakallio S (2005) Practical application of Espoo convention in Finland. Presentation from the Finnish Ministry of the Environment. UNECE Conference, Stockholm

Rio Declaration (1992) Rio Declaration on Environment and Development, UN Doc. A/CONF.151/26 (vol. I), 31 I.L.M. 874 (1992)

SEA Directive (2001) Directive 2001/42/EC of the European Parliament and of the Council of 27 June 2001 on the assessment of the effects of certain plans and programmes on the environment,21 July 2001, 2001 O.J. (L 197/30)

SEA Protocol (2011) Protocol on Strategic Environmental Assessment to the Convention on Environmental Impact Assessment in a Transboundary Context, 21 May 2001, UNECE Document ECE/MP.EIA/2003/2. Entered into force 11 July 2010

Smith A (2008) Impact assessment in the marine environment—the most challenging of all. IAIA08—Impact assessment in the marine environment. <http://www.iaia.org/iaia08perth/pdfs/themeforums/TF4-1_sector_marine_Smith.pdf>. Accessed 8 Feb 2013

Swedish Polar Research Secretariat (2012) Environmental Impact Assessment (EIA) for OATRC 2012, Dnr 2012-160, Swedish Polar Research Secretariat, Stockholm, Sweden

Wang M Overland JE (2009) A sea ice free summer Arctic within 30 years? Geophys Res Lett 36, L07502. doi:10.1029/2009GL037820

Chapter 9
Pan-Arctic Marine Spatial Planning: An Idea Whose Time Has Come

Charles N. Ehler

Abstract Driven by outside economic forces and the effects of climate change, the Arctic, its ecosystems, and its people are all faced with substantial change ranging from the loss of ice-dependent species, more intense human uses of the Arctic, and the loss of natural services provided by Arctic ecosystems. In addition to business opportunities, these changes represent new risks to the Arctic's unique natural environment and to the people who now live and work in the Arctic. Once new human activities begin in the Arctic Ocean, it will be difficult for policymakers and planners to put limits on them. This paper explores a new approach to the integrated management of human activities—marine spatial planning (MSP). MSP seeks to reduce conflicts among human activities and balance the conservation of ecologically important areas with the sustainable development of marine resources in the Arctic. With the exception of Norway, most Arctic governments have been slow to advance marine spatial planning. A way to advance MSP in the Arctic would be to explicitly recognize the importance of moving beyond sole reliance on the initiatives of national governments and towards a pan-Arctic approach to guide the future of the region. Networks and partnerships of non-governmental actors, including indigenous peoples, environmental NGOs, academia, and private industry, all of whom have influence over governmental policies and actions, could be used to initiate MSP across the Arctic.

9.1 Introduction

The Arctic Ocean—all 14 million square kilometres of it—is one of the most pristine, yet vulnerable, ecosystems in the world. Protected by its historic inaccessibility, harsh environment, relatively small human population, and slow rate of economic development, the Arctic Ocean has been relatively less affected by human activity than most other marine regions of the earth.

C. N. Ehler (✉)
Ocean Visions Consulting, Paris, France
e-mail: charles.ehler@mac.com

E. Tedsen et al. (eds.), *Arctic Marine Governance*,
DOI: 10.1007/978-3-642-38595-7_9, © Springer-Verlag Berlin Heidelberg 2014

This is about to change. Driven by outside economic forces and the effects of climate change, the Arctic, its ecosystems, and its people are all faced with substantial change ranging from the loss of ice-dependent species, particularly polar bears and walruses, more intense human uses of the Arctic (e.g., oil and gas exploration and development, commercial fishing, and marine transport), and the loss of natural services provided by Arctic ecosystems. As the Arctic warms, its ice melts, and its ecosystems change, and as technology improves and the demand for natural resources increases, opportunities open up for industry—shorter shipping routes, virgin fishing grounds, new areas of oil and gas exploration and development, and new places for commercial tourism. Several scientists have predicted that the Arctic will be ice-free by the 2030s. As well as business opportunities, these changes represent new risks to the Arctic's unique natural environment and to the people who now live and work in the Arctic. *Once these activities begin in the Arctic Ocean, it will be difficult for policymakers and planners to put limits on them.*

International and national interest in mitigating and adapting to these changes has led to increased calls to manage human activities through an ecosystem-based approach. Marine spatial planning has emerged as an operational approach to translate this relatively vague concept into management practice in many marine areas around the world.

Marine spatial planning is already in place in one Arctic country. In 2006, the Norwegian Parliament approved an integrated management plan, including spatial and temporal management measures, for the Norwegian part of the Barents Sea and the Sea Areas off the Lofoten Islands (Norwegian Ministry of the Environment 2005). The plan integrates previously separate management regimes for fisheries, shipping, oil and gas, and nature conservation. The Barents Sea plan was revised in 2010. The Norwegian Parliament also approved an integrated management plan for the Norwegian Sea in 2010 (Norwegian Ministry of the Environment 2009).

Canada has developed an integrated management plan for its part of the Beaufort Sea that included development of a marine spatial plan as one of its future actions (Beaufort Sea Partnership 2009). However, while the Canadian Beaufort Sea plan was 'supported' by the Minister of Fisheries and Oceans in 2010, no funds have been allocated to its implementation.

In the United States (US), the federal government is currently "implement[ing] comprehensive, integrated, ecosystem-based coastal and marine spatial planning and management in the United States", including its Beaufort, Chukchi, and Bering seas (National Ocean Council 2012). However, planning for the Arctic Ocean lags behind other marine regions of the US.

Russian legislation and regulations currently have no references to integrated management or marine spatial planning. Governmental organizations of the Russian Federation that have responsibilities for marine management operate on a sector-by-sector basis. However, existing legislation does contain areas in which marine spatial planning could be an important instrument for addressing problems of Russian marine management. Marine planning is under discussion now only in the Russian academic community. Russia has identified biologically important areas (Spiridonov et al. 2011). Still, only a few marine areas have been designated as protected areas. For example, Wrangel Island and its surrounding waters between the East Siberian Sea and the Chukchi Sea are designated as a national 'Zapovednik' (strict nature preserve) and listed as the only World Heritage marine site in the entire Arctic Ocean. Some coastal terrestrial areas have been designated as protected areas, a few of which extend a kilometre into the territorial sea.

While the European Union (EU) has no authority for marine (or maritime) spatial planning in the Arctic, it recently committed to pursue its involvement within existing international frameworks on Arctic issues such as biodiversity, ecosystem-based management, and marine protected areas (European Commission and High Representative 2012). The EU has been a strong advocate for maritime spatial planning—one of the pillars of its Maritime Strategy (European Commission 2010). The EU is planning to issue a directive on coastal and maritime spatial planning to Member States during 2013.

9.2 Why is Marine Spatial Planning Needed?

Before the last century, the oceans of the world were used mainly for two purposes: marine transportation and fishing. Conflicts between uses were few and far between except around some ports. Fisheries were managed separately from marine transportation, which over the last century was managed separately from offshore oil and gas development and other emerging marine activities, despite increasing real conflicts between and among these uses.

Single-sector management has often failed to resolve conflicts among users of marine space, rarely dealing explicitly with trade-offs among uses, and even more rarely dealing with conflicts between the cumulative effects of multiple uses and the marine environment. New uses of marine areas including offshore wind, ocean energy, offshore aquaculture, and marine tourism, as well as the demand for new marine protected areas, have only exacerbated the situation. Single-sector management has also tended to reduce and dissipate the effect of enforcement at sea because of the scope and geographic coverage involved and the environmental conditions in which monitoring and enforcement have to operate. In sharp contrast to the land, little 'public policing' of human activities takes place at sea—and is even less likely in the Arctic Ocean due to its inaccessibility.

As a consequence, marine ecosystems around the world are in trouble. Both the severity and scale of impacts on marine ecosystems from overfishing, habitat loss and fragmentation, pollution, invasive species, and climate change are increasing, with virtually no corner of the world left untouched.

Awareness is growing that the ongoing degradation in marine ecosystems is in large part a failure of governance (Crowder et al. 2006). Many scientists and policy analysts have advocated reforms centred on the idea of 'ecosystem-based management' (EBM). To date, however, a practical method for translating this concept into operational management practice has not emerged (Arkema et al. 2006).

Many recommendations for an ecosystem-based approach to marine management in the Arctic already exist. In fact, most Arctic countries are working to implement ecosystem-based management of their marine areas. The Arctic Council has repeatedly called for an ecosystem-based approach to marine management. For example, in its 2004 Arctic Marine Strategic Plan, EBM is defined as an approach that "[...] requires that development activities be coordinated in a way that minimizes their impact on the environment and integrates thinking across environmental, socio-economic, political and sectoral realms" (Arctic Council 2004). The key features of this approach include consideration of multiple scales, a long-term perspective, the recognition that humans are an integral part of ecosystems, an adaptive management perspective, and a concern for sustaining production and consumption potential for goods and services. The Best Practices in Ecosystem-based Oceans Management in the Arctic (BePOMAr) Project of the Arctic Council has summarized the practices that Arctic countries have used to apply an ecosystem-based management approach to marine management (Hoel 2009). The conclusions of that project include:

- Flexible application of effective ecosystem-based oceans management;
- Decision making must be integrated and science based;
- National commitment is required for effective management;
- Area-based approaches and transboundary perspectives are necessary;
- Stakeholder and Arctic resident participation is a key element; and
- Adaptive management is critical.

An integrated, ecosystem-based management approach has been identified in many marine places as an appropriate evolution to address problems caused by today's incremental, single-sector approach to marine management. However, examples of practical applications of an ecosystem-based approach are elusive. How to begin is the first challenge. A step in the right direction is the increasing worldwide interest in 'marine spatial planning' as an operational process through which to implement EBM.

9.3 What is Marine Spatial Planning?

Marine spatial planning (known as *maritime* spatial planning in Member States of the EU) or MSP is a practical way to create and establish a more rational organization of the use of marine space and the interactions between its uses, to balance demands for development with the need to protect marine ecosystems, and to achieve social and economic objectives for marine regions in an open and planned way.

MSP is a public process of analyzing and allocating the spatial and temporal distribution of human activities in marine areas to achieve ecological, economic, and social goals and objectives that are usually specified through a political process (Ehler and Douvere 2007). Its characteristics include:

- *Integrated* across economic sectors and governmental agencies, and among levels of government;

- *Strategic* and *future-oriented,* focused on the long-term;
- *Participatory,* including stakeholders actively in the entire process;
- *Adaptive,* capable of learning by doing;
- *Ecosystem-based,* balancing ecological, economic, social, and cultural goals and objectives toward sustainable development and the maintenance of ecosystem services; and
- *Place-based or area-based,* i.e., integrated management of all human activities within a spatially defined area identified through ecological, socioeconomic, and jurisdictional considerations (Ehler and Douvere 2007).

It's important to remember that we can only plan and manage human activities in marine areas, not marine ecosystems or components of ecosystems. We can allocate human activities to specific marine areas by objective, e.g., development or preservation areas, or by specific uses, e.g., offshore energy, offshore aquaculture, or sand and gravel mining.

9.4 Why is Marine Spatial Planning Needed?

Most countries already designate or zone marine space for a number of human activities such as maritime transportation, oil and gas development, offshore renewable energy, offshore aquaculture, and waste disposal. However, the problem is that usually this is done on a sector-by-sector, project-by-project basis without much consideration of effects either on other human activities or the marine environment. Consequently, this situation has led to two major types of conflict:

- Conflicts among human uses (user–user conflicts); and
- Conflicts between human uses and the marine environment (user-nature conflicts).

These conflicts weaken the ability of the ocean to provide the necessary ecosystem services upon which humans and all other life on Earth depend. Furthermore, decision makers in this situation usually end up only being able to react to events, often when it is already too late, rather than having the choice to plan and shape actions that could lead to a more desirable future of the marine environment.

By contrast, MSP is a future-oriented process. It offers a way to address both these types of conflict and select appropriate management measures to maintain and safeguard necessary ecosystem services. It is the missing piece that can lead to truly integrated planning from coastal watersheds to marine ecosystems.

When effectively put into practice, MSP can be used to:

- *Set priorities*—to enable significant inroads to be made into meeting the development objectives of marine areas in an equitable way, it is necessary to provide a rational basis for setting priorities, and to manage and direct resources to where and when they are needed most;
- *Stimulate opportunities* for new users of marine areas, including ocean energy;

- *Coordinate actions and investments* in space and time to ensure positive returns from those investments, both public and private, and to facilitate complementarities among jurisdictions and institutions;
- *Provide a vision and consistent direction,* not only of what is desirable, but what is possible in marine areas;
- *Protect nature,* that has its own requirements that should be respected if long-term sustainable development is to be achieved and if large-scale environmental degradation is to be avoided or minimized;
- *Reduce fragmentation of marine habitats,* i.e., when ecosystems are divided by human activities and prevented from functioning properly;
- *Avoid duplication of effort* by different public agencies and levels of government in planning, monitoring, and permitting; and
- *Achieve higher quality of service* at all levels of government, e.g., by ensuring that permitting of human activities is streamlined when proposed development is consistent with a comprehensive spatial management plan for the marine area.

9.5 Why are Space and Time Important?

Some areas of the ocean are more important than others—both ecologically and economically (Crowder and Norse 2008). Species, habitats, populations of animals, oil and gas deposits, sand and gravel deposits, and sustained winds or waves—are all distributed in various places and at various times. Successful marine management needs planners and managers that understand how to work with the spatial and temporal diversity of the sea. Understanding these spatial and temporal distributions and mapping them is an important aspect of MSP. Managing human activities to enhance compatible uses and reduce conflicts among uses, as well as to reduce conflicts between human activities and nature, are important outcomes of MSP. Examining how these distributions might change due to climate change and other long-term pressures (e.g., overfishing) on marine systems is another important step of MSP.

9.6 What is the Principal 'Driver' of MSP in the Arctic?

Pressures from human activities have often led to initiatives to better manage marine areas. For example, the Santa Barbara (US) oil spill from an oil well blowout in 1969 resulted in numerous pieces of legislation focused on marine pollution. In the 1970s, the threat of offshore oil and gas development and phosphate mining lead to initial efforts to protect the Great Barrier Reef. More recently, particularly in western Europe, MSP has been driven by national policies to develop offshore wind energy in Belgium, the Netherlands, and Germany (all of whom have developed and implemented marine plans), and the United Kingdom (England began development of marine spatial plans for two sub-regions of its marine area in 2011) and the requirement to designate more marine protected areas under directives of the European Commission. These 'new' uses, including offshore aquaculture, have had to compete with traditional users for scarce ocean space.

However, most of the Arctic Ocean is different. One of the major issues in the Barents Sea was the potential expansion of oil and gas activities into areas of the Barents used by fisheries and living marine resources. MSP is at the core of the plan, identifying particularly valuable and vulnerable areas, either from ecological and/ or human perspectives. Within the plan, access to specific areas for human activities is carefully managed, for example, by moving shipping lanes outside Norwegian territorial waters (12-nautical miles), limiting trawling in sensitive areas, not opening particularly valuable and vulnerable areas to petroleum activities, including the ice edge, and extending marine protected areas and fishery closure areas to protect spawning aggregations, fish eggs and larvae, and juvenile fish and shellfish.

The level of human use of the Arctic marine environment remains relatively undeveloped. This situation is rare for marine areas. Marine spatial planners usually have to fit new uses into a heavily-used mosaic of other uses. The opportunity to shape the future of the Arctic in a socially desirable way is also a principal driver.

9.7 What are the Benefits of Marine Spatial Planning?

When developed effectively, marine spatial planning can have significant economic, social, and environmental benefits. The following table identifies some of the most important benefits of marine spatial planning (Table 9.1).

Table 9.1 Examples of benefits of marine spatial planning

Economic	Greater certainty of access to desirable areas for new private sector investments, frequently amortized over 20–30 years
	Identification and early resolution of conflicts between incompatible uses through planning instead of litigation
	Streamlining and transparency in permit and licensing procedures
	Improved capacity to plan for new and changing human activities, including emerging technologies and their associated effects
Environmental	Identification of biologically and ecologically important areas as a basis for space allocation
	Establish context for planning a network of marine protected areas
	Identification and reduction of the cumulative effects of human activities on marine ecosystems
Social	Improved opportunities for local community and citizen participation
	Identification of effects of decisions on the allocation of ocean space (e.g., closure areas for certain uses, protected areas) on communities
	Identification and preservation of social, cultural, and spiritual values related to use of ocean space
Administrative	Improve speed, quality, accountability, and transparency of decision making, and reduction of regulatory costs
	Improve consistency and compatibility of regulatory decisions
	Improve information collection, storage and retrieval, access, and sharing

9.8 What are the Key Steps of Marine Spatial Planning?

The development and implementation of MSP involves a number of steps, including:

- Identifying need and establishing *authority*;
- Obtaining *financial support*;
- Organizing the process through *pre-planning;*
- Organizing *stakeholder participation;*
- Defining and analyzing *existing conditions;*
- Defining and analyzing *future conditions;*
- Preparing and approving the *spatial management plan;*
- *Implementing and enforcing* the spatial management plan;
- *Monitoring and evaluating performance;* and
- *Adapting* the marine spatial management process (Ehler and Douvere 2009).

These ten steps are not simply a linear process that moves sequentially from step to step. Many feedback loops should be built into the process. For example, goals and objectives identified early in the planning process are likely to be modified as costs and benefits of different management measures are identified later in the planning process. Analyses of existing and future conditions will change as new information is identified and incorporated into the planning process. Stakeholder participation will change the planning process as it develops over time. Planning is a dynamic process and planners and stakeholders have to be open to accommodating changes as the process evolves over time.

Comprehensive MSP provides an integrated framework for management that provides a guide for, but does not replace, single-sector management. For example, MSP can provide important contextual information for guiding marine protected area management or for fisheries management, but does not replace it.

MSP answers four simple questions:

1. *Where are we today?* What are the baseline conditions?
2. *Where do we want to be?* What are the alternative spatial scenarios of the future? What is the desired vision?
3. *How do we get there?* What spatial management measures move us toward the desired future?
4. What *have we accomplished?* Have the spatial management measures moved us in the direction of the desired vision? If not, how should they be adapted in the next round of planning? (Ehler and Douvere 2009)

9.9 What are Examples of MSP Management Measures?

Management measures are at the heart of any management plan. They are the collective actions that will be implemented to achieve the overall management goals and objectives of the plan.

However, marine spatial planning cannot do it all. An integrated management plan for a marine area will have many management measures that will be applied to the important sectors of human activities (e.g., fisheries, marine transport, minerals extraction, and oil and gas) that use the resources of the marine area. There are four categories of management measures:

- *Input management measures*: measures that specify the inputs of human activities in marine areas;
- *Process management measures*: measures that specify the nature of the production process of human activities in marine areas;
- *Output management measures*: measures that specify the outputs of human activities in marine areas; and
- *Spatial/temporal management measures*: measures that specify where in space and when in time human activities can occur in marine areas (see Table 9.2).

Table 9.2 Examples of MSP management measures

Human activity	Spatial/temporal management measures
Oil and gas exploration	Restrict seismic operations when marine mammals are present in the marine area
	Restrict oil and gas activities in areas of subsistence access and harvest
	Identify areas where oil and gas activities should be prohibited at any time
	Identify areas where oil and gas activities should be prohibited by season, e.g., during marine mammal migrations
	Restrict oil and gas activities to winter months to reduce effects on biota and habitats, e.g., ice roads instead of permanent roads
	Select supply routes, frequency, and timing to avoid effects on biota or harvesting of wildlife by residents
	Prohibit discharges of drilling cuttings and produced water in sensitive marine areas
	Require reconditioning or recycling of drilling fluids in sensitive marine areas
	Prohibit discharges of solid waste into marine environment
	Install, operate, and maintain pipelines to minimize disturbance of seafloor habitats and other uses of the seafloor
Marine transportation	Designate ecologically and biologically sensitive Arctic areas as 'special areas' or 'particularly sensitive sea areas' (PSSAs)
	Require double-hulled and ice-strengthened vessels in Arctic waters
	Manage shipping to reduce marine mammal strikes
	Develop a traffic information system to improve monitoring of vessel traffic in Arctic waters
	Improve information on the use of Arctic marine environment by indigenous communities to avoid conflicts with marine transportation
	Identify areas of heightened ecological and cultural significance in light of changing climate conditions and increasing marine use to protect these areas from impacts of Arctic shipping
	Identify 'areas to be avoided' (ATBAs) based on navigation hazards and biological characteristics
	Improve hydrographic, oceanographic, and meteorological information about the Arctic marine environment to support safe navigation

(Continued)

Table 9.2 (Continued)

Human activity	Spatial/temporal management measures
Commercial fishing	Prohibit commercial fishing in Arctic waters until more information is available to support sustainable management (the US and Denmark have employed such a ban, Canada, Russia, and Norway have not)
	Improve baseline data on living marine resources, including potential changes in their composition in the Arctic
	Improve knowledge base of hydrographic and oceanographic features of benthic areas (habitat, communities, and species) and effects of fishing in those areas
	Increase transparency and participation of stakeholders, including indigenous communities, in international fisheries management planning
Tourism	Limit the negative effects of visits to any special area and local cultures
	Ensure the tourism activities do not conflict with nature conservation efforts
	Limit the size of visiting groups according to nature and wildlife vulnerabilities of special areas at any time
Nature conservation	Designate ecologically and biologically sensitive areas (EBSAs), e.g., nesting and feeding habitats of seabirds
	Designate marine protected areas
	Designate marine reserves (no take areas)
Indigenous peoples	Identify and designate subsistence hunting areas
	Identify and designate culturally important areas

9.10 What are the Outputs of Marine Spatial Planning?

The principal output of MSP is a comprehensive spatial plan for a marine area or ecosystem. The plan moves the whole marine system toward a 'vision for the future'. It sets out priorities for the area and—more importantly—defines what these priorities mean in time and space. Typically, a comprehensive spatial management plan has a 10–20-year horizon and reflects political and social priorities for the area. The comprehensive marine spatial plan is often implemented through a zoning map, zoning regulations, and/or a permit system similar to a comprehensive regional plan on land. Individual permit decisions made within individual sectors (for example, the fisheries, oil and gas, or tourism sectors) should then be based on the zoning maps and regulations. A strategic environmental assessment (SEA) or programmatic environmental impact statement (PEIS) is usually required before a marine spatial plan is approved by government (see Chap. 8).

MSP does not replace single-sector planning and decision making. Instead it aims to provide guidance for a range of decision makers responsible for particular sectors, activities, or concerns so that they have the means to make decisions confidently in a more comprehensive, integrated, and complementary way.

9.11 Why is Stakeholder Participation Critical to Marine Spatial Planning?

Involving key stakeholders, including indigenous peoples in the Arctic, in the development of MSP is essential for a number of reasons. Of these, the most important is because MSP aims to achieve multiple objectives (social, economic, and ecological) and should therefore reflect as many expectations, opportunities, or conflicts that are occurring in the MSP area. The scope and extent of stakeholder involvement differs greatly from country to country and is often culturally influenced. The level of stakeholder involvement will largely depend on the legal or cultural requirements for participation that often exist in each country.

Generally speaking, all individuals, groups, or organizations that are in one way or another affected, involved, or interested in MSP can be considered stakeholders. However, involving too many stakeholders at the wrong moment or in the wrong form can be very time consuming and can distract resources from the expected or anticipated result. To involve stakeholders effectively (e.g., leading toward expected results) and efficiently (e.g., producing expected results at least-cost), three questions should be asked:

- *Who* should be involved?
- *When* should stakeholders be involved?
- *How* should stakeholders be involved?

Where no legal obligations exist, it is important to define what type of stakeholder participation will be most suitable for a successful result. For instance, involving indigenous people in MSP efforts may not be a legal requirement, but they could however be greatly affected (positively or negatively) by MSP management measures, and should therefore participate.

Wide-ranging and innovative approaches to stakeholder participation and proactive empowerment should be used in the MSP process. Stakeholder participation and involvement in the process should be early, often, and sustained throughout the process. Stakeholder participation and involvement encourages "ownership" of the plan and can engender trust among the various stakeholders. Different types of stakeholder participation should be encouraged at various stages of the MSP process. The key stages at which stakeholders should be involved in the process include:

- *The planning phase*: Stakeholders need to be involved and contribute to the setting of goals and objectives of MSP. They also need to be involved in the evaluation and choice of specific management measure options and the consequences of these choices on their areas of interest;
- *The implementation phase*: Stakeholders should be involved in the actual implementation of MSP and its management measures. For example, an approach to enforcement may be identified that would involve local communities in the regulatory and enforcement process. When the local communities understand the problems and benefits of taking action—and agree upon the management

measures to be taken—they will be part of the enforcement process, at least to the extent of encouraging compliance; and

- *The monitoring and evaluation (post-implementation) phase*: Stakeholders should be involved in the evaluation of the overall effectiveness of MSP in achieving goals and objectives. The post-evaluation effort should involve all stakeholders in a discussion to identify plan results, evaluate results against objectives, and plan for the next round of planning (Pomeroy and Douvere 2008).

9.12 How Can Pan-Arctic Marine Spatial Planning Advance?

While most existing MSP efforts have been focused at the national level, especially in Norway, several current initiatives have a pan-Arctic perspective. For example, the Arctic Council has several spatial data initiatives that are developing regional or circumpolar datasets or to provide a framework that would allow integration, access, and coordination of spatial data on the Arctic. It has discussed the possibility of developing a common interface for access to spatial data. However, existing efforts of the Arctic Council to address cartography, geographic information systems, and spatial analysis have been conducted in isolation with no attempt at harmonization or integration. The working group on Emergency Prevention, Preparedness, and Response (EPPR), the working group on the Conservation of Arctic Flora and Fauna (CAFF), and the Arctic Monitoring and Assessment Program (AMAP) working group have discussed the possibility of developing a common interface for access to spatial data, the first tentative steps toward providing a framework to allow for data standardization and integration within the Arctic Council.

Preliminary work on MSP has begun through the Arctic Council's Protection of the Arctic Marine Environment (PAME) working group Work Plan for 2009–2011 (PAME 2009). It is developing pilot projects to make the large marine ecosystem (LME) assessment approach operational (in the Canadian/US Beaufort Sea and the US/Russian Federation West Bering Sea). However, no current plan exists to develop an integrated marine spatial plan for the pan-Arctic region. Given the working arrangement for PAME (two meetings a year) and the level of investment in its work plan (about $140,000 per year (see PAME 2009)), little progress on a pan-Arctic approach to MSP can be expected over the next few years without a substantial infusion of new resources.

In May 2011, the Arctic Council Ministers established an Expert Group on Arctic ecosystem-based management, recognizing the "…planning and management of human activities on a cross-sectoral basis can assist in reducing conflict among activities and in supporting the conservation and sustainable use of natural resources" (Arctic Council 2012). The Expert Group has met two times and developed a definition of ecosystem-based management and a set of principles. Its work is now focused on compiling EBM best practices, identifying conservation standards, and identifying ecological objectives. The Expert Group will present its findings to the Arctic Council Ministers in 2013.

The International Union for Conservation of Nature (IUCN) and the Natural Resources Defense Council (NRDC) have completed a cooperative Arctic Marine Ecosystem-Based Management Project to explore ways of advancing implementation of ecosystem-based management, and to begin the process of identifying specific ecologically significant and vulnerable marine areas that should be considered for enhanced protection in any new management arrangements (Speer and Laughlin 2011). A workshop held in November 2010 convened 34 scientists and representatives of indigenous peoples with expertise in various aspects of Arctic marine ecosystems and species to identify biologically or ecologically significant or vulnerable habitats using internationally accepted criteria developed under the auspices of the Convention on Biological Diversity (CBD 1992). The seven CBD criteria were: uniqueness; life history importance; importance to endangered/threatened species; vulnerable/fragile/slow recovery areas; areas of high productivity; areas of high diversity; and 'naturalness'. Importance of an area for subsistence or cultural heritage was also considered.

The workshop produced a set of preliminary maps depicting 77 Arctic marine EBSAs based on the CBD criteria. Thirteen 'Super EBSAs' were also identified, i.e., areas that all or almost all of the CBD criteria at a global level of significance. The area around the Bering Strait, for example, was identified as a 'Super EBSA'. Caveats and limitations of the results, including lack of data on some species, were identified in the report of the workshop.

Another way to advance MSP in the Arctic would explicitly recognize the importance of moving beyond sole reliance on the initiatives of national governments and toward a pan-Arctic approach to guide the future of the development. Networks and partnerships of non-governmental actors including indigenous peoples, environmental NGOs, academia, and private industry, all of which have influence over governmental policies and actions, could be used to initiate MSP.

Precedents exist. In Belgium, the University of Ghent laid the groundwork for MSP that was later implemented by the national government. In the US, NGOs and the private sector, particularly new private users of ocean space, e.g., wind farms and offshore aquaculture, have been particularly influential in the development of the national MSP framework. Local indigenous peoples (Coastal First Nations) have led by example the MSP process in British Columbia from the 'bottom up'.

Similarly, indigenous peoples from the Arctic could take the initiative to develop an Arctic-wide approach to MSP through a network of their organizations including the Aleut International Association, the Arctic Athabaskan Council, the Gwich'in Council International, the Inuit Circumpolar Council, the Russian Association of Indigenous Peoples of the North (RAIPON), and the Sámi Council. If capacity building to begin MSP is needed, technical advice could be sought from the Coastal First Nations Planning Office in Vancouver, British Columbia, and the Beaufort Sea Planning Office in Inuvik, Northwest Territories, Canada. Initially the Arctic Council Indigenous Peoples Secretariat in Copenhagen,

Denmark could provide coordination of this initiative. Alternatively, the initiative could be self organizing, relying on the interests and initiative of a few indigenous organizations and their leadership.

Leadership through a MSP initiative by indigenous peoples could provide the basis for other stakeholders, e.g., the business and NGO communities, to collaborate in the planning process. Eventually the Arctic Council and national governments would participate, particularly in the implementation of many spatial and temporal management measures.

References

Arctic Council (2004) Arctic Marine Strategic Plan. 24 Nov 2004. Available at <http://www.pame.is/images/stories/AMSP_files/AMSP-Nov-2004.pdf>. Accessed 5 Feb 2013

Arctic Council (2012) Report to senior Arctic officials. The 2nd Arctic Council Ecosystem-based Management Experts Group Meeting, Gothenburg, Sweden, 16–18 Apr 2012

Arkema KK, Abramson SC, Dewsbury BM (2006) Marine ecosystem-based management: from characterization to implementation. Front Ecol Environ 4(10):525–532

Beaufort Sea Partnership (2009) Integrated Ocean Management Plan for the Beaufort Sea. Beaufort Sea Planning Office, Inuvik

Convention on Biological Diversity (1992) Convention on Biological Diversity, 5 June 1992, 1760 U.N.T.S. 79. Entered into force 29 Dec 1991

Crowder L, Norse E (2008) Essential ecological insights for marine ecosystem-based management and marine spatial planning. Marine Policy 32:772–778

Crowder LB, Osherenko G, Young OR, Airamé S, Norse EA, Baron N et al (2006) Resolving mismatches in US ocean governance. Science 313(5787):617–618

Ehler C, Douvere F (2007) Visions for a Sea Change. Report of the first international workshop on marine spatial planning. Intergovernmental Oceanographic Commission and Man and the Biosphere Programme. IOC Manual and Guides, 46: ICAM Dossier, 3. UNESCO, Paris

Ehler C, Douvere F (2009) Marine Spatial Planning: a step-by-step approach toward ecosystem-based management. Intergovernmental Oceanographic Commission and Man and the Biosphere Programme. IOC Manual and Guides No. 53, ICAM Dossier No. 6. UNESCO, Paris

European Commission (2010) Communication from the Commission to the European Parliament, the Council, the European Economic and Social Committee and the Committee of the Regions: Maritime Spatial Planning in the EU—Achievements and Future Development, COM (2010) 771 final. 12 Dec 2010

European Commission and High Representative (2012) Joint Communication to the European Parliament and the Council. Developing a European Union Policy towards the Arctic Region: progress since 2008 and next steps. JOIN(2012) 19 final. 26 June 2012

Hoel AH (2009) Best practices in ecosystem-based oceans management in the Arctic. Report No. 129 of the Norwegian Polar Institute, Polar Environmental Centre

National Ocean Council (2012) Draft National Ocean Policy Implementation Plan, 12 Jan 2012. <http://www.whitehouse.gov/sites/default/files/microsites/ceq/national_ocean_policy_draft_ implementation_plan_01-12-12.pdf>. Accessed 5 Feb 2013

Norwegian Ministry of the Environment (2005) Integrated Management of the Marine Environment of the Barents Sea and the Sea Areas Off the Lofoten Islands. Report No. 8 to the Storting

Norwegian Ministry of the Environment (2009) Integrated Management of the Marine Environment of the Norwegian Sea. Report No. 37 to the Storting

PAME (2009) PAME Work Plan 2009–2011. <http://www.pame.is/pame-work-plan/work-plan-2009-2011>. Accessed 5 Feb 2013

Pomeroy R, Douvere F (2008) The engagement of stakeholders in the marine spatial planning process. Marine Policy 32:816–822

Speer L, Laughlin T (2011) IUCN/NRDC Workshop to identify areas of ecological and biological significance or vulnerability in the Arctic marine environment. Workshop report. 7 Apr 2011

Spiridonov VA, Gavrilo MV, Krasnova ED, Nikolaeva NG (2011) Atlas of marine and coastal biological diversity of the Russian Arctic. WWF Russia, Moscow

Chapter 10
Marine Protected Areas as a Tool to Ensure Environmental Protection of the Marine Arctic: Legal Aspects

Ingvild Ulrikke Jakobsen

Abstract The sensitive Arctic marine environment, with its fragile ecosystems and habitats, may be threatened by increasing human activities and melting of sea ice due to climate change. Marine protected areas (MPAs) are an environmental tool that, by ensuring a higher level of protection within a defined geographical area, can benefit conservation and sustainable use of marine biological diversity. To ensure conservation of marine biological diversity, a number of states have committed themselves to establishing a coherent network of MPAs within 2012, an objective endorsed also by the Arctic states. In this chapter, political developments with regard to the establishment of MPAs within and outside of national jurisdiction and legal obligations applicable to the marine Arctic are examined. Despite having both legal obligations and political initiative, there is currently no network of MPAs in the marine Arctic. To ensure implementation of global obligations in the Arctic region, there is a need for further operationalization and clarification of scientific criteria for the selection and designation of sites that may contribute to a future network of MPAs. To fulfill the objective of an ecologically coherent network of MPAs in the marine Arctic, stronger efforts must also be made by the Arctic states through the Arctic Council.

10.1 Introduction

The marine environment in the Arctic is vulnerable with its unique ecosystems and habitats, and numerous rare and threatened species (CAFF 2010). As a result of climate change, increasing sea temperatures and melting sea ice have been witnessed in the Arctic. These climatic changes have extensive consequences for the marine environment and Arctic biodiversity, but also provide increased opportunities for human activities and economic development (ACIA 2005). Due to this

I. U. Jakobsen (✉)
University of Tromsø, Tromsø, Norway
e-mail: ingvild.jakobsen@uit.no

E. Tedsen et al. (eds.), *Arctic Marine Governance*,
DOI: 10.1007/978-3-642-38595-7_10, © Springer-Verlag Berlin Heidelberg 2014

development and the expansion of activities—such as for fishing, shipping, oil and gas, and tourism—in combination with climate change, the sensitive Arctic marine environment may face new and increasing threats.

The Convention on Biological Diversity (CBD 1992), along with a number of other international agreements, obliges states to protect and conserve marine biodiversity (hereby ecosystems, habitats, and species). To help meet this obligation, states have committed themselves to establishing representative networks of marine protected areas (MPAs), consistent with international law and based on scientific information, by 2012 (WSSD 2002, para. 32). With climatic changes and increasing impacts from human activities in the marine Arctic, the need for networks of protective MPAs within the Arctic is increasingly apparent (CAFF 2010).

The aim of this chapter is to provide an introduction to MPAs as a legal environmental management tool to ensure protection of the marine environment and conservation of biological diversity and ecosystems in the marine Arctic. The chapter examines existing global and regional obligations applicable to the marine Arctic on the conservation and sustainable use of biological diversity and ecosystems, to be met through the use of MPAs (see AOR 2011). MPAs have been on the global political agenda for the past two decades, during which time the legal regime for MPAs has further developed through interactions between political and legal instruments. This chapter therefore reviews political developments with regard to MPAs as well.

The maritime areas of the Arctic are subject to different legal regimes, ranging from territorial seas—subjected to the sovereignty of the coastal states -, to Exclusive Economic Zones (EEZs)—where coastal states enjoy sovereign rights over natural resources and other states enjoy the right of navigation -, to the high seas—where the principle of the freedom of the seas applies.[1] Even though most of the marine Arctic is within the maritime zones of the Arctic coastal states, there are four high seas pockets in the marine Arctic.[2] Living resources that are important for the conservation of biological diversity and ecosystems move between areas both within and beyond the limits of national jurisdiction. Analyses here thus include the possibilities of establishing MPAs in areas beyond national jurisdiction.

10.2 MPAs as an Environmental Tool

10.2.1 Defining MPAs

There is no formal definition of 'MPA'. The term is defined in different ways in a variety of international and regional instruments, thus, there are various types as

---

[1] As discussed in Chap. 1, there are many different definitions for areas constitute the marine Arctic. Regarding conservation of biological diversity, there are good reasons for adopting a wide definition of the 'marine Arctic'. In this chapter, the same definition as established in Chap. 1 and throughout this book, which includes the Arctic Ocean and its adjoining seas, is applied.

[2] The 'Banana hole' in the Norwegian Sea, the 'Loop Hole' in the Barents Sea, the 'Donut Hole' in the Bering Sea, and the 'Central Arctic Ocean'.

well as varying definitions of MPAs. The International Union for Conservation of Nature (IUCN) has adopted a definition that is often applied and has served as the basis for definitions of MPAs within certain legal instruments:

> Any area of inter tidal or sub tidal terrain, together with its overlaying water and associated flora and fauna, historical and cultural features, which has been reserved by law or other effective measures to protect part or all of the enclosed environment (IUCN/WCMC 1994).

Under the CBD, 'protected areas' is defined as a "geographically defined area which is designated or regulated and managed to achieve specific conservation objectives" (CBD 1992, art. 2). Under the OSPAR Convention (1992) MPAs are defined as:

> [...] an area within the maritime area for which protective, conservation, restorative or precautionary measures, consistent with international law have been instituted for the purpose of protecting and conserving species, habitats, ecosystems or ecological processes of the marine environment (OSPAR Commission 2003a, para. 1.1.).

Generally, MPAs can be characterized as geographically limited areas that are protected to achieve one or more conservation objectives through regulations and prohibitions on human activities which may threaten or damage the environment (Koivurova and Molenaar 2009). A prohibition on all human activities is not necessary for classification as an MPA. Rather, the level of restrictions or prohibitions that apply in an MPA can vary from a total ban on all human activities to more moderate limitations for sustainable use of natural resources (IUCN 2011).

The focus of this chapter is on integrated, cross-sectoral MPAs. These are MPAs where all activities that may threaten conservation objectives may be regulated or prohibited. Related concepts that do not cover all activities, but aim to protect a specific area against one particular human activity, such as areas closed for fisheries or areas where shipping is regulated (e.g., MARPOL 'special areas' and particularly sensitive sea areas (PSSAs)) are not addressed here (see Tanaka 2012; see also Chap. 6).

10.2.2 Potential Benefits of Using MPAs

MPAs help to ensure a higher level of protection of a defined geographical area from the environmental impacts of human activities, to ensure conservation and sustainable use of ecosystems, habitats, and sensitive and vulnerable marine species. Conservation of Arctic marine biological diversity is important not only for the Arctic region, but also for global biological diversity. Many species, such as migratory sea birds and mammals, use the marine Arctic seasonally, as Arctic habitats provide vital resources (WWF 2012). As explained, climate change is emerging and identified as a significant threat to the Arctic's biodiversity (CAFF 2010). Even though the establishment of MPAs may not address sources of greenhouse gas emissions, MPAs may nonetheless serve as tool for mitigation and adaptation to climate change. Through the establishment of protected areas, the total pressures on ecosystems may be controlled, thus strengthening ecosystems and making them more adaptive and resilient in the face of climate change. MPAs have also been emphasised as a mitigation tool, as MPAs may be designed to protect marine ecosystems that serve as sinks and reservoirs of greenhouse gases. The CBD has emphasised the role of MPAs in this respect.[3]

MPAs are, however, accompanied by certain limitations. Not all activities that cause environmental damage take place within areas that are protected, and therefore these activities may not be regulated within an MPA. For example, pollution from land-based activities is considered a major threat to the marine environment and biological diversity in the Arctic, but may fall outside of protected areas. Additionally, marine species move across the geographical limits of MPAs and between different legal regimes. It is therefore important that MPAs are used in addition to other traditional management structures and regulation of activities that may cause damage to the marine environment. To achieve ecosystem-based management of the marine Arctic, MPAs must be established and managed together with other tools that provide for integrated management of human activities, such as marine spatial planning.

10.3 International Framework for MPAs

10.3.1 Introduction

In this section, the competence to establish MPAs in maritime zones within and beyond national jurisdiction is reviewed. Next, the political development of MPAs is examined, followed by a review of significant global obligations and developments. Finally, regional obligations and cooperation between the Arctic states are investigated.[4]

10.3.2 The Establishment and Management of MPAs Under the LOS Convention

The competence to establish and manage MPAs is based on the United Nations Law of the Sea Convention (LOS Convention 1982). MPA regimes are closely linked to the jurisdiction and powers that states can exercise in maritime zones

---

[3] See CBD COP Decision VII/5, para. 8, Seventh Ordinary Meeting of the Conference of the Parties to the Convention on Biological Diversity, 9–20 Feb 2004, Kuala Lumpur. CBD COP Decision IX/33, para. 8, Ninth meeting of the Conference of the Parties to the Convention on Biological Diversity, 19–30 May 2008, Bonn.

[4] See Jakobsen (2012).

under the LOS Convention. Whereas states enjoy full sovereignty over land-based territory, the situation is different in the sea.

In the territorial sea, coastal states enjoy sovereignty and therefore may, in this zone, adopt any regulations on human activities to protect marine biodiversity, including the creation of MPAs. However, MPAs can only be adopted subject to the rights of other states to enjoy innocent passage (LOS Convention 1982, art. 17). As coastal states enjoy sovereignty in the territorial sea, this provides jurisdictional competence to regulate innocent passage, such as for purposes of environmental protection (LOS Convention 1982, arts. 21 and 22).

In the EEZ, coastal states enjoy sovereign rights over marine resources and jurisdiction over the marine environment (LOS Convention 1982, art. 56(1)(a), (b) (iii)). A coastal state may therefore, for the purposes of conservation of living resources and environmental protection, adopt MPAs within its EEZ.[5] Other states, however, continue to enjoy the freedom of navigation and the navigational rights of other states must be respected. Due to jurisdiction over the marine environment, coastal states have the authority to adopt regulations on shipping for the purpose of protecting the marine environment against marine pollution. These are, pursuant to Article 211(5) of the LOS Convention, limited to regulations that are "...conforming to and giving effect to generally accepted international rules and standards established through the competent international organization...". The competent international organization referred to here is the International Maritime Organization (IMO).

On the continental shelf, coastal states have, as set out in Article 77, sovereign rights "...for the purpose of exploring it and exploiting its natural resources". Natural resources consist of mineral and other non-living resources together with "...living organisms belonging to sedentary species..." (LOS Convention 1982, art. 77(4)).[6] The wording of "exploring and exploiting" is wide enough to include a right for coastal states to adopt measures for management and conservation of natural resources (Frank 2007). Thus, coastal states may establish MPAs to restrict fishing for sedentary species or other activities, such as mining and oil and gas extraction, that may threaten sedentary species (Frank 2007).

Such MPAs may also be adopted for the outer continental shelf, where coastal states have delineated the outer limits on the basis of the requirements of the Commission on the Limits of the Continental shelf (LOS Convention 1982, art. 76(8)). Regulations adopted for MPAs on the continental shelf must not, however, according to Article 78(2), "...infringe or result in any unjustifiable interference with navigation and other rights and freedoms of other States..." provided for by the LOS Convention. In relation to this, it can be questioned to what extent a

---

[5] For a discussion of the competence to establish MPAs in the EEZ, see Frank (2007), Lagoni (2003), and Jakobsen (2010).

[6] Sedentary species are defined in the provision as "...organisms which, at the harvestable stage, either are immobile on or under the seabed or are unable to move except in constant physical contact with these a-bed or the subsoil".

coastal state may regulate fishing activities on the outer continental shelf, where sedentary species are not the target species, but rather where the regulated activity may threaten sedentary species. On the one hand, it is natural that coastal states—when exercising their sovereign rights—may create an MPA where activities that may threaten sedentary species, such as bottom trawling, are restricted. This may, however, easily come into conflict with the freedoms of the high seas. Therefore, to be legitimate, such regulations would have to be based on a careful consideration of these conflicting interests (Frank 2007). A more appropriate way for coastal states may be to approach the relevant Regional Fisheries Management Organization (RFMO) with competence over the destructive fishing activities and call upon the adoption of appropriate regulations.

The principle of freedom of the high seas means that all states enjoy certain freedoms in high sea areas, such as the freedom of navigation, over flight, to lay submarine cables and pipeline, fishing, or scientific research (LOS Convention, Article 87). As no states have sovereignty or sovereign rights over marine resources on the high seas, no states are entitled to unilaterally define and designate high seas areas as MPAs or to adopt regulations on foreign vessels or other activities. To date, there are no legal regimes that provide for MPAs on the high seas.[7] If some states, for instance the Arctic states, were to agree to take measures to conserve an ecosystem in the high seas through the establishment of MPAs, the resulting restrictions could not apply and be binding on states that were non-parties to the agreement. Even so, the creation of MPAs on the high seas is controversial due to the lack of a clear legal basis in international law. To ensure legitimacy of MPAs established on the high seas on the basis of an agreement between states, careful consideration must be given both to the selection of areas and to compatibility with the freedoms of the high seas (Tanaka 2012). In addition to cooperation between states, MPAs on the high seas require both coordination and cooperation between different competent organizations (Molenaar and Oude Elferink 2009).

The 'Area' is defined in the LOS Convention Article 1(1) as the "…seabed and ocean floor and subsoil thereof beyond the limits of national jurisdiction". The Area and its resources are the common heritage of mankind (LOS Convention 1982, art. 136) When an MPA is established in an area beyond national jurisdiction, it must be determined whether an activity that desired for regulation is subject to the legal regime of the high seas, or to the legal regime of the Area. It is, for instance, not clear which legal regime that is applicable to the living resources of the Area. Questions related to the relationship between these two legal regimes will not, however be dealt with any further here.[8]

---

[7] Note that while MPAs on the high seas are established under the OSPAR Convention, none of these are within the maritime areas of the Arctic.

[8] For a discussion of this, see Molenaar and Oude Elferink (2009).

10.3.3 Global Political Developments with Regard to MPAs

Agenda 21 and the WSSD Plan of Implementation

Agenda 21, the United Nations (UN) Programme of Action, was adopted at the Rio Conference in 1992 and encourages states to take a more holistic approach to ocean management that is to be "…integrated in content and are precautionary and anticipatory in ambit" (Agenda 21 1992, para. 17.1). MPAs are emphasised as a significant tool and a core element of this new approach. Agenda 21 stresses the importance of protecting species, habitats, and ecologically sensitive areas, both within and beyond areas of national jurisdiction (Agenda 21 1992, para. 17.46(e) and (f), 17.75(e) and (f)).

The significance of adopting an integrated ecosystem-based management system of the oceans and of establishing MPAs was confirmed in the Word Summit on Sustainable Development (WSSD) Plan of Implementation, adopted in Johannesburg in 2002. In particular, the WSSD Plan of Implementation declares that states should:

> …maintain the productivity and biodiversity of important and vulnerable marine and coastal areas, including in areas within and beyond national jurisdiction.

and

> …develop and facilitate the use of diverse approaches and tools, (…) including (…) the establishment of marine protected areas, consistent with international law and based on scientific information, including representative networks by 2012… (WSSD 2002, para. 32c).

These stated objectives have had significance for the legal development of MPAs. As a response to the Plan's commitment, the objective to establish MPAs has been endorsed in legal instruments such as the CBD and OSPAR Convention, as well as within EU law.[9] Further, to achieve the objective, it has been transformed into more specific goals within the frameworks of instruments such as through the adoption of decisions, recommendations, and guidelines adopted by the Conference of Parties (COP) of CBD and the OSPAR Commission.

United Nations General Assembly

While the LOS Convention does not contain any specific references to the use of MPAs, the UN General Assembly (UNGA) has in recent years addressed the use of MPAs as a tool to protect oceans and to ensure conservation of biological diversity (UNGA 2008, para. 212–135). The UNGA has reaffirmed the need to develop and facilitate tools for conserving and protecting vulnerable marine ecosystems, including the establishment of MPAs and the development of representative networks of

---

[9] See, for instance, the EU's Marine Strategic Framework Directive (2008).

marine protected areas within 2012 (UNGA 2008, para. 134). In the latest resolution of significance here, UNGA encourages states to:

...further progress towards the 2012 target for the establishment of marine protected areas, including representative networks, and calls upon States to further consider options to identify and protect ecologically or biologically significant areas, consistent with international law and on the basis of the best available scientific information (UNGA 2011, para. 178).

The UNGA has also addressed conservation of biological diversity in areas beyond national jurisdiction, reaffirming its central role. At the same time that the UNGA reaffirmed its role in this area, it made note of the work performed under the CBD (UNGA 2011, para. 172). The UNGA's latest resolution recalls that the COP has adopted scientific criteria for identifying ecologically or biologically significant marine areas in need of protection in open-ocean waters and deep-sea habitats in addition to scientific guidance for selecting areas to establish a representative network of marine protected areas, including in open-ocean waters and deep-sea habitats (UNGA 2011, para. 179).

In 2004, the UNGA requested the appointment of an Ad Hoc Open-ended Informal Working Group to study issues relating to the conservation and sustainable use of marine biological diversity. At its fourth meeting in 2011, the Working Group adopted by consensus a set of recommendations, requesting a process to be initiated by the UNGA with a view to the "possible development of a multilateral agreement" under the LOS Convention to ensure conservation and sustainable use of the biological diversity in areas beyond national jurisdiction (IISD 2012). It was also recommended that the process address conservation and sustainable use of biological diversity in areas beyond national jurisdiction (by the use of MPAs, for instance).[10] At the group's fifth meeting, the issue of an implementation agreement under the LOS Convention was discussed. The EU called for an implementing agreement that provides for a global mechanism for establishing MPAs in areas beyond national jurisdiction (IISD 2012). While the Working Group achieved consensus with regard to the recommendation of a "possible development of a multilateral agreement under" the LOS Convention, it did not yet manage to agree on a process forward to develop the legal framework (IISD 2012). At the 2012 Rio + 20 Conference, states gave support to the development of a new implementation agreement and agreed to initiate, as soon as possible, the negotiation of an implementing agreement to the LOS Convention that addressed the conservation and sustainable use of marine biodiversity in areas beyond national jurisdiction (UNGA 2012, para. 181; United Nations 2012).

Following these developments, there appears to be broad support in the international community for creating such an implementation agreement. This is significant for the Arctic Ocean, as parts of the marine Arctic are beyond national jurisdiction. Moreover, with the melting of sea ice, large areas in the Central Arctic Ocean will become ice free, at least in the summer months, and will be part of the high seas. With the prospect of increased human activity in these areas, it

---

[10] These recommendations were endorsed in UNGA (2011) This obligation is reasonable, para. 167.

will be necessary to identify sensitive or valuable areas that need protection and that may constitute a part of a global network of MPAs. This requires clarification of the legal basis and competence to establish and manage MPAs on the high seas.

10.3.4 Global Legal Obligations with Regard to MPAs

The LOS Convention and the Obligation to Protect and Preserve the Marine Environment

The LOS Convention aims to establish a "legal order for the oceans" (preamble). The LOS Convention part XII includes legal obligations to protect and conserve the marine environment that apply to all states in all maritime zones. In the Ilulissat Declaration (2008), the five Arctic coastal states declared that the law of the sea is the relevant legal regime for the management of the Arctic and also acknowledged that the Arctic Ocean is a unique ecosystem that they have a particular responsibility to protect. The states expressed a commitment to:

> ...take steps in accordance with international law both nationally and in cooperation among the five states and other interested parties to ensure protection and preservation of the fragile marine environment of the Arctic Ocean (Ilulissat Declaration 2008).

In the light of these statements and interest in the development of a framework for MPAs, it is of interest to examine whether the general obligations of the LOS Convention also contain obligations to use this particular tool or to provide for its use.

Does the obligation of LOS Convention Article 192 contain a duty for states to adopt MPAs in order to ensure conservation of biological diversity within their maritime zones? It follows from the Article that states "have the obligation to protect and preserve the marine environment". This obligation is reasonable to understand as providing states with the discretion to decide which measures and tools to adopt to comply with it. A legal duty to adopt MPAs would thus mean constraining the discretion of the states when it comes to the choice of means to fulfil this obligation (Jakobsen 2010).

'Biological diversity' and 'ecosystems' are relatively new legal terms which were introduced to international law through the CBD in particular. A relevant question in relation to LOS Convention Article 192 and the use of MPAs is whether the wording "marine environment" also covers the concept of 'biological diversity'. The term 'marine environment' is not defined in the LOS Convention, but is arguably wide enough to also cover biological diversity, e.g., the diversity within species, between species, and of ecosystems.[11] States are also, according to

---

[11] Biological diversity is defined in the CBD as "the variability among living organisms from all sources including, *inter alia,* terrestrial, marine and other aquatic ecosystems and the ecological complexes of which they are part: this includes diversity within species, between species and of ecosystems" (CBD 1992, art. 2).

Article 194(5), required to adopt measures "...necessary to protect and preserve rare or fragile ecosystems as well as the habitat of depleted, threatened or endangered species and other forms of marine life." A contextual interpretation of Article 192 implies that the 'marine environment' includes protection of 'ecosystems', 'fragile habitats', and 'marine life'.

Following the adoption of the LOS Convention, there has been development from the protection of the environment against single activities, such as fishing and pollution, to a broader approach to protection of the environment, whereby ecosystems and biological diversity are conserved and protected (Birnie et al. 2009). Subsequent instruments, such as the CBD and Agenda 21, spell out the need to take a holistic ecosystem-based approach to the regulation of human activities together with an understanding of the natural variations of species and ecosystems. According to the Vienna Convention (1969), the term 'marine environment' must be interpreted in light of these 'new' developments (art. 31(3) (c)). It follows from the Vienna Convention that when a "rule" is interpreted, "any relevant rules of international law applicable in the relations between the parties" shall be applied as a source of interpretation. Such dynamic interpretation also finds support in international case law,[12] although the suggested interpretation must, however, remain within the wording and objective of the convention (Boyle 2006). An interpretation where 'marine environment' covers the newer legal term 'biological diversity' is acceptable within the wording of the LOS Convention.[13]

An MPA is a cross-sectoral measure where all potentially damaging activities may be restricted to protect biological diversity. Related is the question of whether Article 192 requires states to protect the marine environment against all human activities. It can be held that Part XII of the LOS Convention deals primarily with marine pollution. On the other hand, the wording of 'protection' and 'preservation' implies that states must adopt a broader approach in order to comply with the obligation (Nordquist and Rosenne 1991). Article 193 suggests that the conservation of living resources is part of protection of the marine environment, as is also confirmed in international case law.[14] Formulations in the preamble to the LOS Convention such as "the problems of the ocean space are closely interrelated and need to be considered as a whole" and that the aim is to establish "a legal order for the oceans", indicate that Article 192 is an obligation to protect the marine environment against all activities that may cause damage, not only pollution. This could include, for example, a duty to ensure that coral reefs are not damaged by shipping or oil and gas activities.

---

[12] See Namibia Advisory Opinion, *ICJ Reports* 1971, 16: 31; Case concerning the Gabcikovo-Nagymaros Project (Hungary v. Slovakia), *ICJ Reports* 1997, para. 140; and the arbitration case between Belgium and the Netherlands: *The Arbitration Regarding the Iron Rhine*, The Permanent Court of Arbitration, 2005, para. 59.

[13] The relationship between the CBD and the LOS Convention is addressed in CBD Article 22.

[14] Southern bluefin cases (New Zealand v. Japan; Australia v. Japan), 38 *ITLOS Report* 1724 (1999), para. 170.

Consequently, the obligation in Article 192 does not give any direction regarding which tools states shall apply to achieve protection of the marine environment and states are free to choose which measures and tools they want to apply to fulfil the obligation, including MPAs. Article 194(5) specifically deals with measures to protect the marine environment against pollution, although an interpretation of Article 192 in the light of Article 194(5) suggests that there is a duty for states to adopt measures to protect a sensitive ecosystem that is endangered by pollution or other threats. Even though Article 192 does not contain a general duty to establish MPAs, the use of this tool is consistent with legal and political developments, and is an appropriate way to comply with the obligation. Article 192 is not an absolute obligation and it does not require that all activities that may cause damage to the marine environment be prohibited.

The CBD and MPAs

All of the Arctic states are parties to the CBD, with the exception of the United States (US).[15] The objective of the Convention is to ensure the conservation of biological diversity, sustainable use of its components, and the fair and equitable sharing of the benefits arising from genetic resources (CBD 1992, art. 1).

CBD Article 8 includes different measures to ensure in situ conservation of biological diversity. In situ conservation is defined in Article 1 as:

> ...the conservation of ecosystems and natural habitats and the maintenance and recovery of viable populations of species in their natural surroundings and, in the case of domesticated or cultivated species, in the surroundings where they have developed their distinctive properties.

Several of the obligations in Article 8 deal with protected areas (CBD 1992, art. 8(a), (b), (c) and (e)). States are, according to Article 8(a), required to "as far as possible and as appropriate" establish a system of protected areas. The wording "system of protected areas" can be understood as a 'network', and implies that states should establish protected areas in a systematic way, as part of a wider plan for conservation of biodiversity. The obligation to establish protected areas applies, pursuant to Article 4(a), in maritime zones within national jurisdiction.[16] The obligation to establish and manage MPAs must, however, be implemented so that it does not infringe with the rights and duties of other states pursuant to the LOS Convention, as follows from CBD Article 22(2).[17]

It follows from Article 8 that states "shall" establish protected areas. This indicates a legal duty to establish protected areas—not something states may choose to do

---

[15] For an overview of the member states, see <www.cbd.int/convention/parties/list>.

[16] As for areas outside national jurisdiction, it follows from Article 4(b) that the CBD applies in the case of "...processes and activities, regardless of where their effects occur, carried out under its jurisdiction or control, within the area of its national jurisdiction or beyond the limits of national jurisdiction".

[17] Regarding Article 22, see Wolfrum and Matz (2000) and Jakobsen (2010).

when or if they find it "possible" and "appropriate".[18] Arguably, "as far as possible" points to the level of implementation of this duty. States are committed to comply with the obligation in accordance with their conditions and capabilities (Burhenne-Guilmin and Casey-Lefkowitz 1992). This is consistent with the principle of differentiated responsibility.[19] The other element of the clause, "as appropriate", is reasonable to understand as a reference to the manner of implementation, providing states with discretion as to how and where MPAs should be established.[20] This is supported by Article 8(b) which leaves it to the states to develop guidelines for "the selection, establishment and management of protect areas…".

Article 8 does not provide guidance as to what areas that should be selected, what human activities should be prohibited or regulated, or how strict regulations adopted to protect and conserve marine biodiversity should be. Restrictions do apply however, and regulations within protected areas must be designed to meet the objectives of Article 8 and of the CBD.

Requirements as to the content of MPAs may be understood through interpretation of relevant sources, such as the Article's objectives, relevant principles such as the precautionary principle, principle of sustainable development, and the ecosystem approach, and a general legal obligation to ensure conservation and sustainable use of biological diversity (Jakobsen 2010). Work within the CBD regime is also of significance.

The CBD COP has adopted several legally non-binding decisions of importance for MPAs, elaborating on criteria for the selection and management of MPAs.[21] The COP has adopted scientific criteria for identifying ecologically or biologically significant marine areas in need of protection in open ocean waters and deep sea habitats.[22] With regard to management of MPAs, the COP has established that the MPAs should be "…effectively managed, ecologically based […] where human activities are managed […] to maintain the structure and functioning of the full range of marine and coastal ecosystems".[23]

Article 8(l) is also important when defining legal requirements for MPAs. According to Article 8(l) "…where a significant adverse effect on biological diversity has been determined pursuant to Article 7…" the states shall as far as possible and as appropriate "…regulate or manage the relevant processes and categories of activities…". This implies a duty for states to take actions when activities have "significant adverse effects". Within an MPA, which is protected due to its sensitive or vulnerable ecosystems, activities may have a negative effect on

---

[18] The interpretation and application of the clause "as far as possible and as appropriate" is discussed in Jakobsen (2010).

[19] Regarding the principle of differentiated responsibility, see Birnie et al. (2009).

[20] This interpretation is also adopted by the expert committee who proposed the Norwegian Nature Management Act of 2009, see NOU (2004):28, p. 160.

[21] The COP decisions are available at <http://www.cbd.int/>.

[22] CBD COP Decision IX/20. Ninth meeting of the Conference of the Parties, 19–30 May 2008, Bonn.

[23] CBD COP Decision VII/5, *supra* note 3, para. 18.

biological diversity and therefore states should adopt appropriate regulations for these activities. The duty in Article 8(l) can, for instance, imply that a state must approach the IMO for approval to adopt traffic regulation measures such as a sea lane to direct the traffic outside a sensitive area in the EEZ, if shipping is found to have a "significant adverse effect" on biological diversity. This is also set out in a decision by the COP urging states to address threats to achieve an effective management of MPAs.[24]

A reading of Articles 4(b), 8, and 22 suggests that the obligation to establish MPAs under the CBD only applies in maritime zones within national jurisdiction. The need to establish MPAs on the high seas has nevertheless been emphasised under the CBD.[25] The COP acknowledges that the CBD has a key role in supporting the work of the UNGA to ensure conservation of marine biological diversity in areas beyond national jurisdiction.[26]

10.3.5 Regional Obligations and Cooperation Between the Arctic States on MPAs

General

The section analyzes how global obligations to protect and conserve biological diversity through MPAs are implemented in the marine Arctic. These obligations are generally to be implemented by states at the national level; however, marine ecosystems are large and species migrate across the maritime zones of different states. Further, many threats to biological diversity are transboundary in nature. This raises jurisdictional issues and requires that, in order for protection to be successful, states in the Arctic cooperate in establishing MPAs.

There is no comprehensive regional agreement for the marine Arctic. The OSPAR Convention (1992) applies only to the marine environment in the North East Atlantic (art. 1). The Arctic states of Norway, Denmark, Iceland, Sweden, and Finland are together with Belgium, France, Ireland, Luxembourg, the Netherlands, Portugal, Spain, Switzerland, and the United Kingdom, as well as the European Community, contracting parties to the Convention. Russia, however, is not a contracting party, which means that not all of the European marine Arctic is covered.

Without a comprehensive regional agreement in the marine Arctic, and due to the fact that not all of the Arctic states are parties to relevant global conventions, political cooperation on environmental protection becomes especially important. The Arctic Council was established in 1996 with protection of the environment as

---

[24] CBD COP Decision VII/5, ibid, para. 26.

[25] CBD COP Decision VII/24, para. 35–47, Seventh Ordinary Meeting of the Conference of the Parties, 9–20 Feb 2004, Kuala Lumpur.

[26] CBD COP Decision IX/20, *supra* note 23.

one of its main objectives and serves as a forum for high level political cooperation between the Arctic states (Ottawa Declaration 1996). The Arctic Council does
not possess the competence to adopt legally binding regulations and has been
described as a consensus-based and project-driven apparatus, rather than an operational body (Koivurova and Molenaar 2009). In recent years, however, the Arctic
Council has begun contributing to the development of legally binding regulations
and agreements.[27] The Arctic Council also serves a role in implementing the
global obligations of states at the regional level (Henriksen 2010). On this basis,
political cooperation under the Arctic Council to conserve biological diversity
through the use of MPAs is included in this analysis.

The OSPAR Convention

The OSPAR Convention contains principles and obligations to ensure protection of the marine environment and marine biodiversity in the northeast Atlantic.
The Convention applies to parts of the Arctic Ocean both within and beyond
national jurisdiction (OSPAR Convention 1992, art. 1(a)). Under Article 2 of the
Convention, states are under a general obligation to take "all possible steps to prevent and eliminate pollution" and furthermore to take "the necessary measures to
protect the maritime area against the adverse effects of human activities" to safeguard the humans health and to conserve marine ecosystems.

In 1998, OSPAR Convention Annex V on the protection and conservation of the
ecosystems and biological diversity was adopted. Article 2(a), Annex V declares
an obligation whereby states shall take "the necessary measures to protect and
conserve the ecosystems and the biological diversity of the maritime area". This
wording may cover the use of MPAs.

The OSPAR Commission is under a duty "to develop means, consistent with
international law, for instituting protective, conservation, restorative or precautionary
measures related to specific areas or sites or related to particular species or habitats"
(OSPAR Convention 1992, Annex V, art. 3). On this basis, the Commission has
adopted guidelines for the establishment of MPAs, resolving as a goal to establish a
representative network of marine protected areas within 2012 in the OSPAR maritime area (OSPAR Commission 2003/3). The recommendation of the Commission
is consistent with the obligations of CBD Article 8(a) and the objectives of the
WSSD. Even though the recommendation on MPAs is not legally binding, it is significant for the interpretation of Article 2, Annex V.[28] All of the contracting parties to
the OSPAR Convention are also parties to the CBD. It is explicit in Article 2 that
states shall take necessary measures to protect and conserve ecosystems and biological diversity to meet their obligations under the OSPAR Convention, *as well as* to to

---

[27] An example of this is the agreement on search and rescue which is negotiated under the auspices of the Arctic Council (Arctic SAR Agreement 2011). See also Chap. 1.

[28] The Commission has the competence to adopt both legally binding decisions and non-legally
binding recommendations (OSPAR Convention 1992, art. 10(3) and 13).

fulfil their obligations under the CBD to "to develop strategies, plans or programmes for the conservation and sustainable use of biological diversity". Consequently, the obligation in Article 2, Annex V must be interpreted in light of the CBD and also the obligation to adopt protected areas under CBD Article 8(a). It is reasonable to understand the wording of "take the necessary measures" in Article 2, Annex V to allow states to retain discretion in choosing the means of compliance; nevertheless, there is strong political will under OSPAR to establish MPAs that are part of an ecologically coherent network of MPAs in the OSPAR maritime area. Even though Article 2, Annex V does not include a general duty to establish MPAs, MPAs could be considered as part of states' obligations.

The OSPAR Commission has, in order to assist the contracting parties in establishing MPAs in their domestic maritime zones, adopted recommendations on identifying and selecting areas that may be included in the network as well as recommendations on MPA management (OSPAR Commission 2003b, c). Moreover, the Commission has adopted a guidance document for interpreting the concept of a "ecologically coherent network" (OSPAR Commission 2006). The contracting parties may take this document into account when selecting sites for the OSPAR MPA network.

When it comes to the management and regulation of activities within MPAs, the OSPAR Convention takes a broad, holistic approach, as it aims to protect the marine environment and ecosystems against human activities. There are, however, explicit exceptions made for fisheries management and shipping (OSPAR Convention, preamble and Annex V, art. 4). This means that the Commission cannot adopt recommendations and decisions where restricting these activities, as questions relating to the management of fisheries are to be regulated under international and regional agreements dealing specifically with such questions, and authority is similarly ceded to the IMO as the competent international body for international shipping regulation. Nevertheless, these activities are included in assessments of the quality and status of the maritime area (Molenaar and Oude Elferink 2009). Moreover, the precautionary principle is part of the Convention and OSPAR takes an ecosystem approach to the protection of the marine environment, suggesting that when adopting MPAs, states are obliged to cooperate through relevant organizations such as the IMO and RFMOs to ensure appropriate protection of MPAs. One may argue that a duty could be interpreted under the OSPAR Convention for a state to cooperate with the IMO and obtain approval for a traffic regulation, such as a sea lane in the EEZ, to ensure protection of an MPA against shipping activities.[29]

OSPAR has made progress on the establishment of MPAs on the high seas. The recommendations by the Commission on MPAs also address MPAs on the high seas. In 2009, the Commission identified eight areas as potential high seas MPAs and in 2010, six areas were established as MPAs in areas beyond national jurisdiction

---

[29] In the Guidelines for the Management of Marine Protected Areas in the OSPAR Maritime Area, 2003/18, both fisheries and shipping are mentioned as examples of activities that may require regulation.

(OSPAR Commission 2011). None of these MPAs are, however, established within the marine Arctic areas.

Overall, progress on the OSPAR network is underway, but the objective of an ecologically coherent network of MPAs has not yet been achieved.

The Arctic Council

The establishment of a network of protected areas was emphasised at the outset of the Arctic Council through the Arctic Environmental Protection Strategy (AEPS 1991).

The work carried out by Arctic Council's Conservation of Arctic Flora and Fauna (CAFF) working group has provided critical knowledge about biological diversity in the Arctic. On this basis, the Circumpolar Protected Areas Network (CPAN), a strategy and action plan for the establishment of a circumpolar network of MPAs, was adopted in 1998.[30] The use of protected areas is recognized by CPAN as an effective and necessary tool to ensure conservation and sustainable use of biological diversity (CAFF 1996a). The goal of CPAN is to establish "an adequate and well managed network of protected areas that has a high probability of maintaining the dynamic biodiversity of the Arctic region" (CAFF 1996a). The network shall, as far as possible, cover the large variation of ecosystems in the Arctic (CAFF 1996a). CPAN does not include any legal obligations, but aims to provide a common framework for states for the selection and management of protected areas to ensure protection of significant areas at the national, regional, and circumpolar level (CAFF 1996a). Under CPAN, guidelines for the selection and designation of protected areas were adopted in 1996 (CAFF 1996b). It is emphasised in the CPAN strategy and action plan that an additional goal is to contribute to states' obligation pursuant to the CBD Article 8(a).

The Arctic Marine Strategic Plan was developed by the working group Protection of the Marine Environment (PAME) and adopted by the Arctic Council in 2004 (Arctic Council 2004). The objective of the strategy is to protect the marine environment, marine biological diversity, ecosystems, and their functions. The use of MPAs as an environmental tool is recognized and emphasised in the plan. The objective of the WSSD of establishing MPAs is repeated here, stating that the Arctic Council shall:

> Promote WSSD actions related to the marine and coastal environment, including the application of an ecosystem approach and establishment of marine protected areas, including representative networks (Arctic Council 2004).

Work towards establishing protected areas that are part of a circumpolar network is documented through reports from the states to CAFF in 1997 and 2004 (CAFF 1997, 2004). In spite of the legal obligations of the states in the Arctic

---

[30] Information and relevant documents and publications about CPAN are available at <http://www.caff.is/protected-areas-cpan>. See also Koivurova (2009).

and that the significance of establishing protected areas was emphasised as early as 1996 with the establishment of the Arctic Council, few marine protected areas have been adopted in the region.[31] As the current work plans for CAFF and PAME do not include action on MPAs, this indicates that MPAs are not currently a priority area for the Arctic Council (CAFF 2009, 2011; PAME 2011).

A CPAN task at the national level is to identify gaps in national networks of protected areas and select candidate sites for further action (CAFF 1996a). It is also emphasised that the guidelines for selection and identification of protected areas are to be used by each Arctic state within its own legislative framework (CAFF 1996b). The guidelines are broad and do not provide clear criteria for site selection. Although MPAs must be implemented at the national level, this suggests that the Arctic states have not been willing to commit to a common understanding on the selection and management of sites for Arctic network of MPAs. Stronger commitments from the Arctic states when it comes to selection and designation of marine areas would strengthen the chances of achieving the objective of a circumpolar protected area network.

10.4 Conclusions

Despite legal obligations and political initiative at the global and regional levels, there is currently no network of MPAs in the marine Arctic. Given the transboundary character of biological diversity and of environmental threats, states must cooperate in order to successfully protect the marine environment. The work carried out under the CBD in elaborating and operationalizing obligations to establish MPAs is significant for achieving the objective of a global network of MPAs. Moreover, work under the CBD and UNGA to develop an implementation agreement is crucial for creating the necessary legal basis for conservation of marine biodiversity in areas beyond national jurisdiction.

Global obligations must be successfully implemented by the Arctic states at the regional and national levels. To achieve a network of MPAs in the Arctic, as part of a wider global network, there is a need for further operationalization of global and regional obligations by clarifying scientific criteria for the selection and management of areas as potential MPAs in this network. This would assist Arctic states in overcoming jurisdictional and sectoral challenges that may hamper the designation and effective management of MPAs in the region. Given the lack of a comprehensive, overarching regional agreement, it is natural that the Arctic Council would play an active role in this.

No state alone can ensure the conservation of the Arctic marine biodiversity. It is unlikely that the Arctic states will develop and adopt a legally binding instrument on conservation of marine biological diversity and the establishment of

---

[31] Possible explanations are provided in Lalonde (2010).

MPAs, but the efforts made by the Arctic states under the Arctic Council are essential for realizing an ecological network of MPAs in the Arctic.

References

ACIA (2005) Arctic Climate Impact Assessment. Cambridge University Press, New York

AEPS (1991) Arctic Environmental Protection Strategy, 30 I.L.M. 1624 14 Jan 1991

Agenda 21 (1992) Agenda 21: Programme of Action for Sustainable Development. Conference on Environment and Development. U.N. GAOR, 46th Sess., Agenda Item 21, UN Doc A/Conf.151/26

AOR (2011) Phase I Report (2009-2011) of the Arctic Ocean Review (AOR) project. <www.pame.is>. Accessed 1 Feb 2013

Arctic Council (2004) Arctic Marine Strategic Plan. 24 Nov 2004. <http://www.pame.is/images/stories/AMSP_files/AMSP-Nov-2004.pdf>. Accessed 5 Feb 2013

Arctic SAR Agreement (2011) Agreement on Cooperation on Aeronautical and Maritime Search and Rescue in the Arctic, 12 May 2011, 50 I.L.M. 1119 (2011). Entered into force on 19 Jan 2013

Birnie P, Boyle A, Redgwell C (2009) International law and the environment, 3rd edn. Oxford University Press, Oxford

Boyle A (2006) Further development of the 1982 convention on the law of the sea. In: Barnes R, Freestone D, Ong D (eds) The law of the sea: Progress and prospects. Oxford University Press, Oxford

Burhenne-Guilmin F, Casey-Lefkowitz S (1992) The Convention on Biological Diversity: A hard won global achievement. Yearb Int Environ Law 3(1):43–59

CAFF (1996a) Circumpolar Protected Areas Network: Strategy and Action Plan. CAFF Habitat Conservation Report No. 6

CAFF (1996b) Circumpolar Protected Area Network (CPAN): Principles and Guidelines. CAFF Habitat Conservation Report No. 4

CAFF (1997) CPAN Progress Report 1997. CAFF Habitat Conservation Report No. 7

CAFF (2004) CPAN Country Updates Report 2004. CAFF Habitat Conservation Report No. 11

CAFF (2009) CAFF WORK PLAN 2009–2011 Ministerial Period. CAFF International Secretariat, Akureyri

CAFF (2010) Arctic Biodiversity Trends 2010—Selected indicators of change. CAFF International Secretariat, Akureyri. <http://www.arcticbiodiversity.is/images/stories/report/pdf/Arctic_Biodiversity_Trends_Report_2010.pdf>

CAFF (2011) Conservation of Arctic flora and fauna (CAFF) Workplan. CAFF Administrative Series Report No. 1. CAFF International Secretariat, Akureyri

Convention on Biological Diversity (1992) Convention on Biological Diversity, 5 June 1992, 1760 U.N.T.S. 79. Entered into force 29 Dec 1991

Frank V (2007) The European community and marine environmental protection in the international law of the sea: Implementing global obligations at the regional level. Utrecht University, Leiden

Henriksen T (2010) Conservation and sustainable use of Arctic marine biodiversity. Arctic Rev Law Politics 2(1):249–278

Ilulissat Declaration (2008) Arctic Ocean Conference. 48 I.L.M. 382 (2009) Ilulissat, Greenland. 27 May 2008

IISD (2012) Earth negotiations bulletin: Summary of the fifth meeting of the working group on marine biodiversity beyond areas of national jurisdiction, vol 25. P 83. 7–11 May 2012

IUCN (2011) Guidelines for applying the IUCN Protected Area Management Categories to Marine Protected Areas (supplementary to the 2008 guidelines), Second Draft. June 2011

IUCN/WCMC (1994) Guidelines for Protected Area Management Categories. IUCN, Gland and Cambridge

Jakobsen IU (2010) Marine protected areas in international law: A Norwegian perspective. PhD Thesis, University of Tromsø. <http://hdl.handle.net/10037/4409>. Accessed 11 Feb 2013

Jakobsen IU (2012) Marine verneområder i Arktis: Kyststatens forpliktelser og rettigheter. In: Henriksen T, Ravna Ø (eds) Juss i nord: Hav, fisk og urfolk. Gyldendal, Oslo

Koivurova T (2009) Governance of protected areas in the Arctic. Utrecht Law Rev 5(1):44–60

Koivurova T, Molenaar EJ (2009) International governance and regulation of the marine arctic: Overview and Gap Analysis. Report prepared for the WWF International Arctic Programme, Oslo

Lagoni R (2003) Marine protected areas in the exclusive economic zone. In Kirchner A (ed) International marine environmental law: Institutions, implementation and innovations. Kluwer Law International, The Hague

Lalonde S (2010) A Network of Protected Areas in the Arctic: Promises and Challenges. In: Nordquist M, Moore J, Heidar T (eds) Changes in the Arctic Environment and the Law of the Sea. Martinus Nijhoff, Leiden, pp 131–142

LOS Convention (1982) United Nations Convention on the Law of the Sea, 10 Dec 1982, 1833 U.N.T.S. 396. Entered into force 16 Nov 1994

Marine Strategy Framework Directive (2008) Directive 2008/56/EC of the European Parliament and the Council of 17 June 2009 establishing a framework for community action in the field of marine environmental policy, June 17 2009, 2008 O.J. (L 164)

Molenaar EJ, Oude Elferink AG (2009) Marine protected areas in areas beyond national jurisdiction: The pioneering efforts under the OSPAR Convention. Utrecht Law Rev 5(1):5–20

Nordquist MH, Rosenne S (1991) United Nations convention on the Law of the Sea 1982: A commentary, vol IV. Martinus Nijhoff, Dordrecht

OSPAR Convention (1992) Convention for the Protection of the Marine Environment of the North-East Atlantic, 22 Sept 1992, 32 I.L.M. 1072 (1993). Entered into force 25 Mar 1998

OSPAR Commission (2003a) Guidelines for the identification and selection of marine protected areas in the OSPAR maritime area, 2003/17

OSPAR Commission (2003b) Guidelines for the management of marine protected areas in the OSPAR maritime area, 2003/18

OSPAR Commission (2003c) Recommendation on a network of marine protected areas, 2003/3, amended by Recommendation 2010/2

OSPAR Commission (2006) Guidance on developing an ecologically coherent network of OSPAR marine protected areas, 2006/3

OSPAR Commission (2011) Status report on the OSPAR network of marine protected areas

Ottawa Declaration (1996) Declaration on the Establishment of the Arctic Council, 35 I.L.M. 1382 19 Sept 1996

PAME (2011) PAME Work Plan 2011–2013. <http://www.pame.is/images/PAME_NEW/PAME%20Work%20Plan/PAME_Work_Plan_2011-2013.pdf>. Accessed 5 Feb 2013

Tanaka Y (2012) The international law of the sea. Cambridge University Press, Cambridge

UNGA (2008) Oceans and the Law of the Sea, A/Res/63/111

UNGA (2011) Oceans and the Law of the Sea, A/Res./66/231, 2011

UNGA (2012) Oceans and the Law of the Sea, A/Res./67/78

United Nations (2012) The Future We Want: Outcome document adopted at the Rio + 20 Conference. <http://www.un.org/en/sustainablefuture>. Accessed 8 Feb 2013

Vienna Convention (1969) The Vienna Convention on the Law of Treaties, 23 May 1969, 1155 U.N.T.S. 331. Entered into force 27 Jan 1980

Wolfrum R, Matz N (2000) The Interplay of the United Nations Convention the Law of the Sea and the Convention on Biological Diversity. In: Bogdandy AV, Rüdiger W (eds) Max Planck yearbook of United Nations law, Max Planck Institute, The Hague

WSSD (2002) Plan of Implementation of the World summit on Sustainable Development. United Nations World Summit on Sustainable Development, Johannesburg, 26 Aug to Sept 2002

WWF (2012) The Circle. World Wildlife Fund. World Wildlife Magazine 3

Part IV
Opportunities for Transatlantic Cooperation

Chapter 11
EU–US Cooperation to Enhance Arctic Marine Governance

Elizabeth Tedsen and Sandra Cavalieri

Abstract Enhanced transatlantic cooperation can help leverage emerging opportunities to improve protection of the Arctic marine environment. Through both international and domestic action, the European Union (EU) and United States (US) can work together to promote environmental leadership and shape agendas and policy making. This chapter examines the shared objectives and interests of the EU and US in the marine Arctic. With this common ground in mind, the chapter examines areas of opportunity for addressing shortcomings in the international legal and policy framework of the marine Arctic through transatlantic action. The need to support a more integrated, ecosystem-based approach to governance is highlighted as a necessary way forward, although sector-based gaps and policy options for the areas of fisheries, shipping, and offshore hydrocarbon development are also considered. The role of indigenous peoples in Arctic decision making and governance is emphasized. Finally, the chapter considers the need for joint action in tackling Arctic governance's greatest underlying challenge: climate change.

11.1 A Changing Environment and the Need for Policy Action

Climate change is occurring more rapidly in the Arctic than in any other region of the world, with sea ice retreating at a pace that exceeds even the most dramatic predictions of scientists. Access to newly accessible Arctic waters is creating

---

Based on: Best A. et al. (2009a, b).

S. Cavalieri (✉)
Ecologic Institute, Pfalzburger Strasse 43–44, 10717 Berlin, Germany
e-mail: sandra.cavalieri@ecologic.eu

E. Tedsen
Ecologic Institute, 1630 Connecticut Ave NW, Suite 300, 20009 Washington, D.C., USA
e-mail: elizabeth.tedsen@ecologic.eu

E. Tedsen et al. (eds.), *Arctic Marine Governance*,
DOI: 10.1007/978-3-642-38595-7_11, © Springer-Verlag Berlin Heidelberg 2014

economic opportunities for the fishing, shipping, energy, and tourism industries, all of which are expected to expand in both scope and intensity. These changes bring with them new challenges. Increased human activity in the Arctic marine area will require effective policies and international cooperation if the world hopes to protect fragile Arctic ecosystems as well as to safeguard the rights and interests of indigenous peoples.

11.2 Recent Policy Developments

The sense of urgency initiated by rapidly melting Arctic sea ice has contributed to new international action on Arctic governance. Political dynamics are changing nearly as fast as the ice. The world took notice of changing Arctic geopolitics in the summer of 2007 when the Russian flag was planted on the North Pole seabed and the lowest Arctic sea ice extent to that time, since surpassed, was recorded. Since then, Canada, Denmark, Finland, Iceland, Sweden, Russia, the United States (US), and the European Union (EU) have all released new and revised Arctic policies. An increasing number of non-Arctic states and other political entities, such as the EU, have begun developing Arctic policies and seeking a role in Arctic governance, with increasing interest from non-Arctic states and entities in the Arctic Council. Arctic governance is no longer solely of interest to Arctic states, indigenous peoples, academics, and non-governmental organizations (NGOs), but rather, it has become a major item on the policy agenda for a range of international actors.

Emerging global powers are vying to become permanent observers at the Arctic Council and in the process shifting the forum's internal dynamics. A number of non-Arctic states, including China, Italy, Japan, Korea, Singapore, and India, as well as the EU, have applied for permanent observer status with the Arctic Council. The European Commission announced in its November 2008 Arctic Communication (European Commission 2008) that it intended to seek permanent observer status at the Arctic Council, marking a change from its previous policy of participating as an ad-hoc observer, along with the European External Action Service (EEAS) and other EU agencies. At the Arctic Council's April 2009 Ministerial Meeting, a decision on permanent observer status was delayed for all applications, including the European Commission. Although the EU was denied membership—perhaps due to Russia's reluctance as well as conflict with Canada over a proposed seal import ban—the strongest and most comprehensive policy statements from the group of non-Arctic players have come from the EU. The EU updated its application after the Arctic Council's adoption of criteria for observers in May 2011 and reiterated its interest in a Joint Communication in June 2012 (European Commission and High Representative 2012). A decision on observer applications will be taken at the Council's upcoming[1] May 2013 Ministerial Meeting in Kiruna. However, even if the EU is granted observer

---

[1] As of time of writing, March 2013.

status, it remains unclear what role it would actually play in the Arctic Council as the role of observers is currently limited; for example, observers may only submit written statements at ministerial meetings.

Although climate change in the Arctic is indeed an issue of global concern, the Arctic coastal states are reluctant to have the region viewed—and governed—as a global 'commons'. As such, the basis for the EU and other non-Arctic states defining and pursuing interests in the Arctic may not be as self-evident as supposed (Best et al. 2009a, b).

Influenced by the rate of Arctic change, the Arctic Council itself is undergoing shifts so as to better address new governance needs. The Council does not have the competence to impose legally binding obligations of any kind on its members, permanent participants, or observers. The most it can do directly from a governance perspective is to issue policy recommendations, such as the one commissioning the Arctic Climate Impact Assessment (ACIA), and to adopt guidelines and recommendations. The Arctic Council is an influential contributor to policy making in the Arctic, although largely through issuing non-binding guidelines and recommendations. However, the Council's role has recently evolved to serve as a forum under the auspices of which Arctic states negotiate agreements, which are then signed and transposed by the respective national procedures to become legally binding decisions. At the May 2011 Ministerial Meeting, the Arctic coastal states adopted a legally binding Arctic Search and Rescue agreement (Arctic SAR Agreement 2011). An agreement on Marine Oil Pollution Preparedness and Response (Arctic MOPPR Agreement 2013) is scheduled to be signed by the coastal states during the May 2013 Ministerial Meeting. Also, following a recommendation by the Arctic Environmental Ministers, an agreement on reductions of black carbon emissions could be negotiated under the new Canadian chairmanship (Ministry of the Environment 2013). Thus, while the Arctic Council itself remains limited in its mandate and ability to adopted binding measures, it has nonetheless strengthened its role as an institution that both influences and helps support new policies and governance measures.

Alongside these trends, dynamics among the established actors in the Arctic Council have also undergone a shift in recent years. In part, this change was marked by a meeting of the five Arctic Ocean coastal states in Ilulissat, Greenland in May 2008. Perceiving that the Arctic Ocean was on the brink of crossing a significant threshold, they declared that "[b]y virtue of their sovereignty, sovereign rights and jurisdiction in large areas of the Arctic Ocean the five coastal states are in a unique position to address these possibilities and challenges" (Ilulissat Declaration 2008). The Arctic Ocean coastal states also announced their intention to protect the Arctic environment and the interests of indigenous peoples and local inhabitants:

> Climate change and the melting of ice have a potential impact on vulnerable ecosystems, the livelihoods of local inhabitants and indigenous communities.... The Arctic Ocean is a unique ecosystem, which the five coastal states have a stewardship role in protecting. Experience has shown how shipping disasters and subsequent pollution of the marine environment may cause irreversible disturbance of the ecological balance and major harm to the livelihoods of local inhabitants and indigenous communities.

The Arctic Ocean coastal states also expressed their opinion therein that there is "no need to develop a new comprehensive international legal regime to govern the Arctic Ocean" because:

> Notably, the law of the sea provides for important rights and obligations concerning the delineation of the outer limits of the continental shelf, the protection of the marine environment, including ice-covered areas, freedom of navigation, marine scientific research, and other uses of the sea. We remain committed to this legal framework and to the orderly settlement of any possible overlapping claims. This framework provides a solid foundation for responsible management by the five coastal States and other users of this Ocean through national implementation and application of relevant provisions (Ilulissat Declaration 2008).

Despite the fact that Denmark, which is an Arctic state by virtue of Greenland, had earlier insisted that Arctic Ocean coastal state cooperation should not compete with the Arctic Council (SAO 2007), the meeting in Ilulissat produced friction among the Arctic Council members. Iceland expressed the greatest concern among the three non-Arctic Ocean coastal states (the other two being Finland and Sweden). It had already expressed reservations about strengthened Arctic Ocean coastal state cooperation at a 2007 meeting of Senior Arctic Officials (SAOs)[2] and reiterated its concern during the August 2008 Conference of Arctic parliamentarians (Conference of the Parliamentarians 2008). The meeting in Ilulissat in May 2008 also provoked a reaction from one of the strongest Arctic Council permanent participants, the Inuit Circumpolar Council (ICC) and national Inuit leaders, who issued a statement[3] outlining, *inter alia*, their concerns:

> Concern was expressed among us leaders gathered in Kuujjuaq that governments were entering into Arctic sovereignty discussions without the meaningful involvement of Inuit, such as the May, 2008 meeting of five Arctic ministers in Ilulissat, Greenland. The Kuujjuaq summit noted that while the Ilulissat Declaration asserts that it is the coastal nation states that have sovereignty and jurisdiction over the Arctic Ocean, it completely ignores the rights Inuit have gained through international law, land claims, and self-government processes. Further, while the ministers strongly supported the use of international mechanisms and international law to resolve sovereignty disputes, it makes no reference to those international instruments that promote and protect the rights of indigenous peoples (ICC 2008).

---

[2] In the discussions at the Narvik SAO meeting in 2007, Iceland expressed concerns that: "separate meetings of the five Arctic states, Denmark, Norway, US, Russia and Canada, on Arctic issues without the participation of the members of the Arctic Council, Sweden, Finland and Iceland, could create a new process that competes with the objectives of the Arctic Council. If issues of broad concern to all of the Arctic Council Member States, including the effect of climate change, shipping in the Arctic, etc. are to be discussed, Iceland requested that Denmark invite the other Arctic Council states to participate in the ministerial meeting. Permanent participants also requested to participate in the meeting. Denmark responded that the capacity of the venue may be an issue" (SAO 2007).

[3] Similar sentiments were expressed in the ICC's subsequent 2009 'Circumpolar Inuit Declaration on Sovereignty in the Arctic' that referenced the previous statement and Ilulissat meeting (ICC 2009).

But the ICC and the Inuit leaders were also critical of current Arctic governance structures:

> We recognized the value of the work of the Arctic Council…We further noted the meaningful and direct role that indigenous peoples have at the Arctic Council, while at the same time expressing concern that the Council leaves many issues considered sensitive by member states off the table, including security, sovereignty, national legislation relating to marine mammal protection, and commercial fishing (ICC 2008).

Clarifying their position regarding the creation of any new governance arrangements, the ICC made the following statement:

> We called upon Arctic governments to include Inuit as equal partners in any future talks regarding Arctic sovereignty. We insisted that in these talks, Inuit be included in a manner that equals or surpasses the participatory role Inuit play at the Arctic Council through ICC's permanent participant status (ICC 2008).

Several years later, the Illulisat Declaration may continue to be the best indicator of which way the political winds are blowing. A central purpose of the meeting of the five Arctic Ocean coastal states in Greenland was to demonstrate to the international community and the media that there would not be a scramble for resources in the region, but rather an orderly process governed by the law of the sea. They embraced the rhetoric of environmental conservation, stating "the Arctic Ocean is a unique ecosystem, which the five coastal states have a stewardship role in protecting".

11.3 Transatlantic Cooperation and Common Ground

The EU and US hold notably different positions in the marine Arctic—the US, unlike the EU, possesses Arctic coastal territory—however, both share common interests in promoting environmental protection and sustainable development in the Arctic region and, as global leaders, have particular capacity for realizing these aims. Further, while none of the Arctic EU Member States (Denmark, Finland, and Sweden) have direct influence on Arctic waters, the EU is, by virtue of its Member States, among the largest maritime powers in the world and as such can significantly contribute to the discussion on environmental governance in the marine Arctic. In addition, the EU accession of candidate state Iceland might add a territorial perspective to EU marine governance in the Arctic in the mid-term future that goes beyond the already close economic cooperation under the European Economic Area (EEA) agreement.

Both the EU and the US have released policy statements regarding their Arctic policies. In November 2008, the European Commission issued its Arctic Communication (European Commission 2008) which laid out EU policy objectives in a number of different areas, including environmental protection, indigenous peoples, sustainable use of resources, and international governance options. In June 2012, a Joint Communication on the EU's Arctic Policy (European Commission and High Representative 2012) set out the case for a refined policy and increased EU engagement on Arctic issues based on knowledge, responsibility to achieve sustainable development, and engagement with Arctic states, indigenous peoples, and other partners. Like the 2008 Communication, it highlighted the issues of climate change, research, indigenous peoples, maritime safety, sustainable economic development, and multilateral cooperation. The US's Presidential Directive on Arctic Region Policy (NSPD-66 2009) outlined a similar set of issues, with the notable addition of US security interests, focusing on Arctic security needs, protection of the environment and resources, sustainable development, international cooperation, indigenous communities' participation in decision making, and scientific monitoring and research.

These policy statements have a noteworthy level of agreement, with clear areas of opportunity for potential cooperation. Both affirm commitments to the existing law of the sea framework. Both indicate a preference for working within existing institutions and frameworks rather than creating a new overarching governance regime, while indicating a willingness to modify these frameworks to fit unique Arctic conditions. Both highlight the Arctic Council as a forum for continued cooperation in the region. Both recognise the threats posed to indigenous communities by rapid environmental change and poorly regulated economic expansion and the importance of including indigenous people in Arctic decision making. Both indicate a commitment to greater cooperation in scientific research and monitoring. Both point to agreement that Arctic governance should be informed by principles of ecosystem-based management.

Likewise, in their approaches to marine governance, the EU and US also find common ground. The EU has adopted a system of integrated and holistic maritime policies and recognised the need for an integrated policy approach to the Arctic Ocean in its 2007 Integrated Maritime Policy (IMP) (European Commission 2007). The EU's Marine Strategy Framework Directive (2008) states that marine policies will use an ecosystem-based approach to the management of human activities. The US, through its National Ocean Policy (Exec. Order No. 13547 2010), has also adopted an ecosystem-based management approach. The National Ocean Policy's draft implementation plan singles out the Arctic Ocean as a priority area for ocean policy action (National Ocean Council 2012).

Importantly, the EU and US have also both emphasized the need for international cooperation in approaching Arctic environmental protection. The European Commission made international cooperation one of three key policy objectives in its 2008 Communication and cooperation—internationally and with Arctic stakeholders—was again emphasized in the 2012 Joint Communication, including regarding the sustainable management of marine resources. The US's 2009 Arctic Region Policy highlights the need for international cooperation in responding effectively to environmental challenges, as well as for international scientific and shipping cooperation.

With this common basis in mind, transatlantic cooperation between the EU and US can take advantage of emerging opportunities for improving protection of the Arctic marine environment. Through formal cooperation and informal channels, and both international and domestic action, the EU and US can promote environmental leadership and provide resources in shaping agendas and policy making.

11.4 Addressing Governance Shortcomings

The Arctic marine area is governed by a complex array of legal instruments, including bilateral and multilateral agreements, supra-national, national, and sub-national legislation, as well as soft law arrangements. Arctic governance also involves institutions that are national, regional, and global in scope, and that possess mandates ranging from the provision of scientific advice and issuance of recommendations to the prescription of legally binding obligations (see Chap. 3).

The principle challenges with the Arctic's environmental legal and policy framework today are, first, gaps and shortcomings in adherence to global instruments and their implementation; second, insufficient coordination and integration between existing instruments; and third, rapid environmental, socioeconomic, and geopolitical changes that may outpace the ability of current institutions and agreements to adapt to and address new governance challenges.

At the international level, certain challenges stem from the fact that international conventions provide general frameworks that are not specific to Arctic conditions and not all Arctic states are parties to relevant instruments (Best et al. 2009b). At all levels, many of the gaps in Arctic governance today are the result of an incremental, sector-based approach to management that has historically been static in nature and failed to consider interlinkages between systems. Relevant regional data and scientific knowledge suffer from similar gaps, owing both to the complexity of Arctic marine ecosystems and that many scientific efforts to date have been directed toward specific issues, with less attention paid to interdependencies and cause-and-effect relationships present in Arctic ecosystems.

In addition to gaps between different sectoral governance regimes, there are also gaps within these regimes as they apply to the Arctic. Additional discussion on the areas of fishing, shipping, and offshore hydrocarbon extraction can be found in Chaps. 5, 6, and 7.

The changes taking place in the Arctic pose immense challenges for maintaining environmental and cultural sustainability in the region. Many of these are complex, in that they entail a number of ecosystem components that are affected by multiple drivers of change. Additionally, ecosystems sometimes span territorial boundaries and often involve a broad range of stakeholders. Furthermore, approaches to governance in marine environments are often less developed than in terrestrial environments. Implementation of natural resource management in marine ecosystems is arguably more difficult than in terrestrial ecosystems due to the lack of visible boundaries between marine ecosystems and the vast areas of international waters. It is, however, important to understand linkages between terrestrial and marine ecosystems. For instance, the Arctic Ocean receives more river runoff than any other global ocean (AMAP 1998).

To address these challenges, there is a need for flexible and adaptive management approaches in the Arctic that recognize cultural and governmental/legal differences, apply an integrated and interdisciplinary approach to understanding and managing ecosystems, and, ultimately, maintain the resilience of Arctic ecosystems and communities.

Integrated, cross-sectoral governance strategies, taking into account both natural systems and human activities in a holistic and integrated manner, should thus be a key aim of regional and global policies. Cross-sectoral policy options can be distinguished from those of a more narrow focus by their substantive scope and level of participation. An integrated, ecosystem-based management approach is increasingly recognized as a superior approach to marine management, rather than the status quo sectoral approach that is prevalent in the Arctic and elsewhere (see e.g., Ehler 2011). Such an approach can help to maintain ecosystem resilience while also recognizing community and stakeholder needs, maintaining cultural traditions, allowing for economic opportunities, and encouraging flexibility and adaptability. Further, both scientific and traditional knowledge can be incorporated in integrated processes, and multiple interests—e.g., environmental, social, and economic—can be addressed in conjunction. Flexible management approaches can allow governance systems to adapt to changing environmental and climatic conditions as well as new scientific knowledge. As highlighted in EU and US policy statements, the existing governance framework should continue to be modified to better address changing Arctic conditions.

As Arctic challenges are both local and global in nature, there is need for coordination and integrated governance both among Arctic states and at the international level to successfully manage pressures in the Arctic region. Suggestions for improving the current framework range from new sector-based agreements to a comprehensive 'Arctic treaty'; however, as stated in the Ilulissat Declaration, the Arctic coastal states themselves have rejected the idea of a comprehensive new regime.

It is of importance to note that no governing body has a mandate to develop legally binding rules for the entire Arctic region. While it has taken a more active role in policy development, providing a forum for Arctic states to negotiate legally binding instruments, the Arctic Council nonetheless remains limited in its mandate and without regulatory powers of its own: The Ottawa Declaration on the Establishment of the Arctic Council (1996) does not impose binding obligations on its participants and neither is the Arctic Council empowered to do so. The law of the sea provides a general governance framework for the marine environment, which Arctic states agreed to abide by in the Ilulissat Declaration, but relies on additional institutions to implement its provisions.

Recent years have witnessed a surge of interest in Arctic governance and in efforts to improve the existing framework so as to better meet new changes. These include the Arctic SAR agreement (2011), designed to fill gaps in coverage as Arctic marine traffic increases, and the forthcoming Arctic MOPPR Agreement, which was drafted in anticipation of a rise in Arctic offshore hydrocarbon extraction. Will new efforts such as these be sufficient to fill Arctic marine governance gaps? Can they be flexible enough to meet continuing changes? And will they be inclusive in allowing participation by indigenous peoples?

There are several strategies by which a cross-sectoral system of governance in the marine Arctic could be implemented, carrying varying degrees of political support from different Arctic players:

- Relevant actors, including non-Arctic states, indigenous peoples' organizations, industry, and NGOs, could establish *new* issue- or sector-specific instruments and institutions that are complimentary in nature;

- Relevant actors, including non-Arctic states, indigenous peoples' organizations, industry, and NGOs, could participate in multilateral negotiations within the context of *existing* institutions and instruments in order to *modify* them in a coordinated fashion.
- The *Arctic Council* could serve as a forum and coordinator in efforts to *supplement or modify* existing frameworks to function in a more integrated and comprehensive fashion. Observers could play a key role in supporting this work through stronger participation in the Arctic Council working groups and by being given a more active voice in the Council's deliberations. Further, observers could support a strengthened Arctic Council in other fora in which they play a more prominent role themselves.
- State actors, with the involvement of other relevant actors, such as, indigenous peoples' organizations, industry, and NGOs, could negotiate an *overarching*, legally binding regional instrument specifically tailored to address the unique conditions and challenges of the Arctic.

Given the need for a flexible governance regime, and the wide range of sovereign actors involved in the Arctic, the utility of soft law instruments should not be underestimated. Existing international bodies such as the Arctic Council and legal instruments with institutional components may be well situated to create and update guidelines and best practices for the region, although the non-legally binding nature of soft law instruments may produce weaker commitments and less certain outcomes.

11.5 Transatlantic Support for Integrated Management

Regardless of the implementation strategy—or possibly a mix of several strategies—taken, the following outlines certain approaches that could provide a basis for cross-sectoral governance measures and frameworks:

- *Ecosystem-based management* (EBM)—EBM is as a comprehensive, integrated approach to management of human activities that offers flexibility and is widely regarded as a best practice of environmental governance. As illustrated in Chap. 3, EBM has the potential serve as an *organizing framework for integrated management* by balancing a variety of human activities, interests, and priorities while supporting ecosystem sustainability.
- *Marine spatial planning* (MSP)—MSP organizes marine space and uses, balancing development demands with ecosystem protection, to achieve ecological, economic, and social objectives in a planned process. As Chap. 9 describes, MSP is an *operational process through which to implement EBM*. MSP supports managing human activities to enhance compatible uses and reduce conflicts among uses.
- *Marine protected areas* (MPAs)—MPAs often serve as an important *component* of EBM and MSP and can be a helpful tool for implementing the precautionary principle in management by setting aside spaces for protection as conditions

and uses change. As Chap. 10 explains, currently, most protected areas are for terrestrial rather than marine environments and little of the Arctic marine area is designated as MPAs.

EU and the US statements regarding their Arctic policies point to broad areas of agreement, as outlined in Sect. 11.3. In addition to the synergies identified relating to indigenous peoples, the environment, and international cooperation, there also appears to be agreement that marine Arctic governance should be informed by principles of EBM. The EU's 2008 Arctic Communication states that holistic, ecosystem-based management of human activities should complement any efforts to mitigate and adapt to changes in the Arctic caused by climate change, and the 2012 Joint Communication highlights steps for the EU to take on Arctic ecosystem management. Similarly, the US's 2009 Arctic Region Policy states that relevant executive agencies should pursue marine ecosystem-based management in the Arctic.

Both the EU and the US have experience with EBM within their own maritime zones and could push for wider application in transboundary, cross-sectoral Arctic marine governance. The EU has been a strong advocate for MSP, an important component of the IMP, and is planning to issue a directive on coastal and maritime spatial planning in 2013. In the US, the federal government has begun implementing its ocean policy, applying integrated management and MSP, including in Alaska's Beaufort, Chukchi, and Bering Seas. The US National Ocean Policy provides for a strategic plan for EBM in the Arctic, and the draft implementation plan highlights EBM as a guiding theme.

The EU and US support the Arctic Council as a primary forum for cooperation and governance in the Arctic region. As such, supporting the Arctic Council's work may be one of the best areas for EU and US cooperation on promoting integrated management approaches. The Arctic Council has undertaken and continues to lead important work on EBM. As described in Chap. 3, past projects of significance include the PAME (Protection of the Arctic Marine Environment) working group's work on Large Marine Ecosystems (LMEs) and the 2009 Best Practices in Ecosystem-based Oceans Management in the Arctic (BePoMAR) report (Hoel 2009). Currently, an expert group on EBM for the Arctic environment, composed of government experts from Arctic States and representatives from the Arctic Council's permanent participants and working groups, is developing recommendations for advancing EBM in Arctic ecosystems, which will be considered by the Ministers before the end of the Swedish Chairmanship in May 2013.

In practice, EBM approaches are largely implemented at the national level, although regional and international support and coordination are important and can ensure coordination between jurisdictions as well as sharing of best practices. The Arctic Council can help facilitate coordination between states and continue to support the development of EBM principles and practices. In addition to Arctic states, the Arctic Council can also help bring non-governmental actors such as indigenous peoples, NGOs, and industry to the table. Stakeholder participation is an important part of EBM and MSP planning processes and is critical for achieving social, economic, and environmental objectives. The US and EU can support the continuation of this work financially and by appointing experts from multiple cross-cutting agencies to contribute to the working groups and take tasks forward.

Arctic coastal states should designate MPAs in the Arctic, either independently or as part of a larger EBM framework, before a further increase in economic activity might lead to the entrenchment of interests in certain areas. MPAs could help to protect sensitive and unique Arctic ecosystems and increase resilience in the face of changing conditions, as could be especially promoted by a planned network of protected areas, perhaps involving co-management agreements with indigenous communities. MPAs through a network, rather than individually, could cover a wider area, including species' habitat and migratory pathways that cross state boundaries. Thus, working together, states can do more than alone and cooperation can create a transboundary network of protected areas of greater ecological value and resilience. The EU and US could jointly advocate for the creation of an EBM-based network of MPAs.

In its 2012 Joint Communication on the Arctic, the EU outlined the following steps for pursuing Arctic ecosystem management:

- Work through the OSPAR Convention to establish a network of Arctic MPAs and assess measures to manage oil and gas extraction activities in extreme climatic conditions;
- Contribute to work under the Arctic Council's PAME working group; and
- Promote biodiversity protection in areas beyond national jurisdiction in UN bodies such as the LOS Convention.

While the US is neither a contracting party to the OSPAR Convention, nor a party to the LOS Convention, and the EU's application for an observer status at the Arctic Council has yet to be decided upon, the latter two options suggest areas for US as well as EU action. The Arctic Council may present some of the best opportunities for dual action and the EU and US can offer support within the Council, as well as in other international bodies, for environmental priorities and promoting integrated management approaches. Their cooperation in existing regulatory bodies is critical to ensuring that environmental goals remain at the top of the agenda.

11.6 Toward Integrated Management

To successfully implement integrated management approaches in Arctic marine area, key knowledge gaps must first be filled. A commonly identified problem among Arctic policymakers is a lack of information. Good environmental governance must be supported by an understanding of ecosystem conditions, baselines, processes, and changes. Both Arctic and non-Arctic states alike can, via the Arctic Council and other international scientific institutions, continue to improve coordination among scientific research initiatives. The following represent example areas of actions that can support the design, implementation, and adaptation of successful governance practices for the Arctic marine area:

11.6.1 Research and Monitoring

Additional research and data is needed on Arctic systems in order to inform EBM initiatives, as much Arctic research has traditionally had a narrow, issue-based focus. Information is lacking on baseline Arctic ecosystem conditions, thus making the measurement of changes challenging, and on new changes and pressures. Better monitoring and observations are needed to continue to assess changing conditions, activities, interactions, and management needs. To meet adaptation needs in the Arctic, it is necessary to improve observational knowledge and have an understanding of integrated impacts and processes in the region and associated risks (AMAP 2012). Traditional knowledge of indigenous communities should be incorporated into these efforts.

Opportunities for cooperation in Arctic marine science exist between the EU and US, as well as between Arctic and non-Arctic states. Both the EU and US strongly support international scientific cooperation in the Arctic, as noted in their respective Arctic policies. In addition to its 2009 Arctic Policy, the US's Arctic Research and Policy Act (ARPA 1984) provides for a comprehensive national policy dealing with national research needs and objectives in the Arctic. Both the EU and US have devoted significant resources to Arctic research and have helped to promote shared research platforms and collaboration. The US has identified the need to establish a science framework to support science-based EBM implementation as a key action for implementing EBM, noting that "[s]ustainably managing human uses of an ecosystem requires a robust understanding of the nature of the dynamically interacting biological, physical, chemical, and geological components and processes; the effects of human and natural forces; and the results of management efforts" (National Ocean Council 2012).

The Arctic Council's EBM Task Force has recognized the need for supporting EBM through increased knowledge in areas such as ecosystem services, monitoring, data-sharing, and improving understanding of ecosystem interactions of cumulative effects, while, however, noting that "[i]n most cases the missing piece for implementing EBM is not the science but an effective process or organizational structure; without some means to translate the science into a meaningful management approach that meets certain agreed-upon objectives, EBM is just a series of interesting reports" (SAO 2012a, b).

11.6.2 Defining and Assessing Ecosystems

One approach to help distinguish priority areas for policy action is the Large Marine Ecosystem (LME) concept, built on general principles of ecosystem management. LME boundaries are becoming widely used at the international scale to distinguish highly productive areas around the globe for marine ecosystem management.[4] LMEs encompass relatively large areas of approximately 200,000 km^2 or greater and have distinct bathymetry, hydrography, productivity, and trophically-dependent populations (Sherman 1994). They can be evaluated with respect to their productivity, fisheries, pollution, ecosystem health, socioeconomic conditions, and governance (Juda and Hennessey 2001). In addition, LMEs draw attention to the need to understand complex changes in multiple species interactions and the need to manage for resilience rather than composition or structure. LMEs provide a practical basis to evaluate shipping, fishing, and tourism at the regional level.

The Arctic Council PAME working group has developed a map of LMEs in the Arctic that was adopted by the Arctic Council Ministers.[5] Both the Arctic Marine Shipping Assessment (AMSA 2009) and Arctic Oil and Gas Assessment used LMEs as a basis for analysis (IUCN/NRDC 2010), and the Arctic EBM Task Force has suggested using PAME's LME map to "inform EBM implementation" (SAO 2012b). Beyond the Arctic Council, several Arctic Ocean coastal states, notably Canada, Norway, and the US, have organised their national Arctic governance regimes around the concept of LMEs. However, LMEs often cross national borders and there is as yet no established framework for coordinating LME regulatory activities at the bilateral or international level. The EU and US could, for example, continue such efforts through working bilaterally on a comprehensive Arctic Ocean Assessment, drawing on the work of the United Nations (UN) on global reporting and assessment of the marine environment.[6]

An Arctic Ocean Assessment could complement the LME work already taking place and better harmonise governance approaches to issues common to multiple ecosystems. Furthermore, it could build on existing assessments, e.g., environmental impact assessments (EIAs) and strategic environmental assessments (SEAs), to better understand the potential environmental and social impacts of proposed activities and programmes on the Arctic environment. For climate change, EIAs not only help to assess and understand the interactions between a changing environment and human activities, but can also help identify—and mitigate—the climate change impacts of the activity. Both the US and EU are engaged in the UN's global marine assessment and both have their own procedures in place for EIA.

While terrestrial protected areas have increased globally, there remains a need to identify and protect biologically important marine areas, including in the Arctic. An important component of EBM, as well as for the establishment of MPAs, is the identification of ecologically, biologically, or culturally significant and vulnerable areas that should be considered for protection (SAO 2012a; Speer and Laughlin 2010).

---

[4] LMEs are used, among others, by the United Nations Environment Programme, United Nations Development Programme, the World Bank, the US National Oceanic and Atmospheric Association, and the Arctic Council.

[5] See <http://www.pame.is/arctic-large-marine-ecosystems-lme-s>.

[6] See <http://www.un.org/Depts/los/global_reporting/global_reporting.htm>.

These areas can serve as a basis for MPAs as well as for MARPOL 'special areas' or PSSAs (see Sect. 11.8; see also Chap. 6). Efforts have taken place to begin identifying Arctic ecologically and biologically significant areas [EBSAs; as defined by the Convention of Biodiversity (1992)], at the state level - with efforts by Norway, the US, Canada, and Greenland (SAO 2012a), in the AMSA (2009), and by environmental NGOs (Speer and Laughlin 2010).

The identification of EBSAs should be a dynamic process. Many ecologically or biologically and culturally significant areas are tied to Arctic sea ice, which is undergoing dramatic changes (AOR 2012; Speer and Laughlin 2010). The EU and US can support efforts to improve and continue identification of EBSAs as well as to move from the recognition of areas in need of protection to the implementation of protective measures through, *inter alia*, efforts within the IMO, Arctic Council, or bilateral cooperation.

11.7 Transatlantic Policy Options for Fisheries Management

While warmer areas of the Arctic marine area have supported commercial fishing activities for decades, until recently there was little or no major fishing activity in the colder areas of the Arctic, with ice-covered regions completely cutting off access to fishing. The retreat of Arctic sea ice is opening up new parts of the Arctic Ocean to fishing vessels, and there are already signs that certain fish species are migrating north to warming ocean waters.

In light of these changes, gaps exist in both international and national legal and policy frameworks for regulating Arctic fisheries, and the expansion of marine capture fisheries in the Arctic may necessitate adjustments. Any such process would benefit from a needs assessment drawing on basic fisheries research and an evaluation of likely future scenarios regarding, for example, habitats, migration patterns, impacts on target and non-target species, and fishing techniques. For certain Arctic fisheries that have been commercially fished for years, policymakers have access to a wealth of information. In other areas, almost nothing is known. For instance, new fishing opportunities on the Pacific side of the Arctic Ocean may remain located primarily in the maritime zones of coastal states for the near future, whereas fishing opportunities on the Atlantic side may soon extend to areas in the high seas that were previously not fished. There is a pressing lack of scientific data to understanding of Arctic Ocean ecosystems and for use in developing science-based fisheries management.

To address information deficits, individual or collective initiatives geared towards developing mechanisms or procedures similar to an EIA or a SEA for new fisheries in the Arctic marine area could be performed. The EU and US can support further basic research on fisheries, Arctic ecosystems, fish species, and climatic impacts, and for the development of potential scenarios. Assessments could be carried out in the framework of the Arctic Council through its CAFF

(Conservation of Arctic Flora and Fauna) working group or in other forums, such as ICES, where some progress has already been made (see Chap. 5).

Until scientific information and understanding improve, one policy option is to support a freeze on the expansion of commercial fishing in the Arctic, such as the one enacted by the North Pacific Fishery Management Council (NPFMC) in 2009 for fishing in the Alaskan EEZ, until there is sufficient information and scientific understanding of the ecosystem and climatic impacts on fisheries. This freeze followed a 2007 Congressional Joint Resolution (S.J. Res. 17 2007) directing the US to initiate international discussions and take steps to negotiate an agreement for managing migratory and transboundary fish stocks in the Arctic Ocean.

While not having Arctic Ocean coastal waters of its own, the EU likewise "advocate[s] a precautionary approach whereby, prior to the exploitation of any new fishing opportunities, a regulatory framework for the conservation and management of fish stocks should be established for those parts of the Arctic high seas not yet covered by an international conservation and management system" (European Commission and High Representative 2012). Such a statement is made possible since the EU has—according to Articles 3(1)(d) and 38(1) of the Treaty on the Functioning of the European Union (TFEU 2008)—the exclusive competence for the conservation of marine biological resources under its Common Fisheries Policy (CFP), while competence on other fisheries-related issues is shared with EU Member States (however, Member States are only competent to fill gaps in the EU's legislation, see TFEU (2008), art. 2(2)).

Together, both the EU and US could advocate for wider international support of a precautionary position, halting new fisheries activity until scientific understanding and adequate management approaches are established and supporting underlying research and policy measures. Moreover, both could support a declaration that the relevant general principles of the Fish Stocks Agreement (1995), UN Resolutions in relation to vulnerable marine ecosystems and destructive fishing practices (for instance, A/RES/61/105[7]), and relevant conservation and management measures drawn from RFMOs would apply to new and existing fisheries in the Arctic marine area. Following the NPFMC ban, this declaration could stipulate that there shall be no expansion of commercial fishing in the Arctic until adequate assessments of the impacts on target and non-target species and livelihoods of indigenous peoples have been carried out.

Both Arctic Ocean coastal states and other states can adopt individual regulations on fishing activities in the Arctic marine area within their own maritime zones or for their natural and legal persons. The EU and the US could coordinate efforts in this regard, acting independently on the basis of shared concerns, with the EU acting as a flag rather than coastal state in the Arctic Ocean, and thereby expanding the geographic scope and relevance of any adopted regulations. Over time, such transatlantic regulations could serve as a model for international rule making.

---

[7] See <http://www.un.org/Depts/los/general_assembly/general_assembly_resolutions.htm> for more UN Resolutions on sustainable fisheries.

11.8 Transatlantic Policy Options for Arctic Shipping

With sea ice melting, new intra- and trans-Arctic shipping routes are opening to industry and tourism. Trans-Arctic shipping is increasing during summer months already—with an increase from 34 vessels in 2011 to 46 in 2012 (Pettersen 2012)—and it is increasingly important to address the safety and environmental risks associated with a rise in shipping activity.

Significant progress has been made in recent years towards addressing gaps in the international legal and policy framework for Arctic marine shipping. In 2009, for instance, the International Maritime Organisation (IMO) adopted the Polar Shipping Guidelines (2009) and is currently developing new mandatory guidelines for 'Ships Operating in Arctic Ice-Covered Waters', also known as the 'Polar Code'. At the Arctic Council's Seventh Ministerial Meeting, the Arctic SAR Agreement was signed by the Arctic states. In 2013, the new Arctic MOPPR Agreement is planned for signature. Thus, while gaps still remain (see Chap. 6), significant progress has been made in recent years through multilateral cooperation.

For the remaining gaps, there are various options available for modifying the current international framework to account for the risks presented by shipping to Arctic marine ecosystems and human safety. The EU and the US should consider coordinating a joint and harmonised approach towards supporting or initiating various unilateral, regional, and global shipping options. Relevant international bodies in this regard include the IMO, Arctic Council, and the Paris and Tokyo MOUs on port state control. In considering the suitability of regional and global options in the sphere of shipping vis-à-vis individual options, particular account should be taken of the function of competent international organizations like the IMO and the need for uniformity in the international regulation of shipping.

The EU and US share interests in the protection and preservation of the marine environment and marine biodiversity, as well as in the continued exercise of navigational rights and freedoms for their flagged vessels. In recent policy statements both have advocated for work through the IMO on strengthening Arctic navigation standards (such as through the development of the Polar Code), for freedom of navigation in the Arctic, and for promoting navigational safety. Even though the EU cannot act in a capacity comparable to that of an Arctic Ocean coastal state, it can act in a capacity comparable to that of a flag state, a port state, a market state, or with regard to its natural and legal persons. As flag states, and as major shipping powers, or the EU and US could, for instance, impose requirements on vessels that are more stringent than generally accepted international rules and standards (GAIRAS) (e.g., related to special discharge, emission, and ballast water exchange standards), taking forward-looking steps in advance of the forthcoming Polar Code.

In particular, the EU and US can strengthen cooperation to improve shipping regulations by working together in the IMO to support negotiations on the Polar Code and ensure inclusion of stringent standards This could, as suggested in Chap. 6, include new standards such as for ship routeing measures, compulsory pilotage, and ice-breaker or tug assistance. Also through the IMO, the EU and US could jointly propose and support the designation of the marine Arctic (or parts thereof) as a MARPOL special area or PSSA, accompanied by a comprehensive package of associated protective measures (APMs) consisting of one or more of the above standards.

11.9 Transatlantic Policy Options for Offshore Oil and Gas Development

Under scenarios of retreating sea ice, technological advances, and rising oil prices, the exploration, production, and shipping of Arctic oil and gas are increasingly viable and attractive. However, challenges persist under harsh regional conditions, raising the environmental and safety risks of drilling for both the environment and resource-dependent communities. In addition to risks from oil spills, oil and gas development creates significant operational impacts: For example, fuel combustion for onsite power generation, well testing, gas flaring, and operational leaks release black carbon, methane, NOx, SO_2, VOCs, and CO_2 emissions into the Arctic atmosphere.

The positions of the EU and US in relation to offshore hydrocarbons activities in the Arctic are fundamentally different. As a coastal state, the US is directly involved in offshore hydrocarbon extraction, with significant reserves—possibly 30 % of total Arctic reserves—off the coast of Alaska (Bird et al. 2008). By contrast, the EU does not have any coastal state jurisdiction in the Arctic Ocean.

There is no instrument providing comprehensive global regulation of offshore hydrocarbon activities, nor is there any global regulatory or governance body with such a mandate. There are, however, a number of instruments with broader scope that also apply to offshore hydrocarbon activities, including those taking place in the Arctic. Among global instruments, the LOS Convention (1982) sets out the basic rules on access to and control over offshore hydrocarbon resources and the mandate of the International Seabed Authority (ISA). Other instruments with more limited applicability to offshore hydrocarbon activities include MARPOL (1973/1978), OPRC (1990), the OSPAR Convention (1992), and the Espoo Convention (1991).

The Arctic Council's Arctic Offshore Oil and Gas Guidelines (PAME 2009), adopted in 1997 and most recently updated in 2009, can go a long way toward addressing regulatory gaps and deficiencies if put into practice. The Guidelines provide recommendations on standards, technical and environmental best practices, management, impact assessment, emergencies, decommissioning, and regulatory control for Arctic offshore oil and gas operations, and recommend following the precautionary approach, the polluter-pays principle, the principle of sustainable development, and the principle of continuous improvement. While providing an important starting point for regulating offshore oil and gas activities in the Arctic, the Guidelines are, however, non-legally binding and leave coastal states with a wide margin of discretion in their implementation.

The EU's 2008 Arctic Communication stated in its proposals for action to "press for the introduction of binding international standards, building *inter alia* on the guidelines of the Arctic Council and relevant international conventions" (European Commission 2008). Later, in 2009, the 'Council conclusions on Arctic issues' (Council of the European Union 2009), invited the European Commission and Member States to examine the possibilities to endorse the Guidelines. The EU's most recent Arctic policy document, the 2012 Joint Communication, does not address the Guidelines, but references the EU Commission's October 2011 proposed new regulation[8] on the safety of offshore oil and gas prospection, exploration, and production activities. While the proposal is still in inter-institutional procedures, its direct impact on Arctic offshore drilling is doubtful. Norway's offshore activities in Arctic waters would only be covered if the finalized regulation or directive fell inside the scope of the EEA agreement, which has been challenged by Norway (EEA 2012).

It has been suggested that in order to ensure effective protection, the Arctic Offshore Oil and Gas Guidelines should be made legally binding (see Chap. 7). A legally binding agreement, however, could face political challenges from coastal states with established national practices.

Still, the fact that all Arctic eight states have formally endorsed the Guidelines demonstrates at least an initial level of support and agreement on governance of offshore hydrocarbon development. The EU and US can publicly support application of the Guidelines and good management practices. As an overarching recommendation, the Arctic Ocean Review (AOR) draft report has urged for agreement on *non*-binding internationally agreed standards for Arctic offshore oil and gas activity, convened through the Arctic Council (AOR 2012). Generally, Arctic states should work to harmonize and strengthen drilling standards, with involvement from industry and non-Arctic states.

In addition to the Arctic Offshore Oil and Gas Guidelines, the Arctic Council has led other work of significance on Arctic hydrocarbon development and environmental protection. The AMAP (Arctic Monitoring and Assessment Programme) working group provides assessments on petroleum hydrocarbon pollution and oil and gas activities in the Arctic while the EPPR (Emergency Prevention, Preparedness, and Response) working group supports marine oil pollution preparedness and response. The US and EU can provide continuing support for these initiatives, particularly should the EU be granted Arctic Council observer status in 2013.

The OSPAR Commission has developed a comprehensive database for offshore oil and gas installations and a monitoring system whereby the OSPAR Offshore Industry Committee (OIC) collects data on emission and discharges.[9] Additionally, the OSPAR Commission supports integrated environmental assessments and

---

[8] Commission's 2011 'Proposal for a Regulation on safety of offshore oil and gas prospection, exploration and production activities' <http://eur-lex.europa.eu/LexUriServ/LexUriServ.do?uri=CELEX:52011PC0688:EN:NOT>.

[9] Including use and discharge of drilling fluids and cuttings, discharges of oil in produced water, chemicals used and discharged offshore, and accidental spills of oil and chemicals and emissions to air.

implementation of the EU MSFD within its Joint Assessment Monitoring Programme (JAMP), which covers the implementation of OSPAR monitoring and information collection programmes – resulting in a biannual update of the offshore database as well as annual assessment sheets on discharges, emissions, and spills of oil and hazardous substances. To promote monitoring, transparency, and data exchange for Arctic oil and gas operators, such a monitoring system for discharges, emissions, and spills could be replicated more widely in other forums.

A possible way towards this direction can be already seen in the PAME work program for 2011–2013: In a comparison of the existing regulatory structures of the Arctic states, it aims to include aspects such as organisational structures, planning, implementation, and monitoring. The AOR has suggested "[i]dentify[ing] ways for Arctic Ocean coastal states not party to OSPAR to coordinate further with JAMP and OIS, notwithstanding that OSPAR Region I covers only a portion of the Arctic Ocean and excludes the Canadian, Russian, and US marine Arctic" (AOR 2012).

With regard to accidental oil spills, the new legally binding Arctic MOPPR Agreement is being negotiated under the auspices of the Arctic Council. A task force was created to prepare the international instrument on oil pollution preparedness and response, as mandated at the Arctic Council Ministerial Meeting in Nuuk in May 2011 (Arctic Council 2011). The agreement is currently ready for signature and will be presented at the Arctic Council Ministerial Meeting in May 2013. It builds to the OPRC 90 Convention on marine oil pollution and adds a regional approach as foreseen in Article 10. The Arctic MOPPR Agreement sets forth procedures in advance of a spill for clean-up and coordination, including an obligation of mutual assistance, enhancing collaboration, and promoting training and information sharing between response parties in the region.

The (draft[10]) agreement takes significant steps towards preparedness for an Arctic spill incident, yet still leaves certain areas uncovered and does not ensure adequate investments in infrastructure by the parties or appear to provide minimal standards in this regard. Although the binding framework is to be supplemented with non-binding operational guidelines, these are not expected to add substantial standards beyond formal cooperative efforts.

However, there will be opportunities to influence the success of the agreement upon implementation. For example, parties will be required to participate in training exercises, secure equipment, establish programmes to respond to spills, and develop detailed plans and strategies for response. Here, support could be provided in terms of both financial and technical resources. Additionally, private industry owns many of the emergency response resources available in the region, yet the agreement does not harness these, relying solely on governments. Steps should be taken generally to cooperate with private industry in Arctic oil and gas development, building upon cooperative efforts such as the Arctic Oil Spill

---

[10] At the time of writing. Draft agreement for signature on file with the authors.

Response Technology Joint Industry Programme (JIP),[11] and as an important first step, agreements should be undertaken to effectively extend private resources for oil response.

11.10 Indigenous Voice in Arctic Governance

Indigenous peoples have inhabited the Arctic for thousands of years and are not only stakeholders in the Arctic, but also rights holders, and deserve a special status in decision making processes. However, their interests can easily be marginalised or neglected in governance institutions.

Traditionally resilient in the face of change, today Arctic indigenous peoples find themselves unable to fully adapt to the rapid rate of change and the range of external stressors, including climate change. Indigenous communities are extremely vulnerable to climate change due to the dependence of their livelihoods on Arctic ecosystems. To support resilience and adaptation to the impacts of climate change, communities must be empowered and have clear rights and access to participation in resource decision making, including through co-management arrangements. To further ensure that efforts are truly participatory, rather than merely procedural, resources must be available to support participation in meetings and planning processes that are often time and money intensive. For EBM processes, stakeholder involvement, including Arctic indigenous peoples, is critical for achieving and understanding social, economic, and environmental goals. EBM can also utilize traditional, as well as scientific, knowledge.

The EU and US have both recognised the particular vulnerability of indigenous communities as central issues within their Arctic policy statements. To support the adaptation of indigenous communities in the marine Arctic, the best forum for transatlantic efforts is the Arctic Council. The Arctic Council affords indigenous groups special status as permanent participants, empowering them to influence the debate on climate change-related issues and include their perspectives in assessments, such as in the Arctic Climate Impact Assessment (ACIA 2005). The Arctic Council's Arctic Resilience Report (ARR) promises to be another important Arctic Council initiative. Initiated at the SAO's meeting in November 2011,[12] the science-based assessment will analyze changes and drivers across multiple scales, engage stakeholders, and identify policy and management options for strengthening resilience and for adaptation. The EU and US should ensure indigenous participation in the ARR and related future assessments, e.g., by providing adequate financial support to allow experts to attend workshops and meetings and supporting on-site monitoring in local communities.

It is important that the status of Arctic indigenous peoples remains strong under the Arctic Council as it continues to evolve. If this status were lost, it would

---

[11] See <http://www.arcticresponsetechnology.org/>.

[12] See <http://www.arctic-council.org/arr/about/>.

result in significantly less visibility for indigenous peoples' interests. The ICC has emphasized the role of the Arctic Council in exercising rights of self-determination (ICC 2009). The EU and the US should recognise and promote the importance and high-level status of indigenous participation in the Arctic Council as it continues to evolve and in future mechanisms.

The EU and US could also jointly support the creation of an assessment on adaptation in the Arctic to cover issues and challenges associated with the implementation of adaptation policies. Furthermore, across the Arctic, a number of national and sub-national climate change adaptation strategies have been developed. Such strategies have been launched by Canada, the (US) state of Alaska, and Greenland among others. Evaluating existing adaptation strategies and their effectiveness can provide valuable information and best practices for wider use.

11.11 Tackling Climate Change

As highlighted throughout this book, global climate change is driving Arctic change and underlies many of the critical challenges for environmental and cultural sustainability in the region today. While climate change and retreating sea ice may open up new economic opportunities and bring benefits to the region, these will not be without risks to Arctic ecosystems, indigenous communities, and traditional livelihoods. Impacts in the Arctic are predicted to increase faster than elsewhere in the world. For instance, Arctic temperature rise is estimated by models as ranging between 3 and 6 °C by 2080, a greater rate than any other region of the world (AMAP 2011).The rapid rate and extreme level of change is outpacing the ability of management regimes to adapt. Thus, while this book focuses primarily on governance in response to a changed and changing Arctic marine environment, it cannot ignore the forces driving those changes.

Climate change is a global problem in both source and effects, stretching beyond the Arctic region. While climate change has severe impacts within the Arctic, including ocean acidification, permafrost melt, and impacts on biodiversity, these regional impacts also connect to global trends, such as sea level rise, impacts on ocean circulation, and climate feedback loops (see Chap. 2). Approaches in governance have to take both aspects into account—the local and the global drivers.

A required two-fold approach to climate change mitigation should address global warming forcing agents, such as carbon dioxide and methane, as well as mitigation of pollutants that increase the local radiative force within the Arctic, such as black carbon and tropospheric ozone (including its predecessor emissions nitrous oxides and methane). Historically, international efforts have largely focused on creating frameworks for binding commitments to greenhouse gas reductions. Although some progress has been made—particularly by the EU, which has been a climate change leader in international negotiations under the UN

Framework Convention on Climate Change (UNFCCC 1992) and its Kyoto Protocol (1996), as well as internally with the EU's Emissions Trading System and '20–20–20' targets[13]—the potential for a comprehensive global agreement continues to look bleaker and incremental negotiations fail to move at sufficient speed to address the problem. However, new efforts to tackle short-lived climate pollutants (SLCPs) show promise not only for helping to mitigate global climate change, but for the Arctic region in particular. Former US Secretary of State Hillary Clinton launched the Climate and Clean Air Coalition to Reduce Short-Lived Climate Pollutants (CCAC) in February 2012 along with Bangladesh, Canada, Ghana, Mexico, Sweden, and the UN Environment Programme. As of its first anniversary, the CCAC has grown to more than 50 partners, including 27 nations and the European Commission, and has a set of sector-based and cross-cutting initiatives underway aimed at reducing black carbon, methane, and some HFCs (hydrofluoro-carbons). A primary focus of the CCAC has been to highlight the multiple benefits of reducing emissions of SLCPs, especially related to public health and agricultural crop production, thus bringing to the table a wide range of partners motivated by a range of issues for a common goal.

Of the major SLCPs targeted by the CCAC, black carbon is most critical in the Arctic region. The largest black carbon sources have been found in land-based transportation (particularly diesel engines), open biomass burning (including agricultural burning, prescribed forest burning, and wildfires), and residential heating (for instance via wood combustion in stoves and boilers) (SLCF Task Force 2011). Black carbon emissions are also expected to increase with rising shipping activities via Arctic routes and from contributions by gas flaring, for instance from off-shore oil and gas development.

Following the recommendations of the Arctic Council's Short-Lived Climate Forcers Task Force, the Arctic Council Meeting of the Environmental Ministers in February 2013 has (as mentioned in Sect. 11.2) suggested implementing a common monitoring system for Arctic states' black carbon emissions, and possibly negotiating a high-level agreement on reductions of black carbon emissions (Ministry of the Environment 2013). For the future, such an Arctic Council agreement could tackle an important part of the problem—the emissions of local Arctic sources of black carbon have been identified to contribute more radiative forcing per emission unit—while presently, the CCAC demonstrates successful (and necessary) international cooperation. Lessons learned from the willing and positive participation of a complex array of actors should be taken to move toward stronger cooperation for greenhouse gas emissions reduction.

However, SLCP-mitigation actions must the greater challenge of global greenhouse gas emissions reductions. Here, there remains potential for improvement in transatlantic cooperation and efforts. Beyond traditionally taking different approaches towards the UNFCCC regime, the EU and US have differed in other

---

[13] Committing to achieve by 2020 a unilateral reduction in greenhouse gas emissions of at least 20 % from 1990 levels, a 20 % increase in renewable energy share in the EU's energy supply, and a 20 % savings in energy consumption.

mitigation approaches, too, such as, for instance, aviation emissions. The EU has sought to cover aviation emissions—including from foreign carriers—through its ETS, a measure strongly opposed by US and other international airlines. In November 2012, US President Barack Obama signed a bill sheltering US airlines from participating in the programme. Although the EU ETS started coverage of emissions from the aviation sector in the beginning of 2012, the EU Commission proposed a derogation from relevant ETS rules in November 2012—as a "gesture of goodwill"[14]—to prevent action against international aircraft operators for non-compliance with ETS requirements and 'stop the clock' until January 2014 (European Commission 2012). However, this conflict could in fact help to revive cooperative efforts to pursue multilateral solutions through other fora, such as the International Civil Aviation Organization (ICAO) and its upcoming Assembly meetings.

To truly manage the impacts of climate change in the Arctic, as well as the potentially catastrophic global implications, stringent mitigation efforts must be undertaken. The EU and US must, as global leaders and major emitters, find ways to promote cooperation and tackle all sides of the problem.

11.12 Conclusion

Global climate change, transboundary pollution, and global markets all drive Arctic change, underlining the imperative for global action and inclusive governance, with cooperation between Arctic states, non-Arctic states, indigenous communities, industry, and NGOs. Already, important steps have been taken to address marine Arctic governance gaps, but further action is needed to adapt the existing legal and policy framework to meet emerging governance challenges in the marine Arctic. Integrated, flexible management structures are needed in order to effectively manage fast-moving, complex challenges.

While Arctic governance clearly extends beyond the jurisdiction of either the EU or US, transatlantic cooperation is needed. Throughout modern environmental policy, both the US and European countries have often been front-runners in developing environmental policies and instruments, offering political leadership that can assist in addressing the challenges facing the marine Arctic.

On the whole, the Arctic Council continues to be the best forum for transatlantic cooperation as it has a strong record on Arctic environmental issues, an established network of stakeholders, and has successfully brought players together in a flexible forum including now, in its evolving role, helping to facilitate the development of legally binding measures. The Arctic Council has promoted coordination not only between states and indigenous groups, but between governance bodies and regimes (e.g., OSPAR, CBD), and is leading the way on many key policy

---

[14] See <http://ec.europa.eu/clima/policies/transport/aviation/index_en.htm>.

areas such as, for instance, with task forces on EBM and SLCPs. The EU and US should support the continued momentum and success of these efforts, such as by providing resources and sharing knowledge and best practices as models for international action and rule making. A decision on whether the EU will be granted permanent observer status will help determine the potential for transatlantic dialogue and cooperation on the Arctic within this forum.

Enhanced transatlantic cooperation can help leverage emerging opportunities to improve protection of the Arctic marine environment. Through both international and domestic action, the EU and US can promote environmental leadership and provide resources in shaping agendas and policy making. Working together, the EU and US can help ensure that environmental protection measures are on the agenda in international fora, thereby promoting multilateral cooperation on issues of global significance. Working separately, although perhaps in coordination, the two can lead through their own policy measures, setting examples and best practices and serving as a model for international rules and the evolution of future governance frameworks.

References

ACIA (2005) Arctic Climate Impact Assessment. Cambridge University Press, New York

AMAP (1998) AMAP Assessment Report: Arctic Pollution Issues. Arctic Monitoring and Assessment Programme, Oslo

AMAP (2011) Snow, Water, Ice and Permafrost in the Arctic (SWIPA): Climate Change and the Cryosphere. Arctic Monitoring and Assessment Programme, Oslo

AMAP (2012) Arctic Climate Issues 2011: Changes in Arctic Snow, Water, Ice and Permafrost. SWIPA 2011 Overview Report. Oslo: Arctic Monitoring and Assessment Programme

AMSA (2009) Arctic Marine Shipping Assessment 2009 Report. Arctic Council, US

AOR (2012) Draft of 16 Oct 2012 of the Phase II Report of the Arctic Ocean Review Project (doc. AC-SAO-NOV12-4.4a)

Arctic MOPPR Agreement (2013) Agreement on Cooperation on Marine Oil Pollution Preparedness and Response in the Arctic, scheduled to be signed at the Arctic Council's Ministerial Meeting in Kiruna

Arctic SAR Agreement (2011) Agreement on Cooperation on Aeronautical and Maritime Search and Rescue in the Arctic, 12 May 2011, 50 I.L.M. 1119 (2011). Entered into force on 19 Jan 2013

ARPA (1984) Arctic Research and Policy Act of 1984, Pub. L. No. 98–373, amended by Pub. L. No 101–609 (1990)

Best A, Cavalieri S, Jariabka M, Koivurova T, Mehling M, Molenaar EJ (2009a) Transatlantic Policy Options for Supporting Adaptations in the Marine Arctic. Arctic TRANSFORM

Best A, Czarnecki R, Koivurova T, Molenaar EJ (2009b) Comparative policy analyses: U.S., EU and transatlantic Arctic policy. Arctic TRANSFORM

Bird KJ, Charpentier RR, Gautier DL, Houseknecht DW, Klett TR, Pitman JK, Moore TE, Schenk CJ, Tennyson ME, Wandrey CJ (2008) Circum-Arctic Resource Appraisal: Estimates of Undiscovered Oil and Gas North of the Arctic Circle. U.S. Geological Survey, Fact Sheet 2008–3049

Conference of Parliamentarians (2008) Conference Report. The 8th Conference of Parliamentarians of the Arctic Region, Fairbanks, 12–14 Aug 2008

Convention on Biological Diversity (1992) Convention on Biological Diversity, 5 Jun 1992, 1760 U.N.T.S. 79. Entered into force 29 Dec 1991

Council of the European Union (2009) Council conclusions on Arctic issues, 2985th Foreign
    Affairs Council meeting, 8 Dec 2009
EEA (2012) Report on the future of the EU Energy policy and its implications for the EEA.
    European Economic Area Joint Parliamentary Committee, Ref. 1118389, 27 Nov 2012
Ehler C (2011) Marine Spatial Planning in the Arctic: A first step toward ecosystem-based man-
    agement. Aspen Institute
Espoo Convention (1991) Convention on Environmental Impact Assessment in a Transboundary
    Context, 25 Feb 1989, 1989 U.N.T.S. 309. Entered into force 10 Sept 1997
European Commission (2007) Communication from the Commission to the European Parliament,
    The Council, the European Economic and Social Committee and the Committee of the Regions:
    An Integrated Maritime Policy for the European Union, 10 Oct 2007, COM (2007) 575 final
European Commission (2008) Communication from the Commission to the European Parliament
    and the Council: the European Union and the Arctic Region, 20 Nov 2008, COM 763
European Commission (2012) Decision of the European Parliament and of the Council dero-
    gating temporarily from Directive 2003/87/EC of the European Parliament and of the
    Council establishing a scheme for greenhouse gas emission allowance trading within the
    Community, COM (2012) 697. <http://ec.europa.eu/clima/policies/transport/aviation/
    docs/com_2012_697_en.pdf/>
European Commission and High Representative (2012) Joint Communication to the European
    Parliament and the Council. Developing a European Union Policy towards the Arctic region:
    progress since 2008 and next steps. 26 June 2012, JOIN(2012) 19 final
Exec. Order No. 13547 (2010) Stewardship of the Ocean, Our coasts, and the Great Lakes. 10
    July 2010, 75 Fed. Reg. 43,023
Fish Stocks Agreement (1995) Agreement for the Implementation of the Provisions of the United
    Nations Convention on the Law of the Sea of 10 Dec 1982 relating to the Conservation and
    Management of Straddling Fish Stocks and Highly Migratory Fish Stocks, 4 Aug 1995, 2167
    U.N.T.S. 3. Entered into force 11 Dec 2001
Hoel AH (2009) Best practices in ecosystem-based oceans management in the Arctic. report no.
    129 of the Norwegian Polar Institute, Polar Environmental Centre
ICC (2008) Towards an Inuit Declaration on Arctic sovereignty. Inuit Leaders' Summit on Arctic
    sovereignty, Kuujjuaq, 6–7 Nov 2008
ICC (2009) A Circumpolar Inuit Declaration on Sovereignty in the Arctic. Inuit Circumpolar
    Council, US
Ilulissat Declaration (2008) Arctic Ocean Conference. Ilulissat, Greenland. 27 May 2008, 48
    I.L.M. 382 (2009)
IUCN/NRDC (2010) IUCN/NRDC Workshop on Ecosystem-based Management in the Arctic
    Marine Environment. Workshop Report, Washington, D.C. International Union for the
    Conservation of Nature and Natural Resources Defense Council
Juda L, Hennessey T (2001) Governance profiles and the management of the uses of large marine
    ecosystems. Ocean Dev Int Law 32:43–69
Kyoto Protocol to the United Nations Framework Convention on Climate Change (1996) 11 Dec
    1997, 2303 U.N.T.S. 148. Entered into force 16 Feb 2005
LOS Convention (1982) United Nations Convention on the Law of the Sea, 10 Dec 1982, 1833
    U.N.T.S. 396. Entered into force 16 Nov 1994
Marine Strategy Framework Directive (2008) Directive 2008/56/EC of the European Parliament
    and the Council of 17 June 2009 establishing a framework for community action in the field
    of marine environmental policy, June 17 2009, 2008 O.J. (L 164)
MARPOL 73/78 (1973/1978) International Convention for the Prevention of Pollution from
    Ships, 2 Nov 1973, 2 I.L.M. 1319 (1973), as modified by the 1978 Protocol Relating to the
    International Convention for the Prevention of Pollution from Ships (17 Feb 1978, 17 I.L.M.
    546 (1978)) and the 1997 Protocol to Amend the International Convention for the Prevention
    of Pollution from Ships (26 Sept 1997) and as regularly amended
Ministry of the Environment (2013) Arctic Environment ministers agreed to strengthen coopera-
    tion to protect the Arctic environment. Ministery of the Environment, Government Offices of
    Sweden. <http://www.government.se/sb/d/17129/a/208655>. Accessed 10 Feb 2013

National Ocean Council (2012) Draft National Ocean Policy Implementation Plan, 12 Jan 2012. <http://www.whitehouse.gov/sites/default/files/microsites/ceq/national_ocean_policy_draft_implementation_plan_01-12-12.pdf>. Accessed 5 Feb 2013

NSPD-66 (2009) National Security Presidential Directive and Homeland Security Presidential Directive. Arctic Region Policy, 9 Jan 2009, NSPD-66/HSPD-25

OPRC 90 (1990) International Convention on Oil Pollution Preparedness, Response, and Cooperation, 30 Nov 1990. 30 I.L.M. 733 (1991). Entered into force 13 May 1995

OSPAR Convention (1992) Convention for the Protection of the Marine Environment of the North-East Atlantic, 22 Sept 1992, 32 I.L.M. 1072 (1993). Entered into force 25 Mar 1998

Ottawa Declaration (1996) Declaration on the Establishment of the Arctic Council, 19 Sept 1996, 35 I.L.M. 1382

PAME (2009) Arctic Offshore Oil and Gas Guidelines. Last updated 29 Apr 2009

Pettersen T (2012) 46 vessels through Northern Sea Route. Barents Observer. <http://barentsobs erver.com/en/arctic/2012/11/46-vessels-through-northern-sea-route-23-11>. Accessed 10 Feb 2013

Polar Shipping Guidelines (2009) Guidelines for ships operating in polar waters, IMO assembly resolution A.1024(26), 2 Dec 2009

SAO (2007) Meeting of Senior Arctic officials: Final Report. Narvik, 28–29 Nov 2007

SAO (2012a) Arctic Council Ecosystem-Based Management Experts Group Intersessional Report: Knowledge and Process Needs for Arctic EBM. Doc 3.7b. Haparanda, Nov 2012

SAO (2012b) Experts Group on Arctic EBM: Draft Outline of Report to SAOs. Doc 3.7c. Haparanda, Nov 2012

Sherman K (1994) Sustainability, biomass yields, and health of coastal ecosystems: An ecological perspective. Mar Ecol Prog Ser 112:277–301

S.J. Res. 17 (2007) A joint resolution directing the United States to initiate international discussions and take necessary steps with other Nations to negotiate an agreement for managing migratory and transboundary fish stocks in the Arctic Ocean, 110th Cong

SLCF Task Force (2011) Progress Report and Recommendations for Ministers. Arctic Council Task Force on Short-Lived Climate Forcers

Speer L, Laughlin T (2010) IUCN/NRDC Workshop to Identify Areas of Ecological and Biological Significance or Vulnerability in the Arctic Marine Environment. Workshop report, La Jolla

TFEU (2008) Treaty on the Functioning of the European Union. Consolidated version. O.J. C 115/47

United Nations Framework Convention on Climate Change (1992) 1771 U.N.T.S. 107. Entered into force 21 Mar 1994

Annex: List of Relevant Treaties, Instruments, and Agreements

Short title	Full title and citation
AEPS	Arctic Environmental Protection Strategy, 14 January 1991, 30 I.L.M. 1624 (1991)
Antarctic Treaty	The Antarctic Treaty, 1 December 1959, 402 U.N.T.S. 71. Entered into force 23 June 1961
Anti-fouling Convention	International Convention on the Control of Harmful Anti-fouling Systems on Ships, 5 October 2001, IMO Doc. AFS/CONF/26, of 18 October 2001. Entered into force 17 September 2008
Arctic SAR Agreement	Agreement on Cooperation on Aeronautical and Maritime Search and Rescue in the Arctic, 12 May 2011, 50 I.L.M. 1119 (2011). Entered into force on 19 January 2013
Arctic Shipping Guidelines	Guidelines for Ships Operating in Arctic Ice-Covered Waters, IMO MSC/Circ. 1056, MEPC/Circ. 399, 23 December 2002
Basel Convention	Basel Convention on the Control of Transboundary Movements of Hazardous Wastes and their Disposal. 22 March 1989, 1673 U.N.T.S. 57. Entered into force 5 May 1992
BWM Convention	International Convention for the Control and Management of Ships' Ballast Water and Sediments, 13 February 2004, 30 I.L.M. 1455 (1991). Not in force, IMO Doc. BWM/CONF/36, of 16 February 2004
Bunker Oil Convention	International Convention on Civil Liability for Bunker Oil Pollution Damage, 23 March 2001, 402 U.N.T.S. 71. Entered into force 21 November 2008
CBD	Convention on Biological Diversity, 5 June 1992, 1760 U.N.T.S. 79. Entered into force 29 December 1991
CBS Convention	Convention on the Conservation and Management of Pollock Resources in the Central Bering Sea, 16 June 1994, 34 I.L.M. 67 (1995). Entered into force 8 December 1995
CITES	Convention on International Trade in Endangered Species of Wild Fauna and Flora, 3 March 1973, 993 U.N.T.S. 243. Entered into force 1 July 1975
Civil Liability Convention	International Convention on Civil Liability for Oil Pollution Damage, Brussels, 29 November 1969, 9 I.L.M. 45 (1970). Entered into force 19 June 1975. Replaced and entered into force 30 May 1996

(continued)

(continued)

Short title	Full title and citation
CMS / Bonn Convention	Convention on the Conservation of Migratory Species of Wild Animals, 23 June 1979, 1651 U.N.T.S. 33. Entered into force 1 November 1983
COLREG 72	Convention on the International Regulations for Preventing Collisions at Sea, 20 October 1972, 1050 U.N.T.S. 16. Entered into force 15 July 1977
Compliance Agreement	Agreement to Promote Compliance with International Conservation and Management Measures by Fishing Vessels on the High Seas, 24 November 1993, 33 I.L.M. 969 (1994). Entered into force 24 April 2003
Espoo Convention	Convention on Environmental Impact Assessment in a Transboundary Context, 25 February 1989, 1989 U.N.T.S. 309. Entered into force 10 September 1997
Fish Stocks Agreement	Agreement for the Implementation of the Provisions of the United Nations Convention on the Law of the Sea of 10 December 1982 relating to the Conservation and Management of Straddling Fish Stocks and Highly Migratory Fish Stocks, 4 August 1995, 2167 U.N.T.S. 3. Entered into force 11 December 2001
Fund Convention	International Convention on the Establishment of an International Fund for Compensation for Oil Pollution Damage, 18 December 1971, 11 I.L.M. 284 (1972). Entered into force 16 October 1978
Gothenburg Protocol	Protocol to the 1979 Convention on Long-range Transboundary Air Pollution to Abate Acidification, Eutrophication and Ground-level Ozone, 30 November 1999, EB.AIR/1999/1. Entered into force 17 May 2005
Helsinki Protocol	Protocol to the 1979 Convention on Long-Range Transboundary Air pollution on the Reduction of Sulphur Emissions or their Transboundary Fluxes by at least 30 percent, 14 June 1985, 1480 U.N.T.S. 215. Entered into force 2 September 1987
HNS Convention	International Convention on Liability and Compensation for Damage in Connection with the Carriage of Hazardous and Noxious Substances by Sea, 3 May 1996, 35 I.L.M. 1406 (1996). Not in force
HNS Protocol	Protocol on Preparedness, Response and Co-operation to Pollution Incidents by Hazardous and Noxious Substances, 15 March 2000, IMO Doc. HNS-OPRC/CONF/11/Rev.1, of 15 March 2000. Entered into force 14 June 2007
ICCPR	International Covenant on Civil and Political Rights, 16 December 1966, 999 U.N.T.S. 17. Entered into force 23 March 1976
ICRW	International Convention for the Regulation of Whaling, 2 December 1946, 161 U.N.T.S. 72. Entered into force 10 November 1948
ILO Convention No. 169	Convention concerning Indigenous and Tribal Peoples in Independent Countries, 27 June 1989, 28 I.L.M. 1382 (1989). Entered into force 5 September 1991
Ilulissat Declaration	Ilulissat Declaration, Arctic Ocean Conference. Ilulissat, Greenland. 27 May 2008, 48 I.L.M. 382 (2009)

(continued)

(continued)

Short title	Full title and citation
Kyoto Protocol	Kyoto Protocol to the United Nations Framework Convention on Climate Change, 11 December 1997, 2303 U.N.T.S. 148. Entered into force 16 February 2005
London Convention/LC	Convention on the Prevention of Marine Pollution by Dumping of Wastes and Other Matter. 29 December 1972, 1046 U.N.T.S. 120. Entered into force 30 August 1957
London Protocol/LP	Protocol to the Convention on the Prevention of Marine Pollution by Dumping of Wastes and Other Matter, 7 November 1996, 36 I.L.M. 7 (1997). Entered into force 24 March 2006
LOS Convention	United Nations Convention on the Law of the Sea, 10 December 1982, 1833 U.N.T.S. 396. Entered into force 16 November 1994
LRTAP Convention	Convention on Long-range Transboundary Air Pollution, 13 November 1979, 1302 U.N.T.S. 217. Entered into force 16 March 1983
MARPOL 73/78	International Convention for the Prevention of Pollution from Ships, 2 November 1973, 2 I.L.M. 1319 (1973). Entered into force 2 October 1983
Montreal Protocol	Montreal Protocol on Substances that Deplete the Ozone Layer, 16 September 1987, 1522 U.N.T.S. 3. Entered into force 1 January 1989
Murmansk Treaty	Treaty between the Kingdom of Norway and the Russian Federation concerning Maritime Delimitation and Cooperation in the Barents Sea and the Arctic Ocean, 15 September 2010, U.N.T.S. Reg. No. 49095. Entered into force 7 July 2011
NAFO Convention	Convention on Future Multilateral Cooperation in the Northwest Atlantic Fisheries, 24 October 1978, 1135 U.N.T.S. 369. Entered into force 1 January 1979
NASCO Convention	Convention for the Conservation of Salmon in the North Atlantic Ocean, 2 March 1982, 1338 U.N.T.S. 33 (1983). Entered into force 1 October 1983
NEAFC Convention	Convention on Future Multilateral Cooperation in the North-East Atlantic Fisheries, 18 November 1980, 1285 U.N.T.S. 129. Entered into force 17 March 1982
NPAFC	Convention for the Conservation of Anadromous Stocks in the North Pacific Ocean, 11 February 1992. 22 Law of the Sea Bulletin 21 (1993). Entered into force 16 February 1993
OPRC Convention/OPRC 90	International Convention on Oil Pollution Preparedness, Response, and Cooperation, 30 November 1990. 30 I.L.M. 733 (1991). Entered into force 13 May 1995
OSPAR Convention	Convention for the Protection of the Marine Environment of the North-East Atlantic, 22 September 1992, 32 I.L.M. 1072 (1993). Entered into force 25 March 1998
Ottawa Declaration	Declaration on the Establishment of the Arctic Council, 19 September 1996, 35 I.L.M. 1382 (1996)
Part XI Deep-Sea Mining Agreement	Agreement relating to the Implementation of Part XI of the United Nations Convention on the Law of the Sea of 10 December 1982, 28 July 1994. 1836 U.N.T.S. 3. Entered into force 28 July 1996

(continued)

(continued)

Short title	Full title and citation
Polar Bear Agreement	International Agreement on the Conservation of Polar Bears, 15 November 1973, 13 I.L.M. 3 (1974). Entered into force 26 May 1976
Polar Shipping Guidelines	Guidelines for ships operating in polar waters, IMO Assembly Resolution A.1024(26), 2 December 2009
POPs Protocol	Protocol to the 1979 Convention on LRTAP on Persistent Organic Pollutants (POPs), 24 June 1998, 2230 U.N.T.S. 79. Entered into force 23 October 2003
Port State Measures Agreement	Agreement on Port State Measures to Prevent, Deter and Eliminate Illegal, Unreported and Unregulated Fishing, 22 November 2009. Not in force (at 27 November 2012)
Ramsar Convention	Convention on Wetlands of International Importance Especially as Waterfowl Habitat, 2 February 1971, 996 U.N.T.S. 245. Entered into force 21 December 1975
Rovaniemi Declaration	Rovaniemi Declaration (1991). Declaration on the Protection of the Arctic Environment, 14 June 1991, 30 I.L.M. 1624 (1991)
SAR Convention	International Convention on Maritime Search and Rescue, 27 April 1979. 1405 U.N.T.S. 118. Entered into force 22 June 1985
Stockholm Convention	Stockholm Convention on Persistent Organic Pollutants, 22 May 2001, 2256 U.N.T.S. 119. Entered into force 17 May 2004
SEA Protocol	Protocol on Strategic Environmental Assessment to the Convention on Environmental Impact Assessment in a Transboundary Context, 21 May 2001, UNECE Document ECE/MP.EIA/2003/2. Entered into force 11 July 2010
Sofia Protocol	Protocol to the 1979 Convention on Long-Range Transboundary Air Pollution Concerning the Control of Emissions of Nitrogen Oxides or their Transboundary Fluxes, 31 October 1988, 28 I.L.M. 212 (1989). Entered into force 14 February 1991
SOLAS	International Convention for the Safety of Life at Sea, 1 November 1974, 1184 U.N.T.S. 278. Entered into force 25 May 1980
Spitsbergen Treaty	Treaty Concerning the Archipelago of Spitsbergen, 9 February 1920, 2 L.N.T.S. 7. Entered into force 14 August 1925
STCW Convention	International Convention on Standards of Training, Certification and Watchkeeping for Seafarers, 1 December 1978, 1361 U.N.T.S. 2 Entered into force 28 April 1984. As amended and modified by the 1995 Protocol
UNDRIP	United Nations Declaration on the Rights of Indigenous Peoples. General Assembly Resolution. New York, 13 September 2007. A/RES/61/295
UNESCO World Heritage Convention	Convention Concerning the Protection of the World Cultural and Natural Heritage, 16 November 1972, 1037 U.N.T.S. 151. Entered into force 17 December 1975
UNFCCC	United Nations Framework Convention on Climate Change, 9 May 1992, 1771 U.N.T.S. 107. Entered into force 21 March 1994

(continued)

(continued)

Short title	Full title and citation
Vienna Convention	Vienna Convention for the Protection of the Ozone Layer, 22 March 1985, 1513 U.N.T.S. 293. Entered into force 22 September 1988
VOC Protocol	Protocol to the 1979 Convention on LRTAP concerning the Control of Emissions from Volatile Organic Compounds or their Transboundary Fluxes, 18 November 1991, 2001 U.N.T.S. 187. Entered into force 29 September 1997